OTHER VOLUMES IN THIS SERIES

I

Tables of the cumulative non-central chi-square distribution
by G. E. HAYNAM, Z. GOVINDARAJULU and F. C. LEONE
Table I (Power of the chi-square test)
Table II (Non-centrality parameter)

Tables of the exact sampling distribution of the two-sample Kolmogorov-Smirnov criterion $D_{mn} (m \leq n)$ by P. J. KIM and R. I. JENNRICH
Table I (Upper tail areas)
Table II (Critical values)

Critical values and probability levels for the Wilcoxon Rank Sum Test and the Wilcoxon Signed Rank Test
by FRANK WILCOXON, S. K. KATTI and ROBERTA A. WILCOX
Table I (Critical values and probability levels for the Wilcoxon Rank Sum Test)
Table II (Probability levels for the Wilcoxon Signed Rank Test)

The null distribution of the first three product-moment statistics for exponential, half-gamma, and normal scores by P. A. W. LEWIS and A. S. GOODMAN
Tables 1–3 (Normal with lag 1, 2, 3)
Tables 4–6 (Exponential with lag 1, 2, 3)
Tables 7–9 (Half-gamma with lag 1, 2, 3)

Tables to facilitate the use of orthogonal polynomials for two types of error structures
by KIRKLAND B. STEWART
Table I (Independent error model)
Table II (Cumulative error model)

II

Probability integral of the doubly noncentral *t*-distribution with degrees of freedom n and non-centrality parameters δ and λ by WILLIAM G. BULGREN

Doubly noncentral *F* distribution–Tables and applications by M. L. TIKU

Tables of expected sample size for curtailed fixed sample size tests of a Bernoulli parameter
by COLIN R. BLYTH and DAVID HUTCHINSON

Zonal polynomials of order 1 through 12 by A. M. PARKHURST and A. T. JAMES

III

Tables of the two factor and three factor generalized incomplete modified Bessel distributions
by BERNARD HARRIS and ANDREW P. SOMS

Sample size requirement: Single and double classification experiments
by KIMIKO O. BOWMAN and MARVIN A. KASTENBAUM

Passage time distributions for Gaussian Markov (Ornstein-Uhlenbeck) statistical processes
by J. KEILSON and H. F. ROSS

Exact probability levels for the Kruskal-Wallis test
by RONALD L. IMAN, DANA QUADE and DOUGLAS A. ALEXANDER

Tables of confidence limits for linear functions of the normal mean and variance
by C. E. LAND

IV

Dirichlet distribution–Type 1
by MILTON SOBEL, V. R. R. UPPULURI and K. FRANKOWSKI

V

Variances and covariances of the normal order statistics for sample sizes 2 to 50
by G. L. TIETJEN, D. K. KAHANER and R. J. BECKMAN

Means, variances and covariances of the normal order statistics in the presence of an outlier
by H. A. DAVID, W. J. KENNEDY and R. D. KNIGHT

Tables for obtaining optimal confidence intervals involving the chi-square distribution
by G. RANDALL MURDOCK and WILLIAM O. WILLIFORD

OTHER VOLUMES IN THIS SERIES

VI

The distribution of the size of the maximum cluster of points on a line
by NORMAN D. NEFF and JOSEPH I. NAUS

VII

The product of two normally distributed random variables
by WILLIAM W. MEEKER, JR., LARRY W. CORWELL and LEO A. AROIAN
Table I (Power of the chi-square test). Table II (Non-centrality parameter).
Table I (Upper tail areas). Table II (Critical values).
Table I (Independent error model). Table II (Cumulative error model).

Selected Tables in Mathematical Statistics

Volume VIII

Edited by the Institute of Mathematical Statistics

Coeditors
W. J. Kennedy
Iowa State University
and
R. E. Odeh
University of Victoria

Managing Editor
J. M. Davenport
Texas Tech University

AMERICAN MATHEMATICAL SOCIETY
PROVIDENCE, RHODE ISLAND

1980 *Mathematics Subject Classification*
Primary 62Q05; Secondary 62C15, 62E30, 62F07, 62G30, 62J15, 62K05, 62K10.

International Standard Serial Number 0094-8837
International Standard Book Number 0-8218-1908-9
Library of Congress Card Number 74-6283

Table of Contents

Preface

This volume of mathematical tables has been prepared under the aegis of the Institute of Mathematical Statistics. The Institute of Mathematical Statistics is a professional society for mathematically oriented statisticians. The purpose of the Institute is to encourage the development, dissemination, and application of mathematical statistics. The Committee on Mathematical Tables of the Institute of Mathematical Statistics is responsible for preparing and editing this series of tables. The Institute of Mathematical Statistics has entered into an agreement with the American Mathematical Society to jointly publish this series of volumes. At the time of this writing, submissions for future volumes are being solicited. No set number has been established for this series. The editors will consider publishing as many volumes as are necessary to disseminate meritorious material.

Potential authors should consider the following rules when submitting material.

1. The manuscript must be prepared by the author in a form acceptable for photo-offset. The author should assume that nothing will be set in type although the editors reserve the right to make editorial changes. This includes both the introductory material and the tables. A computer tape of the tables will be required for the checking process *and* the final printing (assuming it is accepted). The authors should contact the editors prior to submission concerning the requirements for the computer tape.

2. While there are no fixed upper and lower limits on the length of tables, authors should be aware that the purpose of this series is to provide an outlet for tables of high quality and utility which are too long to be accepted by a technical journal but too short for separate publication in book form.

3. The author must, whenever applicable, include in his introduction the following:

(a) He should give the formula used in the calculation, and the computational procedure (or algorithm) used to generate his tables. Generally speaking FORTRAN or ALGOL programs will not be included but the description of the algorithm used should be complete enough that such programs can be easily prepared.

(b) A recommendation for interpolation in the tables should be given. The author should give the number of figures of accuracy which can be obtained with linear (and higher degree) interpolation.

(c) Adequate references must be given.

(d) The author should give the accuracy of the table and his method of rounding.

(e) In considering possible formats for his tables, the author should attempt to give as much information as possible in as little space as possible. Generally speaking, critical values of a distribution convey more information than the distribution itself, but each case must be judged on its own merits. The text portion of the tables (including column headings, titles, etc.) must be proportional to the size 5–1/4″ by 8–1/4″. Tables may be printed proportional to the size 8–1/4″ by 5–1/4″ (i.e., turned sideways on the page) when absolutely necessary; but this should be avoided and every attempt made to orient the tables in a vertical manner.

(f) The table should adequately cover the entire function. Asymptotic results should be given and tabulated if informative.

(g) Examples of the use of the tables must be included.

4. The author should submit as accurate a tabulation as he/she can. The table will be checked before publication, and any excess of errors will be considered grounds for rejection. The manuscript introduction will be subjected to refereeing. Since an inadequate introduction may lead to rejection, the author should strive for an informative manuscript, which not only establishes a need for the tables, but also explains in detail how to use the tables.

5. Authors having tables they wish to submit should send two copies to:

Dr. Robert E. Odeh, Coeditor
Department of Mathematics
Univerity of Victoria
Victoria, B. C., Canada V8W 2Y2

At the same time, a third copy should be sent to:

James M. Davenport, Coeditor
Department of Mathematics
Texas Tech University
P. O. Box 4319
Lubbock, Texas 79409

Additional copies may be required, as needed for the editorial process. After the editorial process is complete, a camera-ready copy must be prepared for the publisher.

Authors should check several current issues of *The Institute of Mathematical Statistics Bulletin* and *The AMSTAT News* for any up-to-date announcements about submissions to this series.

Acknowledgments

The tables included in the present volume were checked at the Universiy of Victoria. Dr. R. E. Odeh arranged for, and directed this checking with the assistance of Mr. Bruce Wilson. The editors and the Institute of Mathematical Statistics wish to express their great appreciation for this invaluable assistance. So many other people have contributed to the instigation and preparation of this volume that it would be impossible to record their names here. To all these people, who will remain anonymous, the editors and the Institute also wish to express their thanks.

Selected Tables in Mathematical Statistics
Volume 8, 1985

EXPECTED SIZES OF A SELECTED SUBSET IN PAIRED COMPARISON EXPERIMENTS

B. J. Trawinski
University of Alabama in Birmingham

ABSTRACT

Tables are provided for the design and analysis of a balanced paired comparison experiment intended for selecting from a set T of treatments a subset S containing the best treatment. The principal criteria for designing the experiment are a restriction on the expected size of S and a specified probability of including in S the best treatment of T. Selected examples of applications range from a design of a paired comparison experiment, through an exploration of experiment sizes, to data analysis in selection and screening problems.

1. INTRODUCTION AND SUMMARY

The method of paired comparisons is applicable to comparative studies of units which may be objects, treatments, and the like. It is especially useful when meaningful absolute measurements are difficult or impossible to obtain, as is the case in subjective studies. The method involves a judgment of "preference" expressed for one of the units in each pair considered by a panelist or judge. By concentrating on one pair of units at a time, the judge

---

Received by the editors December 1978 and in revised form April 1979 and March 1980.
1980 Mathematics Subject Classification: Primary 62Q05; Secondary 62F07,62J15.
Research supported by Health Special Research Resources Grant FR-00145. Funding for completion of computer work provided by the University of Alabama in Birmingham Graduate School.

0094-8837/85 $1.00 + $.25 per page

is freed of confusing effects such as fatigue or lack of recall. In a number of situations, unit pairs occur naturally (for an example, see (ii) below).

In paired comparison experiments, the preferences are based on assessments of a well defined common characteristic for the units in a set, and a homogeneous perception of this characteristic when several judges are involved. The latter is a requirement ordinarily fulfilled by proper training of qualified personnel.

The following are some examples of areas of applicability of the method: (i) the food industry where samples are compared for taste or other qualities, (ii) optical research where contact lenses are tested for least irritability, (iii) rehabilitation where treatments for the improvement of patient performance are compared, (iv) histological studies where staining procedures are compared for effectiveness, (v) tournaments where performance is evaluated on the basis of competition between members of pairs. Paired comparisons have also been used extensively in behavioral research.

The theory and methodology dealing with paired comparisons is extensive (see David [2], and for a bibliography up to 1976, see Davidson and Farquhar [3]). The procedure for the inclusion of the best unit of a set in a selected subset is formulated and discussed by Trawinski and David [9]; Trawinski [7] develops asymptotic formulae for the expected size of the subset and shows [8] that the discriminant relation of the procedure (see Section 3) is consistent for the inclusion of the best unit in a selected subset. The tables of this paper are based on both the exact and asymptotic developments of expected sizes of a selected subset; these tables are applicable in the design of experiments and supportive of data analysis requiring the inclusion, with probability not smaller than a given number, of the best unit of a set in a subset whose size can be controlled. The tables can also be used in screening of units from a fixed set.

In accordance with the customary terminology of design and analysis of experiments, the units subjected to paired comparisons are referred to throughout this discussion as treatments. Notation, definitions, data collection methods, and probability distribution are discussed in Section 2; included in this section is a subsection on linear models. Except for a brief introduction to functions which govern the methodology for the selection of a subset containing the best treatment (material connected with (3.1) through (3.3)), Section 3 is devoted to quantitative discussion of properties and structure of the selection procedure, and methods of applications following therefrom. Section 4 is concerned with computation of table entries, and Section 5 deals with structure of tables and interpolation.

2. PRELIMINARIES

DEFINITIONS AND DISTRIBUTION

A balanced paired comparison experiment consists of $\frac{1}{2}t(t-1)$ preferences, each pertaining to one member of a pair (T_r,T_s) formed from the set $T = \{T_1,\cdots,T_t\}$ of t treatments. Throughout this paper balanced paired comparison experiments of n independent replications are considered.

To each pair (T_r,T_s) $(r\neq s;\ r,s=1,\cdots,t)$ of replication γ $(\gamma=1,\cdots,n)$, there corresponds a characteristic random variable

$$X_{rs\gamma} = 1, \text{ if } T_r \text{ is preferred to } T_s$$
$$= 0, \text{ if } T_s \text{ is preferred to } T_r,$$

and a preference probability p_{rs} defined by

$$P[X_{rs\gamma} = 1] = p_{rs},\ P[X_{rs\gamma} = 0] = 1-p_{rs} = p_{sr}.$$

For further reference a configuration of all $\frac{1}{2}t(t-1)$ preference probabilities is denoted by $C\{p_{rs}\}$.

In terms of the realizations $x_{rs\gamma}$ of $X_{rs\gamma}$, the number of times T_r is preferred to each of the remaining treatments in the n(t-1) comparisons is given by $x_r = \Sigma_{\gamma=1}^{n} x_{r\cdot\gamma}$, where $x_{r\cdot\gamma}$ stands for the sum of the $x_{rs\gamma}$ over the index s, excluding s=r. Thus to the set of treatments $T = \{T_1,\cdots,T_t\}$ there corresponds a vector of preference scores $x = (x_1,\cdots,x_t)$ obtained from a balanced paired comparison experiment of n replications. The components X_r of the vector random variable X, of which x is an outcome, have the same structure as the x_r.

Below is an illustration of a method of preference scoring when $T = \{T_1,T_2,T_3\}$ and $n = 2$. For example, the first scoring table indicates that T_1 is preferred to T_3, T_2 to T_1, and T_2 to T_3.

The vector of realizations for the set T due to the first replication is in this case (1,2,0); the vector of realizations for the set T due to both replications is x = (2,3,1).

	x_{rs1}					x_{rs2}			
	T_1	T_2	T_3	$x_{r\cdot 1}$		T_1	T_2	T_3	$x_{r\cdot 2}$
T_1	-	0	1	1	T_1	-	1	0	1
T_2	1	-	1	2	T_2	0	-	1	1
T_3	0	0	-	0	T_3	1	0	-	1

Alternatively, if α_{rs} denotes the number of times T_r of (T_r,T_s) is preferred to T_s in n replications, then the components of the vector x can be written as

$$x_r = n(t-r) + \Sigma_{u=1}^{r-1}\alpha_{ru} - \Sigma_{v=r+1}^{t}\alpha_{vr}(r=1,\cdots,t).$$

Let the above t relations for the x_r $(r=1,\cdots,t)$ be denoted by R(n). The probability function for the vector of preference scores, x, has the form

$$f(x;C\{p_{rs}\}) = \Sigma_{R(n)}\Pi_{r>s}B(n,\alpha_{rs})p_{rs}^{\alpha_{rs}}(1 - p_{rs})^{n-\alpha_{rs}} \tag{2.1}$$

where $B(n,\alpha_{rs})$ denotes the binomial coefficient.

The following pertains to an ordering in T and is intended to set up notation. Denote by $T_{(t)}$ and $T_{(t-1)}$ respectively the best and the second best treatments in T (criteria for such designations vary with analytical procedures, and one criterion appropriate for the inclusion of $T_{(t)}$ in a selected subset is given at the beginning of Section 3). To simplify notation, the parentheses on subscripts of the preference probabilities, or on the subscripts of other quantities corresponding to a theoretical ordering of the $T_1,\cdots,T_t$, will be omitted. Thus, for example, the preference probability that $T_{(t)}$ is preferred to $T_{(t-1)}$ will still be denoted by $p_{t,t-1}$; cases where such simplifications are used will be clear from the context.

LINEAR MODELS

The definition of a common characteristic C in T carries the connotation that the treatments $T_1,\cdots,T_t$ have unknown true merits or values $V_1,\cdots,V_t$ which are points on some continuous numerical scale V. If instead of expressing preference for one member of the pair (T_r,T_s), one could observe the values V_r of T_r and V_s of T_s in V by means of outcomes of the continuous random variables Y_r and Y_s, then X_{rs} (of the preceding subsection, with replication subscript suppressed) would be the indicator random variable of the event $A_{rs} = [Y_r > Y_s]$, taking on the value 1 if A_{rs} occurs, and 0 otherwise. It follows that the theoretical frequency p_{rs} with which T_r is preferred to T_s is a function of values on a scale V.

The preference probabilities of $C\{p_{rs}\}$ are said to satisfy a "linear model" if each p_{rs} is related to the corresponding V_r,V_s through a continuous function H, so that

$$p_{rs} = H(V_r - V_s). \tag{2.2}$$

In (2.2) H has the properties of a distribution function whose derivative is symmetric about zero.

A consequence of (2.2) is that $p_{rs} > \frac{1}{2}, = \frac{1}{2}, < \frac{1}{2}$ according as $V_r > V_s, = V_s, < V_s$, and that the values $V_1,\cdots,V_t$ of the respective treatments $T_1,\cdots,T_t$ can be represented by t points on a linear scale with arbitrary origin, hence the term "linear".

Assessments of treatments with respect to a multicomponent characteristic $C = (C_1,\cdots,C_k)$ are not, in general, conducive to linear models, unless the resultant of the valuation of T_r can be defined, for example by $V_r = \Sigma_{j=1}^{k} V_{rj}$ for $r = 1,\cdots,t$, where V_{rj} is the value of T_r in relation to C_j.

The important feature of a linear model is that the ordered values, $V_{(1)} \le \cdots \le V_{(t)}$, of $V_1,\cdots,V_t$ permit a theoretical ranking of the treatments $T_1,\cdots,T_t$ (not just an ordering which may be partial). Thus associated with $V_{(r)}$ is the r-th ranked treatment $T_{(r)}$ to which there corresponds the random number of preferences $X_{(r)}$, a component of the vector random variable X defined in the preceding subsection.

ILLUSTRATION 2.1. In a speech therapy investigation involving patients who stutter, the characteristic may be limited in its definition to facial contortion of subjects when pronouncing certain words and treated as a one component characteristic. Filmed patients' responses due to different treatments (randomized and properly spaced in time) can be presented in pairs in random succession to experienced speech therapy workers. A linear model can be justified if the valuation of patients' performance under treatment can be explained in terms of points on the real line, and the degree of preference for one member of each (T_r,T_s) over the other can be perceived on the basis of value difference between T_r and T_s. If for instance T_1, T_2, and T_3 represent treatments which correspond respectively to slight, substantial, and moderate improvement in patients' performance and the pair values V_1 and V_3, and V_3 and V_2, are approximately equidistant, then on the average T_2 should be preferred to T_3 with about the same frequency as T_3 to T_1. In this case $V_2 - V_1 = H^{-1}(p_{31}) + H^{-1}(p_{23}) \cong 2H^{-1}(p_{23})$ and $p_{21} = H(V_2 - V_1) \cong H(2H^{-1}(p_{23}))$ for some continuous distribution function H whose derivative is symmetric about zero. For another set of treatments $\{T_1,T_2,T_3\}$, the values might be such that V_3 is slightly higher than V_1, while V_2 is considerably greater than V_3; this illustrates a case where T_1 and T_3 are about equivalent and T_2 is superior to both. In terms of H, the following relations hold: $V_2-V_1 = H^{-1}(p_{23}) + H^{-1}(p_{31})$ and $p_{21} \cong H(H^{-1}(p_{23}) + H^{-1}(\frac{1}{2})) = H(H^{-1}(p_{23})) = p_{23}$.

In general, for any set of treatments the relative value difference $V_r - V_s$ for (T_r,T_s) may range in absolute value from zero to a very large number (see [9] for two important special cases, the Thurstone-Mosteller model and the Bradley-Terry model). Whenever two or more such differences are equal or nearly equal, the frequencies of preferences for one treatment over the other in each of the corresponding pairs should be nearly equal, when n is sufficiently large.

For the application of the tables, a linear model need not hold (see the following section). If it does, the paired comparison experiment need only be compatible with the general aspects of a linear model without the requirement of specifying a particular form of H.

3. USE OF TABLES

The tables are applicable in the design of experiments and/or data analysis intended for the selection from T of a subset S containing the best treatment $T_{(t)}$ of T. With respect to the procedure for such a selection, $T_{(t)}$ is defined as the treatment in T having the largest total, $p_t = \Sigma_{s \neq t}\, p_{ts}$, of the preference probabilities of $T_{(t)}$ over each T_s of T, and $T_{(t-1)}$ is the treatment in T having the second largest total, $p_{t-1} = \Sigma_{s \neq t-1}\, p_{t-1,s}$, of the preference probabilities of $T_{(t-1)}$ over each T_s of T. In general p_{t-1} is not less than any p_r $(r=1,\cdots,t-2)$ defined similarly.

Let $X_{(t)}$ denote the component of $X = (X_1,\cdots,X_t)$ corresponding to $T_{(t)}$ in T, and let X_{max} stand for the random variable $\max_{1 \leq r \leq t} X_r$ of X. Let ν be a non-negative integer not greater than $n(t-1)$. The procedure for selecting S depends on the discriminant relation $X_r \geq X_{max} - \nu$ operating on the components of X, together with the probability

$$P[X_{(t)} \geq X_{max} - \nu;\ t,n,\ C\{p_{rs}\}] \tag{3.1}$$

denoted by $P_{(t),n}(C\{p_{rs}\},\nu)$, of correctly including in S the best treatment $T_{(t)}$. The size of S is a random variable.

Let $E_{t,n}(C\{p_{rs}\},\nu)$ denote the expected size of S. In addition to the assurance of correct inclusion of $T_{(t)}$ in S with probability (3.1), the random size of S can be controlled through the assignment of a value to

$$E_{t,n}(C\{p_{rs}\},\nu) = \Sigma_{r=1}^{t}\ P[X_r \geq X_{max} - \nu;\ t,n,\ C\{p_{rs}\}] \tag{3.2}$$

and a subsequent solution for a parameter such as, for example, the number n of replications.

Tables 1 and 2 contain respectively values of the functions (3.1), and (3.2) scaled by t^{-1}, when the configuration $C\{p_{rs}\}$ written as

$$\begin{matrix} \tfrac{1}{2}+\tau_{21} & & & & \\ \tfrac{1}{2}+\tau_{31} & \tfrac{1}{2}+\tau_{32} & & & \\ \vdots & \vdots & \ddots & & \\ \tfrac{1}{2}+\tau_{t-1,1} & \tfrac{1}{2}+\tau_{t-1,2} & \cdots & \tfrac{1}{2}+\tau_{t-1,t-2} & \\ p+\tau_{t,1} & p+\tau_{t,2} & \cdots & p+\tau_{t,t-2} & p \quad (p=p_{t,t-1}) \end{matrix} \tag{3.3}$$

is given in forms of statistical models customary in analysis of experimental data. Two cases are of particular utility.

The configuration $C\{\tfrac{1}{2},p\}$, which results from (3.3) when $\tau_{rs} = 0$ for all pairs r,s, is an idealized representation of (3.3) when the τ_{rs} are not too large. Such a configuration of preference probabilities corresponds to the important case of a single outlier, and is compatible with the procedure of including in a selected subset S a strictly better treatment $T_{(t)}$ of T.

The configuration $C\{\frac{1}{2}\}$, which is the same as $C\{\frac{1}{2},p\}$ when p assumes the minimum in the interval $[\frac{1}{2},1]$, denotes the idealized case when all treatments are equivalent with respect to a characteristic defined in T (by tagging treatment $T_{(t)}$, the random variable $X_{(t)}$ is still defined).

As a function of the configuration $C\{\frac{1}{2},p\}$ and any p' in the interval $[\frac{1}{2},1)$, the probability (3.1) has the property

$$P_{(t),n}(C\{\tfrac{1}{2},p\},v) \geq P_{(t),n}(C\{\tfrac{1}{2},p'\},v) \tag{3.4}$$

whenever $p > p'$. Also, for configurations $C\{p_{rs}\}$ in which the preference probabilities satisfy linear models, the inequality relation

$$P_{(t),n}(C\{p_{rs}\},v) \geq P_{(t),n}(C\{\tfrac{1}{2}\},v)$$

holds (see Section 4.4 of [9]). Accordingly v in $X_r \geq X_{max} - v$ is taken as the smallest integer satisfying the relation

$$P_{(t),n}(C\{\tfrac{1}{2}\},v) \geq P^{\circ} \tag{3.5}$$

where P° is a specified number.

In terms of the quantities defined, the expectation (3.2) has the following properties.

$$tP_{(t),n}(C\{\tfrac{1}{2}\},v) = E_{t,n}(C\{\tfrac{1}{2}\},v) \tag{3.6}$$

$$\geq E_{t,n}(C\{\tfrac{1}{2},p'\},v), \quad p' > \tfrac{1}{2}$$

$$\geq E_{t,n}(C\{\tfrac{1}{2},p\},v), \quad p > p'. \tag{3.7}$$

To simplify notation, the probabilities of (3.5) and (3.4) will be written as $P_{(t),n}(\frac{1}{2},v)$ and $P_{(t),n}(p,v)$, and the expectations in (3.6) and (3.7) will be written as $E_{t,n}(\frac{1}{2},v)$ and $E_{t,n}(p,v)$. When no ambiguity arises, the expected size of S will be denoted by $E_{t,n}$.

For an observed vector of preference scores, $x = (x_1, \cdots, x_t)$, S will consist of treatments whose scores are highest and not less than $x_{max} - v$, where $x_{max} = \max_{1 \leq r \leq t} x_r$ and where v, computed according to (3.5), appears in Table 2 above each block of entries of the $t^{-1}E_{t,n}$ for numerous combinations of P°, t and n. The notation (v) or v signifies the respective use of exact or asymptotic theory in the computation of the corresponding blocks of entries (see Section 5, subsection on Structure of Tables).

Although the observed outcome of the random size of S is an integer in $\{1,2,\cdots,t\}$, because of continuity of $E_{t,n}(p,v)$ in p, the expected size of S takes on values in the interval [1,t]. Therefore restrictions related to problems of design or analysis, such as

$$E_{t,n}(p,v) \leq 1 + \varepsilon \tag{3.8}$$

or

$$E_{t,n}(p,v) \cong 1 + \varepsilon \tag{3.9}$$

where ε is a real number in the interval (0,t-1], are plausible, and from the viewpoint of averages have a well founded meaning. For example, the expectation $E_{3,n}(p,v) = 1.01$ may reflect the situation in which in a fixed number of experiments, say 100, ninety-nine result in the size of S equal to 1, and one results in the size of S equal to 2.

When a strictly better treatment is present in T, it can always be included in a subset S of expected size not greater than $1 + \varepsilon$ (ε small), in accordance with $x_r \geq x_{max} - v$, and with probability at least P°, by finding an appropriate value of n. In this connection restriction (3.8) is always desirable and leads to an optimal solution when the smallest number of replications, n, is sought. Examples 3.1 and 3.2 illustrate cases where the sequences of consecutive values of $t^{-1}E_{t,n}$ in Table 2 covering $t^{-1}(1 + \varepsilon)$, are non-increasing for increasing n.

EXAMPLE 3.1. In connection with the speech therapy study of Illustration 2.1 involving three treatments, the design of experiment size can be handled as follows. Suppose the investigator wants to ensure with probability not less than P°=0.975 the inclusion of the superior treatment $T_{(3)}$ in a subset S in such a way that in expectation the size of S should not exceed 1.05 whenever the probability of preference between $T_{(3)}$ and $T_{(2)}$ is at least p = 0.90. Converting the upper bound on the expectation to the proportion 3^{-1}x1.05 = 0.350, one finds by inspection in Table 2, under the heading $P° = 0.975$ and the column for $t = 3$, the pair values

n	$3^{-1}E_{3,n}(0.90,v)$
14	0.353
15	0.344
16	0.344
17	0.339

covering 0.350. Here the design condition

$$3^{-1}E_{3,n}(0.90,v) \leq 0.350$$

is satisfied with $n = 15$ (smallest value of n). In this case the discriminant relation becomes $x_r \geq x_{max} - 10$. The designed paired comparison experiment with three treatments and 15 replications resulted in $\alpha_{21} = 14$, $\alpha_{31} = 9$, $\alpha_{23} = 14$, and therefore $x = (7,28,10)$. Consequently the relation $x_r \geq 28\text{-}10=18$ determined the selected subset $S = \{T_2\}$.

A comment regarding the value of p is in order. Under the assumption of a strictly better treatment $T_{(t)}$ in T, the difference $p_t - p_{t-1}$ is positive, and in terms of the τ_{rs} of (3.3),

$$p > \tfrac{1}{2} + t^{-1} \Sigma_{s=1}^{t-2} (\tau_{t-1,s} - \tau_{t,s}).$$

As the $\tau_{rs} \to 0$, (3.3) becomes $C\{\tfrac{1}{2},p\}$. The design of an experiment with n replications depends on the experimenter's assignment of a value to p, which may be based on some prior information or on an educated assessment of the qualities of the $T_1,\cdots,T_t$. After completion of the designed experiment, the assigned value of p may be compared with the relative frequency with which $T_{(t)}$ is preferred to $T_{(t-1)}$ in n replications. According to Section 2, it can be seen that $n^{-1}\alpha_{t,t-1}$ is such a relative frequency, and therefore an estimator for p. In Example 3.1, where the paired comparison experiment was performed to conform with a linear model, and turned out, at the same time, to be an approximation (3.3) to the configuration $C\{\tfrac{1}{2},p\}$ for a single outlier, the observed value of this estimator is $14/15 \cong 0.93$.

For small or relatively small values of n, because of larger random fluctuations in the components of X, the treatment with the highest score has a lesser chance of being best. However the probability of including the best treatment in the selected subset S increases as the size of S becomes larger. Two numerical illustrations of this relationship follow from Table 2.

P^o:	0.750,	0.900,	0.950,	0.975,	0.990
$E_{20,3}(0.80,v)$:	2.20,	4.40,	5.52,	8.22,	11.3
$E_{20,14}(0.60,v)$:	6.82,	10.7,	12.6,	14.4,	16.5

This property of the selection procedure is particularly useful in problems in which some reduction in the size of T is of primary interest.

EXAMPLE 3.2. In a screening problem of a set of 20 treatments, one is able to assert that a lower bound for the preference probability between the strictly better treatment, $T_{(20)}$, and $T_{(19)}$ can be set at p = 0.65. How many replications of the experiment are needed to reduce T to a subset S of expected size not more than eight, so that S contains the superior treatment with probability not less than P^o= 0.990? To find n, first convert $1 + \varepsilon = 8$ to the proportion $20^{-1}x8 = 0.400$. A scan of the column in Table 2 reveals that n cannot be less than or equal to 8 for the design condition, $20^{-1}E_{20,n}(0.65,v) \leq 0.400$, to be satisfied. For increasing n in the region of interest, the tabulated upper and lower bounds $20^{-1}E_{20,n}(0.60,v)$ and $20^{-1}E_{20,n}(0.70,v)$ for the unknown $20^{-1}E_{20,n}(0.65,v)$ form decreasing sequences of entries; in Example 5.3 the value of $E_{20,14}(0.65,38)$ is found to be 0.389. Thus the sought value of n must not be greater than 14. To check whether it is equal to 13, extend the tabulated entries (see Section 5) for t = 20, n = 13 to obtain the pair values

p	$20^{-1}E_{20,13}(p,36)$
0.50	0.990
0.60	0.831
0.70	0.123
0.80	0.050
0.90	0.050
1.00	0.050.

Newton's Forward Interpolation Formula with initial value 0.990 (obtained from Table 1) and the differences -0.159, -0.549, 1.184, -1.746, 2.235 result in 0.405 for $E_{20,13}(0.65,36)$. Hence the sought value of n is 14.

There may be situations where, for economic or other reasons, (3.9) is acceptable, in which case the calculated number of replications, n, will be somewhat smaller than that satisfying (3.8), and the lower bound p for the preference probability between $T_{(t)}$

and $T_{(t-1)}$ will be automatically increased by a corresponding number δ.

EXAMPLE 3.3. Consider the design restriction

$$E_{3,n}(0.80,v) \leq 1.44 \tag{3.10}$$

for the solution of n, when $P^o = 0.975$. Here $3^{-1} x 1.44 = 0.480$, and for the P^o one obtains by inspection from Table 2 the sequence $(n, 3^{-1}E_{3,n}) = (14,0.523), (15,0.481), (16,0.490), (17,0.455)$ covering 0.480. This sequence suggests that perhaps n could be taken equal to 15 rather than 17. However this choice of n yields

$$0.480 = 3^{-1}E_{3,15}(0.80+\delta,10) < 3^{-1}E_{3,15}(0.80,10),$$

where δ can be shown to have the approximate value 10^{-3}. In the case when the deviation $\delta = 10^{-3}$ from $p = 0.80$ is acceptable, (3.9) would replace the specification (3.10), with $p = 0.80$ and $\varepsilon = 0.44$, resulting in a solution $n = 15$.

The tables are also applicable in selection analysis when data are obtained from undesigned balanced paired comparison experiments.

EXAMPLE 3.4. Consider the data of Example 3.1 and the single requirement that $T_{(3)}$ is to be included in a selected subset S with probability not less than $P^o = 0.990$. Entering Table 2 at the given P^o, one finds at $t = 3$ and $n = 15$ the integer $v = 12$. In this case the discriminant relation results in $x_r \geq 28 - 12 = 16$ and $S = \{T_2\}$. The analysis can be carried further. Corresponding to the estimate 0.93 for the preference probability between $T_{(3)}$ and $T_{(2)}$, one finds the estimate 0.337 for $3^{-1}E_{3,15}(15^{-1}\alpha_{2,3}, 12)$ by applying Newton's Backward Interpolation Formula to the tabulation at $t = 3$, $n = 15$ extended by 0.993 (from Table 1) at $p = \frac{1}{2}$ and 3^{-1} at $p = 1.00$. This yields 1.01 as an estimate for the expected size of S; given sufficient accuracy, this number is greater than the corresponding estimate which would result from the same data under the requirement that $P^o = 0.975$.

The final example explores design conditions with a view toward an optimal solution for experiment size with respect to cost.

EXAMPLE 3.5. In an exploratory study between design restrictions and sizes of paired comparison experiments involving four treatments, the following questions can be considered.

(i) How many replications are needed to include $T_{(4)}$ in S with probability not less than $P^o = 0.990$ so that $E_{4,n}(p,\nu) \leq 1.08$, when p is equal to 0.70?

(ii) If the n in (i) is too large, how can one reduce it by retaining $E_{4,n}(0.70,\nu) \leq 1.08$ with some decrease in the level of P^o?

To answer question (i), convert $1 + \varepsilon = 1.08$ to $4^{-1}\text{x}1.08 = 0.270$. In this case n = 65, a result which is established in Example 5.6 with the corresponding value $4^{-1}E_{4,65}(0.70,31) = 0.269$. To deal with question (ii), refer to Section 5, Interpolation at n, with fixed ν and observe that for exploratory purposes a convenient starting value for ν is 21. This leads to the tabled set of triples $(n,4^{-1}E_{4,n}(0.70,21),P^o)$ = (30,0.525,0.990), (40,0.329, 0.975), (50,0.268,0.950), (70,0.251,0.900). Evidently both graphs defined by the points $(n,4^{-1}E_{4,n}(0.70,21))$, and (n,P^o), are monotonically decreasing. Hence the unknown n is not larger than 50, and P^o is not less than 0.950. The application of the four-point Lagrange interpolation formula results in 0.270 for $4^{-1}E_{4,n}(0.70, 21)$ when n = 49. At this value of n, P^o is taken as 0.952 (by linear interpolation, one finds $P_{(4),49}(\frac{1}{2},21)$ approximately equal to 0.9525). This illustrates a method of computation of n and P^o. A comprehensive exploration of possibilities should in addition include computations with other values of ν.

The structure of probabilities in (3.2) and their continuity in the p_{rs} are sufficient to conclude that for relatively small deviations from the $\tau_{rs} = 0$ in (3.3), the expectation $E_{t,n}(C\{p_{rs}\},\nu)$ deviates negligibly from $E_{t,n}(p,\nu)$. When departures from the $\tau_{rs} = 0$ in (3.3) are not small, the solution for n satisfying (3.9) may be too conservative for the design restriction to absorb a possible, not too negligible, expected increase in $E_{t,n}$. The properties of the selection functions, which are monotonicity as given by (3.4) and (3.7) and continuity in the p_{rs} of the functions of (3.1) and (3.2), suggest the following method as a guide to the solution of n. For a given t and design specifications P^o,p,ε, choose a suitable value p' (p'< p), and find the corresponding integer n in accordance with (3.8). This n should satisfy

$$1 + \varepsilon \geq E_{t,n}(p',v)$$
$$\geq E_{t,n}(p,v), \quad p' < p,$$

and

$$1 + \varepsilon \cong E_{t,n}(C\{p_{rs}\},v), \quad p_{t,t-1} = p,$$

for some $C\{p_{rs}\}$ expected to produce an anticipated increment of approximate magnitude $E_{t,n}(p',v)-E_{t,n}(p,v)$ over the value $E_{t,n}(p,v)$. To illustrate this, suppose that $t = 4$, $p = 0.80$, and that $T_{(4)}$ is to be included in S with probability not less than $P^o = 0.900$ so that (3.9) is satisfied with $1 + \varepsilon = 2$. If $p' = 0.70$ is chosen, then the corresponding number of replications n is 15 (instead of $n = 7$ for $p = 0.80$), and $2 > E_{4,15}(0.70,9) > E_{4,15}(0.80,9)$. This choice of n allows for any configuration $C\{p_{rs}\}$ which is expected to produce an anticipated increase of about 0.9 over the value $E_{4,15}(0.80,9) = 1.10$. In this case the approximation

$$2 \cong E_{4,15}(C\{p_{rs}\},9)$$

should hold.

In the special case when the p_{rs} of $C\{p_{rs}\}$ satisfy a linear model, a heuristic argument can be proposed (by examining the probabilities in (3.2) with respect to departures from $\tau_{rs} = 0$ of (3.3)) to give support to the inequality

$$E_{t,n}(p,v) \geq E_{t,n}(C\{p_{rs}\},v), \qquad p_{t,t-1} = p. \tag{3.11}$$

The following are illustrations of (3.11) for twelve different configurations, $C\{p_{rs}\}$, when $t = 3$, $n = 1$, and H is the uniform distribution function on the interval (-1,1). These illustrations are grouped in sets of six and ordered within each set according to increasing departures of $C\{p_{rs}\}$ from the corresponding configuration $C\{\frac{1}{2},p_{3,2}\}$.

	$E_{3,1}(C\{p_{rs}\},v)$			
	v=0		v=1	
$\tau_{2,1} \backslash p_{3,2}$	0.65	0.85	0.65	0.85
0	1.4550	1.2550	2.2275	2.1275
0.01	1.4518	1.2478	2.2259	2.1239
0.02	1.4482	1.2402	2.2241	2.1201
0.05	1.4350	1.2150	2.2175	2.1075
0.10	1.4050	1.1650	2.2025	2.0825
0.15	1.3650	1.1050	2.1825	2.0525

It should be noted that in the case of $t = 3$, assignment of values to $p_{3,2}$ and $\tau_{2,1}$ of (3.1) specifies the three probabilities of $C\{p_{rs}\}$. For instance when $p_{3,2} = 0.65$ and $\tau_{2,1} = 0.10$, then $p_{2,1} = 0.60$ and $p_{3,1} = H(V_3 - V_1) = H(H^{-1}(p_{3,2}) + H^{-1}(p_{2,1})) = H\ (0.50) = 0.75$.

4. COMPUTATION OF TABLE ENTRIES

EXACT FORMULAE

When each preference probability is equal to ½, (2.1) becomes

$$f(x;C\{\tfrac{1}{2}\}) = 2^{-\frac{1}{2}nt(t-1)}g(x;n),$$

where $g(x;n) = \Sigma_{R(n)}\Pi_{r>s}B(n,\alpha_{rs})$. The function g gives the number of ways in which the vector x of observed preferences can be realized; $2^{-\frac{1}{2}nt(t-1)}$ is the total frequency in the sample space of a paired comparison experiment involving t treatments and n replications.

Let m_j stand for the number of components in x of constant magnitude x_j, and let [x], with ordered components, be a representative element of the set of vectors obtained under permutations of the components of a fixed vector x. The element [x] will be referred to as a partition vector of the sample space containing all possible outcomes x of X. Because of symmetry of g in the $x_1,\cdots,x_t$, the total frequency for all vectors of preferences whose components are permutations of a fixed vector x is

$$t!g(x;n) \ / \ \Pi_{j=1}^{k}m_j! = G([x];n).$$

The following is an illustration of the sample space, expressed in terms of the partition vectors [x], and the corresponding frequencies.

[x]	G([x];2)
[0,2,4]	6
[1,1,4]	6
[0,3,3]	6
[1,2,3]	36
[2,2,2]	10

For each (t,n): t = 3, n = 1(1)20; t = 4, n = 1(1)7; t = 5, n = 1(1)3, reference [6] lists the sequences {[x]} and the corresponding sets of values of G. This reference can be obtained on request.

The determination of the integers, v, satisfying (3.5) for some preassigned values of P^o, requires the computation of the $P_{(t),n}(\frac{1}{2},v)$. The procedure for obtaining $P_{(t),n}(\frac{1}{2},v)$ can be summarized thus.

(a) For each [x]

(i) find the number $s(1 \leq s \leq t)$ of components in [x] which are greater than or equal to $x_{max} - v$, and denote this number by $s([x],v)$;

(ii) form

$$Q([x];n,v) = t^{-1}s([x];v)G([x];n).$$

(b) Compute

$$P_{(t),n}(\tfrac{1}{2},v) = 2^{-\frac{1}{2}nt(t-1)}\Sigma_{[x]}Q([x];n,v).$$

Below is an illustration of the procedure when t = 3, n = 2, v = 1,2,3. For example corresponding to P^o = 0.750 and P^o = 0.975, the respective values of v chosen in accordance with (3.5) are 2 and 4. The latter follows from the fact that the given P^o = 0.975 is greater than $P_{(3),2}(\frac{1}{2},3)$, and for t = 3, n = 2, the integer 4 is the largest permissible value for v (see beginning of Section 3).

Illustration of the Procedure for Determining ν

	$s([x];\nu)$			$Q([x];2,\nu)$		
$[x]$ \ ν	1	2	3	1	2	3
[0,2,4]	1	2	2	2	4	4
[1,1,4]	1	1	3	2	2	6
[0,3,3]	2	2	3	4	4	6
[1,2,3]	2	3	3	24	36	36
[2,2,2]	3	3	3	10	10	10
			$P_{(3),2}(\frac{1}{2},\nu)$:	0.65625	0.87500	0.96875

The following is a procedure for obtaining exact values of $E_{t,n}(p,\nu)$.

(a) For each $[x]$

(i) find the number $s(1 \leq s \leq t)$ of components in $[x]$ which are greater than or equal to $x_{max} - \nu$, and denote the partition element $[x]$ by $[x]_s$;

(ii) using the integers $m_1,\cdots,m_k$ (defined earlier), form

$$m([x]_s;p,\nu) = t^{-1}G([x]_s;n)\Sigma_{j=1}^{k} m_j p^{x_j}(1-p)^{n(t-1)-x_j}.$$

(b) For $s=1,\cdots,t$, compute

$$M(s;p,\nu) = \Sigma_{[x]_s} m([x]_s;p,\nu)$$

and

$$E_{t,n}(p,\nu) = 2^{-\frac{1}{2}n(t-1)(t-2)}\Sigma_{s=1}^{t} sM(s;p,\nu). \tag{4.1}$$

For example, in the special case when $t=3$, $n=2$, and $\nu=2$, the procedure yields

$$E_{3,2}(p,2) = 1+(0.5)p^4+7p^3(1-p)+(11.5)p^2(1-p)^2+6p(1-p)^3+(1-p)^4.$$

When $t=2$, (4.1) can be written as

$$E_{2,n}(p,\nu) = 1 + \Sigma_{|d|\leq\nu} b(\tfrac{1}{2}(n+d);n,p),$$

where $d = x_2-x_1$, and where b stands for the binomial probability function.

ASYMPTOTIC FORMULAE

The extent of computation needed to generate frequencies $G([x];n)$ for experiment sizes (t,n) greater than those given in [6] made it necessary to consider approximations to the probability of correct inclusion of $T_{(t)}$ in S, and to the expected size of S. Both approximations depend on the multivariate central limit theorem.

In [9] it is shown that, asymptotically, $P_{(t),n}(\frac{1}{2},\nu)$ can be replaced by the integral

$$\int_{-\infty}^{\infty} [\Phi(u_t+w)]^{t-1}\, d\Phi(u_t) \tag{4.2}$$

where $w = 2(\nu+\frac{1}{2})(nt)^{-\frac{1}{2}}$, and where Φ is the distribution function of the standard normal random variable. For certain values of P^o assigned to the integral (4.2), the integer ν can be calculated from a tabulation, as given for example by Bechhofer [1]. Section 5 contains values of w for all cases of P^o and t which appear in Table 2. Examples involving evaluation of ν can also be found in Section 5.

The approximation to $E_{t,n}(p,\nu)$ is developed in [7], where it is shown that this expected size of S can be replaced asymptotically by

$$\int_{-\infty}^{\infty} \{(t-1)[\Phi(\zeta_1)]^{t-2}\Phi(\zeta_2) + [\Phi(\zeta)]^{t-1}\}d\Phi(z). \tag{4.3}$$

The symbols in (4.3) are

$$\zeta_1 = z + 2^{\frac{1}{2}}(\nu+\tfrac{1}{2})/\sigma,\quad \zeta_2 = (2^{-\frac{1}{2}}\sigma z+\nu+\tfrac{1}{2}-\delta)c_t^{-\frac{1}{2}},\quad \zeta = (c_t^{\frac{1}{2}}z+\nu+\tfrac{1}{2}+\delta)\ (\sigma_t^2-c_t)^{-\frac{1}{2}},$$

where $\delta = nt(p-\frac{1}{2})$, and

$$\sigma^2 = n[2p(1-p)+\tfrac{1}{2}(t-1)],\quad \sigma_t^2 = n[(t+2)p(1-p)+\tfrac{1}{4}(t-2)],$$

$$c_t = n[(t+1)p(1-p)-\tfrac{1}{4}].$$

The dependence on $c_t^{\frac{1}{2}}$ of the integral in (4.3) restricts its usage to values of p less than $\frac{1}{2} + \frac{1}{2}[t/(t+1)]^{\frac{1}{2}}$. However this is not a stringent restriction, for this integral already admits values of p close to 0.9330, when t=3(the smallest size of T in the tables).

Asymptotic approximations to $E_{t,n}$ expressed as proportions of t in Table 2 were computed by means of Simpson's One-Third Rule applied to (4.3) on the interval $I = [-6-\Delta z, 6+\Delta z]$. Let G(z) and H(z) stand respectively for $[\Phi(\zeta_1)]^{t-2}\Phi(\zeta_2)\phi(z)$ and $[\Phi(\zeta)]^{t-1}\phi(z)$, where ϕ denotes the derivative of Φ. If the interval I is partitioned into subintervals of length $12/(s-2) = \Delta z (s > 2)$, then on I, (4.3) is given by

$$\Sigma_{j=0}^{s+1} \int_{a(j)}^{b(j)} \{(t-1)G(t) + H(t)\}dz \tag{4.4}$$

where $a(j) = -6 + (j-1)\Delta z$, $b(j) = -6 + j\Delta z$. Since $\Phi(-6-\Delta z) = 1 - \Phi(6+\Delta z) < 10^{-9}$ and $\phi(-6-\Delta z) = \phi(6+\Delta z) < 6.1\text{x}10^{-9}$ (see, e.g., [5, p.110]), the value of (4.4) is equal to that of (4.3) if, for example, seven place accuracy is retained for both. An application of Simpson's One-Third Rule to each integral of the sum in (4.4) results in the approximation

$$A_{t,n}(p,v) = A_1(t,n,p,v) + A_2(t,n,p,v) \tag{4.5}$$

to (4.3), where $A_1(t,n,p,v)$ and $A_2(t,n,p,v)$ denote respectively

$$\begin{aligned}(t-1)(\Delta z/6)[G(a(0)) &+ 2\Sigma_{j=0}^{s} G(b(j)) \\ &+ 4\Sigma_{j=0}^{s+1} G(-6+(2j-1)\Delta z/2) + G(-a(0))]\end{aligned} \tag{4.6}$$

and

$$\begin{aligned}(\Delta z/6)[H(a(0)) &+ 2\Sigma_{j=0}^{s} H(b(j)) \\ &+ 4\Sigma_{j=0}^{s+1} H(-6+(2j-1)\Delta z/2 + H(-a(0))].\end{aligned} \tag{4.7}$$

The choice $\Delta z = 0.40$ was made after an examination of numerous computations of $A_{t,n}(p,v)$ with trial increments $\Delta z = 1.00, 0.50, 0.40, 0.25, 0.20, 0.10$, at parameter values t,n,p,v for which (4.3) is known exactly to any number of decimal places, or accurately to 5 decimal places. An example of the former case is the set $p = \frac{1}{2}$, $v = -\frac{1}{2}$, and any t, for which the integral (4.3) is equal to 1 independently of n; in this case the values of $A_{t,n}(\frac{1}{2}, -\frac{1}{2})$ with $\Delta z \leq 0.40$ were found accurate to as many decimal places, less one, as the accuracy with which $\Phi(\cdot)$ was evaluated. An example of the latter case is the set n = 60, t = 10, $p = \frac{1}{2}$ and $v = 41.36423$ for which the integral (4.3) is equal to 0.95000; in this case $A_{t,n}(p,v)$, with $\Delta z = 0.40$, resulted in 0.9500018. Further discussion on the accuracy of the asymptotic approximation can be found in [7], with an overall conclusion that the discrepancies between $t^{-1}A_{t,n}(p,v)$, with $\Delta z = 0.40$, and $t^{-1}E_{t,n}(p,v)$, should be almost entirely due to the multivariate central limit theorem.

The computation of the asymptotic entries in the tables was carried out by a single program for the evaluation of $t^{-1}A_{t,n}(p,v)$, utilizing a subroutine for $\Phi(\cdot)$, $\phi(\cdot)$. The input consisted of the values $w(t,P^o)$ of Section 5, and parameters specifying minima and maxima for t,n,p with appropriate increments to generate entries in the tables as shown.

Individual values of $A_{t,n}(p,v)$ can be obtained in the following manner. For a given t,n,p, and P^o,

(a) solve $w = w(t,P^o)$ for v by rounding upward to the nearest integer the solution $\frac{1}{2}(w(nt)^{\frac{1}{2}} - 1)$;

(b) evaluate $A_1(t,n,p,v)$ and $A_2(t,n,p,v)$ of (4.5) by using a subroutine for $\Phi(\cdot)$ and $\phi(\cdot)$ (see e.g., [4, p.78]) to obtain the terms $G(\cdot)$ and $H(\cdot)$ as given in (4.6) and (4.7); the computation procedure of $A_{t,n}(p,v)$ should involve checks on the magnitudes of ζ, ζ_1, ζ_2 against the values $-6-\Delta z$ and $6+\Delta z$, and utilize the decision $\Phi(\cdot) = 0$ if the argument of Φ is not greater than $-6-\Delta z$, and $\Phi(\cdot) = 1$ if the argument of Φ is not less than $6+\Delta z$.

When p = 1, the probabilities of (3.2) defining $E_{t,n}(1,v)$ lend themselves to simplification so that asymptotically

$$E_{t,n}(1,v) = 1 + (t-1)\Phi(u),$$

where $u = (2v+1-nt)/n^{\frac{1}{2}}(t-2)^{\frac{1}{2}}$.

5. TABLE FORMAT AND INTERPOLATION

STRUCTURE OF TABLES

The main body of Table 2 consists of values of $E_{t,n}$ expressed as proportions of t. Thus each entry provides for a three digit value of $E_{t,n}$ uniformly throughout the table.

The integer v of the discriminant relation $x_r \geq x_{max} - v$ appears at the top of each block of entries, $t^{-1}E_{t,n}(p,v)$ $(p=0.60(0.10) \leq 1.00)$. For example, when $P^o = 0.975$ and $t = 4$, n=4, the corresponding ν is 6, and, together with the values of $4^{-1}E_{4,4}(p,6)$, appears in Table 2 as

(6)
0.963
0.916
0.810
0.622
0.358.

In this case the integer ν(namely 6) is placed in parenthesis to signify that these entries are obtained from exact formulae. Entries headed by ν(instead of (ν)) are computed in accordance with the approximating functions of Section 4.

For each triple P^o,t,n, the corresponding block of entries can be supplemented with the value of $P_{(t),n}(\frac{1}{2},\nu)$ from Table 1 (see (3.6)). Such a value may be needed when interpolating in Table 2 for $E_{t,n}(p,\nu)$ at some p. Additionally, blocks of entries terminating at some $p < 1.00$ can be supplemented with values of t^{-1}; these are the cases where the expected size of S is equal to 1.00, accurately to two decimal places.

In the illustrations below, specific cases from Table 2 are reproduced with the supplemented values underlined. The extended tabulations in the illustrations are used in the examples on interpolation.

ILLUSTRATION 5.1		ILLUSTRATION 5.2		ILLUSTRATION 5.3	
P^o=0.975, t=3, n=15		P^o=0.750, t=3, n=19		P^o=0.990, t=20, n=14	
p	(10)	p	(5)	p	38
0.50	<u>0.976</u>	0.50	<u>0.756</u>	0.50	<u>0.992</u>
0.60	0.929	0.60	0.626	0.60	0.823
0.70	0.749	0.70	0.411	0.70	0.108
0.80	0.481	0.80	0.338	0.80	<u>0.050</u>
0.90	0.344	0.90	<u>0.333</u>	0.90	<u>0.050</u>
1.00	<u>0.333</u>	1.00	<u>0.333</u>	1.00	<u>0.050</u>

Interpolation formulae applied in the sequel are of the standard type; these can be found in most books on mathematical computations or methods of numerical analysis.

INTERPOLATION AT p

For each triple P^o,t,n, the number pairs $(p,t^{-1}E_{t,n}(p,\nu))$, p=0.50(0.10)1.00, are points of a graph of a continuous non-increasing function of the interval $[\frac{1}{2},1]$. The problem of finding the value of $t^{-1}E_{t,n}$ at $p \neq 0.50(0.10)1.00$ can be resolved by the application of a properly chosen interpolation formula.

EXAMPLE 5.1. Consider the problem of finding $3^{-1}E_{3,19}(0.85,5)$ in Table 2 when $P^{\circ} = 0.750$. The relative flatness of the graph of $3^{-1}E_{3,19}(p,5)$ in the interval [0.80,0.90] (see Illustration 5.2) suggests that linear interpolation, which gives the value of 0.335, could be accurate enough. The exact value of $3^{-1}E_{3,19}(0.85,5)$, accurate to three decimal places, is 0.334.

EXAMPLE 5.2. In Table 2, $P^{\circ} = 0.975$, the Lagrange interpolation formula applied to the first five points $(p, 3^{-1}E_{3,15}(p,10))$ of Illustration 5.1 results in 0.965 for $3^{-1}E_{3,15}(0.55,10)$; this is the same as the exact value, accurate to three decimal places. It should be noted that the points of Illustration 5.1, connected by a smooth curve on graph paper which allows for a read-off accurate to two decimal places, produced 0.97 for the value of $3^{-1}E_{3,15}(0.55,10)$.

EXAMPLE 5.3 Consider the interpolation for $t^{-1}E_{t,n}(0.65,v)$ in Table 2, when $P^{\circ} = 0.990$, $t=20$, $n=14$. A sketch of the graph of $20^{-1}E_{20,14}(p,38)$ on the interval $[\frac{1}{2},1]$ through the points of Illustration 5.3 shows that linear interpolation would give much too high a value, while graphical interpolation would not be advisable because of the degree of arbitrariness in drawing the function in the interval (0.60,0.70). In addition a sketch without the initial point (0.50,0.992) would differ appreciably on (0.60,0.70) from the one which utilizes all points. The application of the Lagrange interpolation formula to the six points of Illustration 5.3 results in 0.389 for $20^{-1}E_{20,14}(0.65,38)$.

If the interpolation for $t^{-1}E_{t,n}$ is to be done at a p in the middle part of the interval $[\frac{1}{2},1]$, Bessel's formula can be used.

EXAMPLE 5.4. Utilizing the entire difference table computed from the extended tabulation of Illustration 5.2, Bessel's interpolation formula with initial value 0.411 results in 0.358 for $3^{-1}E_{3,19}(0.75,5)$ when $P^{\circ}=0.750$. This is also the exact value, accurate to three decimal places.

INTERPOLATION AT n

Interpolation for $t^{-1}E_{t,n}(p,v)$ at an $n = n_o$ can be carried out when P°,t,p (interpolation within a table), or when v,t,p (interpolation between tables), are fixed. For a given P°,t,p, let Γ

denote the graph through a sequence of points, $\{(n, t^{-1}E_{t,n}(p,\nu))\}$, which can be obtained from the tables. Two related matters need to be discussed first.

The integer valued ν introduces irregularities in an otherwise continuous graph, Γ_c, which would result from $t^{-1}E_{t,n}(p,\nu)$ as a function of n, if the ν were real numbers satisfying the equality in the relation (3.5) and n were a continuous variable. Such an irregular graph with respect to Γ_c is Γ, as defined above. In this sense Γ_c is the greatest lower bound to all graphs, including Γ, which would result from $t^{-1}E_{t,n}(p,\nu)$ as functions of n, if the ν were any real numbers satisfying (3.5). The irregularities in Γ smooth out with increasing n.

Two additional smoothing properties in the Γ hold which are independent of the smoothing with increasing n. Considered as a sequence of graphs, $\{\Gamma_t\}$, indexed by t, the Γ_t become smooth with increasing t. Also, considered as a sequence, $\{\Gamma_p\}$, indexed by values of p, the Γ_p become smooth with increasing p; for $p \geq 0.70$ most of the graphs can be taken as montonically decreasing and convex on the support [20,100], where interpolation may be needed at a value of n $(20 < n < 100, n \neq 25(5)50(10)90)$.

In certain cases it is suggested that a sketch of the graph Γ in the region of interpolation be constructed. This preliminary task should help in deciding whether an interpolation should be carried out, or whether the required value of $E_{t,n}(p,\nu)$ at a given $n = n_0$ should be computed by the methods of Section 4.

The other matter relates to the computation of the integer ν in $x_r \geq x_{max} - \nu$, once the determination of an entry $t^{-1}E_{t,n}(p,\nu)$ at a new n is made, or in cases where the value of ν is omitted from Table 2. The accompanying tabulation lists values $w = w(t,P^o)$ of the integral (4.2) as pertaining to Table 2. For a given P^o, t, and n, the value ν is calculated from $w = 2(\nu+\frac{1}{2})(nt)^{-\frac{1}{2}}$ by rounding upward to the nearest integer the solution $\frac{1}{2}(w(nt)^{\frac{1}{2}} - 1)$.

VALUES OF w

t \ P^o	0.750	0.900	0.950	0.975	0.990
3	1.4338	2.2302	2.7101	3.1284	3.6173
4	1.6822	2.4516	2.9162	3.3220	3.7970
5	1.8463	2.5997	3.0552	3.4532	3.9196
6	1.9674	2.7100	3.1591	3.5517	4.0121
7	2.0626	2.7972	3.2416	3.6303	4.0860
8	2.1407	2.8691	3.3099	3.6953	4.1475
9	2.2067	2.9301	3.3679	3.7507	4.1999
10	2.2637	2.9829	3.4182	3.7989	4.2456
11	2.3137	3.0294	3.4625	3.8414	4.2859
12	2.3582	3.0709	3.5022	3.8794	4.3221
13	2.3982	3.1083	3.5380	3.9138	4.3548
14	2.4345	3.1423	3.5705	3.9451	4.3846
15	2.4678	3.1734	3.6004	3.9738	4.4121
16	2.4984	3.2021	3.6280	4.0004	4.4374
17	2.5267	3.2288	3.6536	4.0250	4.4610
18	2.5531	3.2536	3.6774	4.0480	4.4830
19	2.5777	3.2768	3.6998	4.0696	4.5036
20	2.6009	3.2986	3.7207	4.0899	4.5230

EXAMPLE 5.5. In Table 2, for $P^o = 0.750$, the value of $18^{-1}E_{18,75}(0.60,\nu)$ is (obviously) $18^{-1} = 0.056$. The corresponding integer ν needed for the selection of S is 47, which is obtained from $\frac{1}{2}(2.553(75\text{x}18)^{\frac{1}{2}}-1) = 46.40$.

EXAMPLE 5.6. Suppose one seeks the value of $4^{-1}E_{4,65}(0.70,\nu)$ in Table 2 when $P^o = 0.990$. Newton's Forward Interpolation Formula, utilizing the leading differences of the difference table constructed from the points $(n,4^{-1}E_{4,n}(0.70,\nu)) = (60,0.277)$, $(70,0.263)$, $(80,0.255)$, $(90,0.252)$, $(100,0.251)$, results in 0.269 for $4^{-1}E_{4,65}(0.70,\nu)$. For $n = 65$, the integer ν calculated from the value of w, at $P^o = 0.990$ and $t = 4$, is 31.

The method of interpolation for $t^{-1}E_{t,n}(p,\nu)$ at an n between tables must necessarily involve P^o as a variable, and the remaining parameter ν in (3.5) fixed, at some value $\nu = \nu_o$. The general procedure for finding $t^{-1}E_{t,n}(p,\nu_o)$ at an $n = n_o$ consists of constructing from the tables a sequence of triples $\{(n, t^{-1}E_{t,n}(p,\nu_o), P^o)\}$, where t and p are fixed numbers. Then $t^{-1}E_{t,n_o}(p,\nu_o)$ is obtained from the sequence $\{(n, t^{-1}E_{t,n}(p,\nu_o)\}$ by means of a properly chosen interpolation method. The corresponding value of P^o at $n = n_o$ is found by an interpolation using the sequence $\{(n,P^o)\}$. Example 3.5 utilizes such a method in an exploratory study between design restrictions and experiment sizes.

ACKNOWLEDGMENT

The author expresses his appreciation to Mr. Paul Shoemaker of the Rust Computing Center for the adaptation of the computation programs to the present IBM unit, for computation of tables, and for programming and execution of the tabulations. The finalization of this research would not have been possible without the support and cooperation of other personnel of the Rust Computing Center.

REFERENCES

[1] Bechhofer, Robert E. (1954). A single-sample mutliple decision procedure for ranking means of normal populations with known variances. Ann. Math. Statist., Vol. 25, pp. 16-39.

[2] David, H. A. (1963). The Method of Paired Comparisons. Charles Griffin & Company Limited, London.

[3] Davidson, Roger R. and Farquhar, Peter H. (1976). A bibliography on the method of paired comparisons. Biometrics, Vol. 32, No. 2, pp. 241-252.

[4] I.B.M. Application Program (1968). System/360 Scientific Subroutine Package. H20-0205-3 Fourth Edition.

[5] Pearson, E. S. and Hartley, H. O. (1956). Biometrika Tables For Statisticians, Vol. 1. Cambridge University Press.

[6] Trawinski, B. J. (1961). Research On Order Statistics And The Design of Experiments. Technical Report No. 52, Virginia Polytechnic Institute and State University, Blackburg, Va.

[7] Trawinski, B. J. (1969). Asymptotic approximation to the expected size of a selected subset. Biometrika, Vol. 56, No. 1, pp. 207-213.

[8] Trawinski, B. J. (1980). Consistency of decision functions in experiments involving subjective judgment. Inst. Math. Statist. Bul., Vol. 9, No. 3.

[9] Trawinski, B. J. and David, H. A. (1963). Selection of the best treatment in a paired-comparison experiment. Ann. Math. Statist., Vol. 34, No. 1, pp. 75-91.

TABLE 1

The values of $P_{(t),n}(\frac{1}{2},\nu)$ of (3.5) are given in the table for each combination of t and n (t = 3(1)20, n = 1(1)20(5)50(10)100), at levels

$$P^o = 0.750,\ 0.900,\ 0.950,\ 0.975,\ 0.990.$$

The corresponding integers ν can be found in Table 2; consistent with the notation (ν) or ν of Table 2, the respective values $P_{(t),n}(\frac{1}{2},\nu)$ are exact or asymptotic.

TABLE 2

The proportions $t^{-1}E_{t,n}(p,\nu)$ are tabulated for the following cases:

$$t = 3(1)20,\ n = 1(1)20(5)50(10)100,\quad p = 0.60(0.10)1.00,$$

at each level

$$P^o = 0.750,\ 0.900,\ 0.950,\ 0.975,\ 0.990.$$

Entries equal to t^{-1} (accurate to three decimal places) are omitted.

TABLE 1

$P_{(t),n}(\frac{1}{2},\nu)$

n \ t	3	4	5	6	7	8	9	10	11	12	13	14	15	16	17	18	19	20
									$P^0=0.750$									
1	.750	.906	.844	.769	.876	.830	.783	.878	.845	.810	.776	.866	.840	.814	.787	.761	.848	.827
2	.875	.769	.834	.764	.831	.778	.841	.800	.758	.823	.789	.756	.818	.790	.762	.821	.797	.774
3	.805	.825	.856	.788	.830	.777	.822	.779	.823	.787	.751	.799	.768	.812	.785	.759	.803	.780
4	.753	.764	.791	.817	.754	.790	.822	.778	.812	.775	.810	.778	.812	.784	.755	.792	.766	.801
5	.831	.825	.832	.761	.784	.807	.758	.786	.812	.775	.803	.770	.799	.770	.798	.772	.800	.777
6	.799	.786	.790	.799	.812	.756	.778	.798	.757	.781	.803	.770	.794	.764	.789	.762	.787	.763
7	.771	.754	.753	.760	.771	.784	.797	.752	.771	.789	.754	.774	.794	.764	.785	.758	.780	.755
8	.832	.811	.801	.799	.802	.808	.759	.772	.786	.800	.764	.781	.797	.768	.785	.758	.777	.752
9	.811	.786	.773	.769	.770	.775	.783	.791	.801	.763	.776	.789	.758	.773	.787	.761	.777	.752
10	.793	.764	.815	.805	.802	.802	.752	.760	.769	.778	.788	.754	.767	.779	.751	.765	.779	.754
11	.776	.814	.793	.782	.776	.775	.777	.781	.787	.793	.757	.767	.777	.787	.759	.771	.782	.758
12	.761	.796	.773	.759	.753	.751	.752	.755	.760	.766	.772	.780	.787	.758	.767	.777	.751	.763
13	.823	.779	.754	.795	.785	.780	.777	.777	.779	.782	.787	.753	.761	.768	.776	.784	.759	.768
14	.810	.764	.795	.777	.765	.759	.755	.755	.756	.759	.763	.768	.773	.779	.750	.758	.766	.774
15	.797	.809	.779	.759	.795	.786	.780	.777	.776	.777	.779	.781	.750	.755	.761	.767	.774	.780
16	.786	.796	.764	.794	.778	.768	.761	.758	.756	.756	.758	.760	.763	.767	.772	.776	.751	.757
17	.775	.783	.802	.778	.762	.751	.785	.780	.776	.774	.774	.775	.777	.779	.751	.755	.760	.765
18	.765	.771	.788	.764	.791	.778	.769	.762	.758	.756	.756	.756	.757	.760	.762	.766	.770	.774
19	.756	.760	.776	.750	.777	.763	.753	.784	.778	.774	.772	.771	.771	.772	.774	.776	.750	.754
20	.805	.802	.764	.783	.763	.789	.777	.768	.762	.758	.756	.754	.754	.755	.756	.758	.761	.763
25	.766	.754	.758	.769	.783	.763	.783	.770	.760	.752	.776	.771	.767	.764	.762	.761	.760	.760
30	.784	.763	.758	.762	.770	.780	.762	.777	.764	.754	.772	.765	.758	.753	.773	.769	.767	.764
35	.757	.773	.761	.759	.762	.768	.777	.759	.772	.759	.773	.764	.755	.771	.765	.760	.756	.752
40	.778	.783	.766	.759	.758	.760	.766	.773	.756	.766	.754	.766	.756	.769	.762	.755	.769	.764
45	.757	.758	.772	.761	.756	.755	.758	.762	.768	.752	.761	.770	.759	.770	.761	.753	.765	.759
50	.778	.771	.778	.764	.756	.752	.752	.755	.758	.764	.769	.756	.764	.753	.762	.754	.764	.757
60	.781	.765	.765	.772	.759	.751	.769	.767	.766	.767	.750	.754	.759	.764	.753	.759	.766	.757
70	.754	.763	.756	.759	.766	.754	.767	.761	.758	.756	.756	.757	.759	.761	.764	.753	.757	.762
80	.762	.763	.751	.772	.752	.759	.768	.759	.753	.767	.764	.762	.762	.762	.763	.751	.754	.757
90	.771	.764	.772	.764	.763	.765	.752	.760	.752	.762	.758	.754	.752	.751	.751	.752	.753	.754
100	.752	.767	.769	.759	.754	.754	.757	.762	.752	.760	.754	.764	.760	.757	.755	.754	.754	.754
									$P^0=0.900$									
1	1.000	.906	.969	.919	.962	.938	.910	.955	.936	.915	.956	.941	.925	.907	.949	.936	.923	.908
2	.969	.905	.935	.951	.919	.945	.918	.944	.923	.947	.930	.910	.937	.922	.904	.932	.918	.903
3	.914	.922	.933	.942	.906	.925	.941	.917	.935	.913	.933	.914	.933	.917	.935	.921	.905	.925
4	.940	.937	.937	.941	.905	.918	.930	.903	.919	.932	.912	.927	.908	.924	.907	.923	.907	.923
5	.909	.902	.900	.904	.911	.918	.926	.934	.910	.921	.931	.912	.924	.906	.919	.902	.916	.900
6	.943	.926	.918	.916	.918	.922	.926	.931	.907	.916	.924	.903	.913	.923	.906	.916	.900	.911
7	.923	.901	.933	.928	.926	.927	.928	.901	.907	.913	.919	.926	.907	.915	.922	.906	.915	.923
8	.903	.925	.912	.905	.901	.901	.902	.905	.909	.913	.918	.922	.903	.910	.916	.922	.907	.914
9	.933	.907	.929	.919	.914	.911	.910	.910	.912	.915	.918	.921	.902	.907	.912	.918	.902	.908
10	.918	.929	.912	.900	.925	.920	.917	.916	.916	.917	.919	.921	.902	.906	.910	.914	.919	.904
11	.905	.914	.929	.916	.907	.901	.924	.922	.921	.920	.921	.900	.903	.906	.909	.912	.916	.901
12	.931	.900	.915	.900	.920	.913	.908	.904	.902	.902	.902	.903	.905	.907	.909	.912	.914	.917
13	.920	.923	.901	.916	.905	.923	.917	.912	.909	.907	.907	.907	.907	.908	.910	.912	.914	.916
14	.910	.912	.919	.903	.918	.908	.901	.920	.916	.913	.911	.910	.910	.910	.911	.912	.914	.915
15	.900	.900	.908	.918	.905	.919	.911	.905	.901	.918	.916	.914	.913	.913	.913	.913	.914	.915
16	.927	.922	.925	.906	.918	.907	.921	.914	.909	.905	.902	.918	.917	.916	.915	.915	.915	.900
17	.919	.912	.915	.921	.906	.918	.909	.901	.916	.911	.908	.905	.903	.902	.901	.901	.901	.901
18	.911	.903	.905	.911	.919	.907	.918	.910	.904	.918	.913	.910	.908	.906	.904	.904	.903	.903
19	.903	.923	.922	.901	.909	.918	.907	.919	.912	.906	.901	.915	.912	.910	.908	.907	.906	.905
20	.927	.916	.913	.916	.921	.908	.917	.908	.901	.913	.908	.904	.900	.914	.911	.910	.908	.908
25	.924	.907	.922	.919	.900	.904	.910	.916	.907	.915	.908	.902	.911	.907	.903	.900	.911	.908
30	.923	.902	.913	.907	.905	.905	.907	.911	.915	.906	.912	.904	.911	.906	.901	.909	.905	.901
35	.902	.921	.906	.916	.911	.908	.908	.909	.911	.900	.905	.909	.902	.908	.901	.908	.903	.910
40	.907	.920	.902	.909	.902	.913	.910	.910	.910	.911	.901	.904	.908	.900	.905	.909	.904	.909
45	.912	.901	.900	.904	.911	.904	.900	.911	.910	.910	.911	.901	.903	.906	.909	.902	.906	.900
50	.917	.903	.916	.901	.905	.911	.906	.902	.912	.910	.910	.910	.901	.902	.904	.907	.900	.903
60	.910	.909	.915	.911	.911	.901	.906	.911	.907	.903	.901	.910	.909	.908	.909	.900	.902	.903
70	.906	.915	.903	.910	.907	.907	.908	.900	.905	.900	.906	.903	.901	.908	.907	.906	.906	.906
80	.904	.909	.907	.910	.905	.902	.902	.903	.906	.909	.903	.908	.905	.902	.908	.906	.905	.904
90	.903	.903	.912	.900	.904	.900	.908	.907	.908	.900	.903	.906	.902	.906	.903	.900	.905	.904
100	.903	.913	.905	.904	.905	.909	.905	.903	.902	.903	.904	.906	.900	.903	.907	.903	.901	.905
									$P^0=0.950$									
1	1.000	1.000	.969	.980	.962	.983	.971	.955	.979	.968	.956	.978	.970	.960	.979	.973	.965	.957
2	.969	.975	.982	.951	.967	.977	.963	.975	.962	.975	.964	.952	.967	.957	.971	.962	.953	.967
3	.973	.972	.974	.974	.953	.962	.970	.954	.965	.950	.961	.970	.960	.969	.960	.970	.961	.952
4	.981	.975	.970	.971	.973	.954	.960	.966	.950	.959	.965	.953	.961	.968	.959	.966	.957	.965
5	.963	.951	.972	.971	.971	.951	.955	.959	.964	.968	.955	.961	.966	.955	.961	.951	.958	.964
6	.975	.963	.954	.951	.950	.951	.953	.956	.959	.962	.965	.953	.957	.962	.951	.956	.961	.952
7	.961	.971	.962	.956	.954	.953	.953	.954	.956	.958	.960	.963	.951	.955	.959	.962	.952	.957
8	.973	.957	.968	.962	.958	.955	.954	.954	.955	.956	.957	.959	.961	.963	.953	.956	.959	.962
9	.963	.967	.956	.967	.962	.958	.956	.955	.955	.955	.956	.957	.958	.960	.962	.951	.954	.957
10	.953	.957	.964	.955	.966	.962	.959	.957	.955	.955	.955	.955	.956	.957	.959	.960	.950	.952
11	.968	.967	.954	.962	.955	.965	.962	.959	.957	.955	.955	.955	.955	.955	.956	.957	.959	.960
12	.960	.959	.963	.952	.961	.955	.950	.961	.958	.957	.955	.955	.954	.954	.955	.955	.956	.957
13	.952	.968	.954	.960	.951	.960	.954	.950	.960	.958	.956	.955	.954	.954	.954	.954	.955	.955
14	.966	.961	.963	.950	.957	.964	.958	.954	.950	.960	.958	.956	.955	.954	.954	.953	.953	.954
15	.960	.954	.955	.958	.963	.955	.962	.957	.953	.950	.959	.957	.956	.954	.954	.953	.953	.953
16	.953	.965	.964	.950	.955	.961	.954	.961	.957	.953	.961	.958	.957	.955	.954	.953	.952	.952

TABLE 1

$P_{(t),n}(\tfrac{1}{2}, V)$

n \ t	3	4	5	6	7	8	9	10	11	12	13	14	15	16	17	18	19	20
$p^0 = 0.950$																		
17	.966	.959	.957	.958	.961	.953	.959	.953	.960	.956	.952	.960	.958	.956	.954	.953	.952	.952
18	.961	.952	.950	.951	.954	.958	.951	.957	.952	.958	.955	.952	.959	.957	.955	.954	.953	.952
19	.956	.963	.959	.959	.961	.951	.956	.961	.955	.951	.957	.954	.951	.958	.956	.955	.953	.952
20	.950	.958	.954	.953	.954	.957	.960	.954	.959	.954	.960	.956	.953	.950	.957	.955	.954	.953
25	.962	.963	.956	.952	.951	.951	.953	.955	.958	.952	.956	.951	.956	.952	.957	.954	.951	.957
30	.960	.957	.960	.954	.950	.959	.958	.958	.950	.952	.955	.957	.953	.956	.952	.956	.952	.956
35	.959	.953	.954	.957	.951	.958	.956	.955	.954	.955	.956	.957	.952	.954	.956	.952	.955	.952
40	.959	.951	.960	.950	.954	.958	.955	.953	.951	.951	.951	.952	.953	.954	.956	.951	.953	.955
45	.959	.960	.956	.955	.957	.951	.955	.952	.957	.956	.955	.955	.955	.955	.956	.951	.952	.954
50	.960	.959	.953	.951	.952	.953	.956	.952	.956	.954	.953	.952	.951	.951	.951	.952	.953	.954
60	.954	.959	.950	.954	.952	.952	.953	.954	.950	.953	.951	.954	.953	.952	.951	.950	.950	.950
70	.959	.951	.958	.951	.955	.952	.952	.952	.953	.954	.951	.953	.951	.954	.952	.951	.950	.954
80	.955	.954	.957	.956	.951	.954	.952	.951	.951	.951	.952	.954	.951	.953	.951	.953	.952	.950
90	.952	.957	.951	.955	.955	.951	.953	.951	.950	.955	.950	.951	.952	.953	.950	.952	.950	.952
100	.950	.953	.952	.955	.954	.954	.955	.952	.950	.954	.953	.953	.953	.954	.951	.952	.953	.951
$p^0 = 0.975$																		
1	1.000	1.000	1.000	.980	.991	.983	.993	.987	.975	.990	.985	.978	.989	.985	.979	.990	.986	.982
2	1.000	.975	.982	.983	.988	.977	.985	.975	.984	.975	.983	.976	.984	.978	.985	.980	.987	.983
3	.996	.993	.991	.990	.979	.983	.986	.977	.982	.986	.979	.984	.977	.983	.977	.982	.977	.982
4	.981	.975	.987	.987	.987	.976	.979	.982	.984	.976	.980	.983	.976	.981	.984	.979	.982	.977
5	.987	.979	.987	.986	.985	.985	.986	.976	.978	.980	.983	.975	.978	.981	.984	.978	.981	.975
6	.975	.983	.976	.986	.984	.983	.983	.983	.984	.975	.977	.979	.981	.983	.977	.979	.981	.976
7	.983	.982	.979	.975	.984	.982	.982	.981	.981	.982	.982	.983	.976	.978	.980	.981	.975	.977
8	.989	.977	.982	.978	.985	.982	.981	.980	.980	.980	.980	.980	.981	.982	.975	.977	.978	.980
9	.982	.982	.985	.980	.976	.983	.981	.979	.979	.978	.978	.978	.978	.979	.979	.980	.981	.975
10	.976	.986	.979	.983	.978	.984	.981	.979	.978	.977	.977	.976	.976	.977	.977	.977	.978	.979
11	.983	.981	.982	.976	.981	.977	.982	.980	.978	.977	.976	.975	.982	.981	.981	.975	.976	.976
12	.978	.985	.976	.979	.983	.979	.975	.980	.978	.977	.975	.981	.981	.980	.980	.980	.980	.980
13	.985	.981	.981	.982	.977	.981	.977	.981	.979	.977	.975	.981	.980	.979	.979	.978	.978	.978
14	.980	.976	.976	.977	.980	.982	.979	.975	.980	.978	.976	.981	.979	.978	.978	.977	.977	.977
15	.976	.982	.980	.981	.982	.977	.980	.977	.981	.978	.976	.981	.979	.978	.977	.976	.976	.975
16	.983	.977	.976	.976	.977	.980	.975	.978	.975	.979	.977	.981	.979	.978	.977	.976	.975	.980
17	.980	.983	.980	.980	.980	.982	.977	.980	.977	.980	.978	.975	.979	.978	.977	.976	.980	.979
18	.976	.979	.976	.975	.976	.977	.979	.975	.978	.975	.978	.976	.980	.978	.977	.975	.979	.978
19	.983	.975	.981	.979	.979	.980	.981	.977	.980	.976	.979	.977	.980	.978	.977	.975	.979	.978
20	.980	.981	.977	.975	.975	.976	.977	.979	.975	.978	.980	.978	.975	.979	.977	.976	.979	.978
25	.983	.982	.977	.980	.978	.977	.977	.978	.979	.980	.976	.978	.975	.977	.975	.978	.976	.978
30	.981	.977	.978	.979	.976	.980	.979	.978	.978	.978	.979	.975	.977	.978	.975	.977	.979	.977
35	.979	.981	.979	.980	.976	.978	.976	.980	.975	.978	.978	.978	.979	.976	.977	.978	.975	.976
40	.978	.978	.976	.975	.976	.977	.979	.977	.976	.975	.979	.978	.978	.978	.975	.976	.976	.977
45	.977	.976	.978	.977	.976	.977	.978	.976	.978	.977	.976	.979	.978	.978	.978	.978	.975	.975
50	.977	.975	.976	.979	.978	.977	.978	.975	.977	.979	.977	.976	.975	.978	.978	.977	.977	.978
60	.978	.979	.978	.979	.976	.975	.979	.979	.976	.977	.978	.976	.978	.977	.976	.978	.977	.977
70	.980	.979	.976	.975	.976	.978	.977	.976	.976	.976	.976	.977	.975	.977	.975	.977	.976	.978
80	.976	.979	.979	.977	.977	.978	.976	.978	.977	.976	.976	.976	.977	.977	.976	.977	.975	.977
90	.979	.979	.979	.976	.978	.978	.975	.977	.975	.977	.977	.976	.976	.977	.977	.975	.976	.977
100	.977	.976	.978	.975	.977	.976	.976	.976	.977	.976	.977	.977	.976	.976	.976	.976	.977	.975
$p^0 = 0.990$																		
1	1.000	1.000	1.000	.997	.991	.996	.993	.997	.994	.990	.996	.993	.997	.995	.993	.997	.995	.993
2	1.000	.997	.997	.995	.996	.992	.995	.990	.993	.996	.993	.995	.993	.995	.993	.990	.994	.991
3	.996	.993	.992	.996	.991	.993	.994	.995	.991	.993	.995	.992	.994	.991	.993	.990	.993	.994
4	.996	.992	.995	.995	.994	.995	.995	.991	.992	.993	.994	.991	.992	.993	.991	.992	.994	.991
5	.996	.992	.995	.993	.993	.992	.992	.993	.993	.994	.994	.991	.992	.993	.994	.991	.992	.993
6	.990	.993	.995	.993	.992	.991	.990	.990	.990	.991	.991	.992	.992	.993	.994	.991	.992	.993
7	.994	.994	.995	.993	.991	.994	.994	.993	.993	.993	.993	.993	.993	.990	.991	.991	.992	.992
8	.995	.994	.991	.993	.991	.994	.993	.992	.992	.991	.991	.991	.991	.991	.992	.992	.992	.993
9	.992	.991	.992	.994	.991	.994	.993	.992	.991	.990	.993	.993	.993	.993	.993	.993	.990	.991
10	.995	.993	.993	.994	.992	.994	.992	.991	.990	.993	.992	.992	.992	.991	.991	.991	.991	.992
11	.992	.994	.994	.991	.992	.990	.992	.991	.993	.992	.992	.991	.991	.990	.990	.990	.990	.990
12	.995	.992	.992	.992	.993	.991	.992	.991	.993	.992	.991	.990	.993	.992	.992	.992	.991	.991
13	.992	.994	.993	.993	.990	.991	.993	.991	.993	.992	.991	.993	.992	.991	.991	.991	.990	.990
14	.990	.991	.991	.991	.991	.992	.993	.991	.993	.992	.990	.992	.992	.991	.990	.992	.992	.992
15	.993	.993	.992	.992	.992	.993	.990	.992	.993	.992	.990	.992	.991	.990	.992	.992	.991	.991
16	.991	.991	.994	.993	.993	.990	.991	.992	.990	.992	.990	.992	.991	.990	.992	.991	.991	.990
17	.993	.993	.992	.991	.991	.991	.992	.992	.991	.992	.990	.992	.991	.992	.991	.991	.990	.992
18	.992	.992	.993	.992	.992	.992	.992	.990	.991	.992	.991	.992	.991	.992	.991	.990	.992	.991
19	.990	.994	.992	.990	.993	.993	.990	.991	.991	.992	.991	.992	.991	.992	.991	.990	.992	.991
20	.993	.992	.993	.992	.991	.991	.991	.991	.992	.990	.991	.992	.991	.992	.991	.992	.991	.991
25	.993	.992	.992	.993	.991	.991	.990	.992	.992	.990	.991	.991	.992	.990	.991	.992	.991	.992
30	.991	.992	.992	.992	.992	.991	.990	.992	.991	.991	.991	.991	.991	.992	.990	.991	.991	.990
35	.993	.990	.992	.991	.991	.992	.991	.992	.991	.990	.990	.992	.992	.990	.990	.990	.991	.991
40	.992	.991	.992	.991	.991	.991	.991	.990	.991	.990	.992	.991	.991	.991	.991	.991	.991	.991
45	.992	.990	.990	.991	.991	.990	.991	.991	.992	.991	.991	.991	.990	.991	.991	.991	.991	.991
50	.991	.992	.991	.992	.991	.990	.992	.990	.990	.991	.990	.991	.990	.991	.991	.990	.990	.991
60	.991	.990	.991	.991	.991	.990	.991	.991	.991	.991	.991	.991	.990	.991	.990	.991	.991	.990
70	.991	.992	.992	.991	.991	.991	.991	.991	.990	.991	.991	.991	.990	.990	.991	.991	.990	.991
80	.991	.991	.991	.991	.990	.990	.990	.991	.991	.991	.990	.990	.990	.990	.990	.991	.991	.990
90	.992	.991	.992	.991	.990	.991	.991	.991	.990	.991	.990	.991	.991	.990	.990	.990	.990	.991
100	.990	.991	.991	.991	.991	.991	.991	.991	.991	.991	.990	.991	.990	.991	.991	.990	.990	.990

TABLE 2

$$t^{-1}E_{t,n}(p,v)$$

$p^0 = 0.750$

p	n \ t	3	4	5	6	7	8	9	10	11	12	13	14	15	16	17	18	19	20
	1	(1)	(2)	(2)	2	3	3	3	4	4	4	4	5	5	5	5	5	6	6
0.60		.747	.902	.836	.761	.869	.821	.772	.869	.834	.798	.762	.855	.827	.799	.771	.744	.833	.810
0.70		.737	.891	.812	.735	.846	.790	.734	.836	.794	.751	.709	.808	.773	.738	.704	.670	.767	.737
0.80		.720	.872	.768	.684	.800	.727	.657	.768	.709	.651	.595	.703	.653	.604	.557	.512	.612	.569
0.90		.697	.846	.699	.593	.720	.617	.523	.643	.557	.478	.408	.514	.445	.383	.328	.280	.362	.312
1.00		.667	.813	.600	.424	.571	.424	.311	.426	.320	.240	.182	.251	.191	.148	.116	.093	.122	.097
	2	(2)	(2)	(3)	3	4	4	5	5	5	6	6	6	7	7	7	8	8	8
0.60		.863	.752	.816	.744	.811	.754	.818	.773	.728	.794	.757	.721	.784	.753	.722	.783	.756	.729
0.70		.823	.695	.756	.675	.738	.667	.730	.669	.611	.676	.625	.576	.639	.594	.551	.610	.570	.532
0.80		.752	.597	.645	.547	.599	.504	.556	.473	.401	.450	.385	.328	.371	.318	.273	.309	.266	.229
0.90		.646	.460	.483	.366	.393	.293	.318	.241	.185	.202	.157	.125	.135	.108	.089	.095	.079	.068
1.00		.500	.297	.287	.199	.192	.144	.138	.111	.095	.090	.079	.072	.068					
	3	(2)	(3)	(4)	4	5	5	6	6	7	7	7	8	8	9	9	9	10	10
0.60		.783	.800	.829	.757	.797	.738	.782	.733	.778	.736	.695	.743	.707	.753	.721	.689	.735	.706
0.70		.716	.718	.738	.647	.678	.597	.632	.563	.600	.538	.482	.519	.467	.502	.454	.410	.443	.402
0.80		.607	.582	.577	.463	.470	.372	.383	.306	.318	.256	.207	.217	.177	.185	.152	.126	.132	.110
0.90		.471	.413	.377	.264	.247	.180	.170	.132	.125	.102	.087	.082	.072	.069	.062	.057	.055	.051
	4	(2)	(3)	4	5	5	6	7	7	8	8	9	9	10	10	10	11	11	12
0.60		.724	.728	.752	.774	.703	.736	.766	.714	.747	.702	.736	.697	.731	.695	.660	.695	.663	.698
0.70		.638	.617	.624	.627	.532	.546	.560	.486	.502	.438	.456	.399	.416	.365	.320	.336	.296	.311
0.80		.514	.456	.429	.402	.301	.289	.278	.214	.208	.163	.159	.128	.125	.102	.086	.084	.072	.070
0.90		.395	.312	.254	.216	.163	.143	.128	.107	.097	.086	.079	.072	.068					
	5	(3)	(4)	5	5	6	7	7	8	9	9	10	10	11	11	12	12	13	13
0.60		.795	.781	.784	.702	.720	.738	.679	.704	.727	.680	.705	.663	.690	.651	.678	.643	.669	.636
0.70		.691	.646	.628	.515	.509	.505	.423	.425	.427	.362	.366	.312	.317	.271	.275	.236	.241	.207
0.80		.540	.456	.403	.290	.260	.235	.178	.164	.151	.120	.112	.093	.087	.075	.071	.063	.060	.054
0.90		.399	.301	.235	.178	.151	.131	.113	.102	.092	.084								
	6	(3)	(4)	5	6	7	7	8	9	9	10	11	11	12	12	13	13	14	14
0.60		.755	.731	.728	.729	.735	.667	.682	.698	.645	.664	.682	.637	.657	.616	.637	.599	.620	.585
0.70		.632	.571	.540	.510	.488	.397	.386	.376	.310	.304	.299	.249	.245	.206	.203	.171	.170	.144
0.80		.477	.382	.317	.268	.230	.173	.152	.136	.111	.100	.092	.080	.074	.066	.062	.057	.054	.051
0.90		.364	.270	.210	.172	.146	.126	.112											
	7	(3)	(4)	5	6	7	8	9	9	10	11	11	12	13	13	14	14	15	15
0.60		.721	.689	.679	.675	.676	.680	.686	.627	.639	.650	.602	.616	.629	.587	.602	.563	.578	.542
0.70		.584	.512	.468	.430	.400	.375	.354	.285	.271	.259	.211	.203	.195	.161	.155	.129	.125	.106
0.80		.433	.334	.266	.219	.184	.158	.137	.113	.101	.091	.081	.075	.069	.064	.060			
0.90		.348	.258	.202	.168														
	8	(4)	5	6	7	8	9	9	10	11	12	12	13	14	14	15	15	16	16
0.60		.776	.741	.717	.702	.695	.691	.625	.629	.634	.639	.589	.597	.605	.561	.571	.531	.541	.504
0.70		.622	.544	.479	.429	.389	.355	.280	.259	.240	.224	.180	.169	.159	.130	.123	.103	.097	.083
0.80		.448	.336	.262	.211	.174	.147	.121	.107	.096	.087	.079	.073	.068					
0.90		.349	.254	.202															
	9	(4)	5	6	7	8	9	10	11	12	12	13	14	14	15	16	16	17	17
0.60		.747	.706	.676	.657	.645	.638	.633	.631	.630	.575	.577	.580	.533	.538	.543	.502	.508	.470
0.70		.581	.492	.423	.369	.327	.292	.262	.237	.215	.171	.157	.144	.118	.109	.102	.086	.080	.069
0.80		.415	.304	.235	.189	.158	.135	.117	.104	.094	.084	.078	.072						
0.90		.341	.251																
	10	(4)	5	7	8	9	10	10	11	12	13	14	14	15	16	16	17	18	18
0.60		.721	.674	.710	.684	.665	.651	.582	.576	.572	.569	.567	.516	.517	.517	.474	.476	.478	.439
0.70		.545	.450	.433	.370	.320	.280	.216	.192	.171	.154	.139	.113	.103	.095	.081	.075	.070	.062
0.80		.391	.283	.233	.186	.154	.132	.113	.101	.092	.084								
0.90		.337																	
	11	(4)	6	7	8	9	10	11	12	13	14	14	15	16	17	17	18	19	19
0.60		.698	.717	.675	.645	.622	.605	.591	.580	.571	.563	.509	.505	.501	.498	.453	.452	.450	.411
0.70		.515	.470	.387	.324	.276	.237	.205	.179	.158	.140	.114	.103	.093	.085	.073	.068	.063	.057
0.80		.374	.285	.219	.177	.148	.128	.113	.101										
0.90		.335																	
	12	(4)	6	7	8	9	10	11	12	13	14	15	16	17	17	18	19	19	20
0.60		.676	.688	.642	.608	.583	.563	.546	.533	.521	.511	.502	.494	.487	.439	.434	.430	.389	.386
0.70		.489	.433	.350	.288	.242	.205	.176	.153	.134	.118	.105	.095	.086	.074	.069	.064	.058	.054
0.80		.362	.272	.211	.172	.145	.126	.112											
0.90		.334																	
	13	(5)	6	7	9	10	11	12	13	14	15	16	16	17	18	19	20	20	21
0.60		.730	.662	.612	.634	.603	.577	.556	.538	.522	.508	.495	.442	.433	.425	.417	.409	.368	.362
0.70		.519	.402	.320	.290	.239	.200	.170	.146	.127	.112	.099	.085	.077	.071	.065	.061	.056	.052
0.80		.368	.264	.206	.171	.145	.126												
0.90		.334																	

TABLE 2

$t^{-1}E_{t,n}(p,v)$

$p^0 = 0.750$

p	n \ t	3	4	5	6	7	8	9	10	11	12	13	14	15	16	17	18	19	20
	14	(5)	6	8	9	10	11	12	13	14	15	16	17	18	19	19	20	21	22
0.60		.710	.637	.644	.602	.567	.539	.516	.495	.478	.462	.447	.434	.422	.411	.366	.357	.349	.341
0.70		.494	.376	.328	.263	.215	.179	.152	.131	.114	.101	.090	.081	.074	.068	.062	.058	.055	.051
0.80		.358	.259	.206	.169	.144													
0.90		.334																	
	15	(5)	7	8	9	11	12	13	14	15	16	17	18	18	19	20	21	22	23
0.60		.691	.676	.616	.571	.586	.554	.525	.501	.479	.460	.442	.426	.376	.364	.352	.341	.331	.321
0.70		.472	.391	.303	.241	.214	.176	.148	.127	.110	.097	.087	.079	.071	.065	.061	.057	.054	.051
0.80		.351	.259	.203	.168	.144													
0.90		.334																	
	16	(5)	7	8	10	11	12	13	14	15	16	17	18	19	20	21	22	22	23
0.60		.673	.653	.590	.595	.554	.519	.489	.463	.440	.419	.401	.384	.368	.353	.339	.326	.288	.278
0.70		.454	.368	.282	.243	.196	.162	.137	.118	.103	.092	.083	.076	.070	.065	.060	.057		
0.80		.346	.256	.202	.168														
	17	(5)	7	9	10	11	12	14	15	16	17	18	19	20	21	21	22	23	24
0.60		.656	.631	.620	.567	.523	.487	.498	.469	.442	.419	.397	.377	.359	.342	.300	.286	.274	.262
0.70		.437	.348	.288	.226	.182	.151	.134	.116	.101	.090	.082	.075	.069	.064	.060			
0.80		.342	.254	.202															
	18	(5)	7	9	10	12	13	14	15	16	17	18	19	20	21	22	23	24	25
0.60		.641	.610	.595	.540	.541	.500	.465	.435	.407	.383	.361	.341	.322	.305	.289	.275	.261	.248
0.70		.423	.331	.271	.212	.182	.150	.127	.110	.098	.088	.080	.073	.068					
0.80		.340	.252	.201															
	19	(5)	7	9	10	12	13	14	16	17	18	19	20	21	22	23	24	24	25
0.60		.626	.591	.572	.515	.513	.471	.435	.440	.410	.383	.358	.336	.315	.297	.280	.264	.229	.216
0.70		.411	.318	.257	.202	.172	.143	.122	.109	.097	.087	.079	.073	.068					
0.80		.338	.251	.201															
	20	(6)	8	9	11	12	14	15	16	17	18	19	20	21	22	23	24	25	26
0.60		.667	.625	.550	.537	.486	.483	.444	.409	.378	.351	.326	.304	.284	.265	.248	.233	.219	.205
0.70		.426	.327	.246	.203	.164	.142	.121	.106	.095	.085	.078	.072						
0.80		.339	.252																
	25	6	8	10	12	14	15	17	18	19	20	22	23	24	25	26	27	28	29
0.60		.603	.540	.499	.470	.447	.395	.381	.341	.306	.275	.269	.243	.220	.200	.182	.166	.152	.139
0.70		.378	.281	.221	.180	.152	.129	.114	.101	.092	.084								
0.80		.334																	
	30	7	9	11	13	15	17	18	20	21	22	24	25	26	27	29	30	31	32
0.60		.592	.513	.459	.417	.384	.356	.308	.289	.252	.221	.210	.185	.164	.146	.139	.125	.112	.101
0.70		.364	.267	.209	.172	.146	.126	.112											
0.80		.334																	
	35	7	10	12	14	16	18	20	21	23	24	26	27	28	30	31	32	33	34
0.60		.546	.490	.425	.375	.335	.302	.274	.233	.213	.183	.169	.147	.129	.120	.106	.094	.084	.076
0.70		.348	.260	.204	.169	.144													
	40	8	11	13	15	17	19	21	23	24	26	27	29	30	32	33	34	36	37
0.60		.540	.470	.396	.341	.296	.260	.230	.205	.173	.156	.134	.122	.106	.097	.086	.077	.072	.065
0.70		.344	.255	.202															
	45	8	11	14	16	18	20	22	24	26	27	29	31	32	34	35	36	38	39
0.60		.504	.426	.372	.312	.266	.229	.199	.174	.153	.130	.116	.105	.092	.084	.075	.067	.063	.058
0.70		.339	.252	.201															
	50	9	12	15	17	19	21	23	25	27	29	31	32	34	35	37	38	40	41
0.60		.500	.412	.352	.289	.242	.205	.175	.152	.133	.117	.105	.091	.083	.074	.068	.062	.058	.054
0.70		.337	.251																
	60	10	13	16	19	21	23	26	28	30	32	33	35	37	39	40	42	44	45
0.60		.470	.371	.303	.254	.207	.172	.151	.129	.113	.099	.087	.079	.072	.067	.062	.058	.054	.051
0.70		.335																	
	70	10	14	17	20	23	25	28	30	32	34	36	38	40	42	44			
0.60		.430	.341	.271	.221	.185	.153	.133	.115	.101	.090	.081	.074	.069	.064	.060			
0.70		.334																	
	80	11	15	18	22	24	27	30	32	34	37	39	41						
0.60		.414	.318	.249	.206	.167	.142	.123	.107	.095	.086	.079	.073						
	90	12	16	20	23	26	29	31	34	36	39	41	43						
0.60		.400	.302	.239	.191	.159	.135	.117	.104	.093	.085	.078	.072						
	100	12	17	21	24	27	30	33	36	38	41								
0.60		.381	.289	.227	.182	.152	.130	.114	.102	.092	.084								

TABLE 2

$t^{-1}E_{t,n}(p,v)$

p	n \ t	3	4	5	6	7	8	9	10	11	12	13	14	15	16	17	18	19	20
										p^0=0.900									
	1	(2)	(2)	(3)	3	4	4	4	5	5	5	6	6	6	6	7	7	7	7
0.60		1.000	.902	.966	.915	.959	.933	.903	.950	.929	.906	.950	.934	.916	.897	.941	.927	.912	.896
0.70		1.000	.891	.959	.899	.947	.915	.877	.932	.905	.874	.927	.904	.879	.852	.908	.887	.865	.841
0.80		1.000	.872	.947	.869	.926	.879	.825	.893	.850	.803	.871	.832	.790	.746	.820	.782	.742	.702
0.90		1.000	.846	.928	.823	.893	.817	.731	.818	.744	.665	.756	.686	.615	.545	.638	.573	.509	.450
1.00		1.000	.813	.900	.743	.841	.701	.556	.674	.545	.428	.538	.430	.337	.261	.344	.270	.210	.163
	2	(3)	(3)	(4)	5	5	6	6	7	7	8	8	8	9	9	9	10	10	10
0.60		.965	.893	.924	.942	.905	.932	.901	.930	.904	.932	.911	.887	.918	.898	.877	.909	.892	.873
0.70		.951	.853	.886	.908	.852	.884	.835	.870	.826	.862	.822	.780	.821	.783	.744	.786	.750	.713
0.80		.927	.778	.808	.836	.743	.776	.690	.727	.648	.687	.613	.542	.583	.517	.455	.496	.438	.384
0.90		.889	.659	.676	.704	.553	.578	.450	.478	.372	.399	.311	.241	.262	.204	.160	.175	.138	.110
1.00		.833	.484	.475	.468	.290	.294	.192	.195	.136	.137	.102	.083	.081	.069	.062	.059	.054	.051
	3	(3)	(4)	(5)	6	6	7	8	8	9	9	10	10	11	11	12	12	12	13
0.60		.899	.904	.915	.924	.880	.901	.918	.887	.908	.879	.902	.876	.899	.876	.898	.877	.855	.880
0.70		.848	.843	.848	.857	.783	.801	.818	.757	.779	.722	.746	.692	.717	.666	.692	.644	.596	.624
0.80		.756	.728	.713	.714	.590	.593	.596	.495	.503	.416	.427	.352	.363	.300	.310	.256	.211	.220
0.90		.614	.553	.507	.482	.333	.315	.299	.212	.204	.150	.144	.111	.107	.086	.083	.069	.060	.058
1.00		.417	.332	.272	.229	.160	.140	.124	.104	.094	.084	.078	.072						
	4	(4)	(5)	6	7	7	8	9	9	10	11	11	12	12	13	13	14	14	15
0.60		.923	.916	.913	.916	.869	.882	.894	.857	.874	.889	.859	.876	.848	.866	.840	.859	.835	.854
0.70		.865	.840	.827	.817	.729	.732	.735	.660	.669	.677	.611	.622	.561	.573	.516	.529	.476	.490
0.80		.757	.693	.655	.614	.476	.454	.434	.337	.325	.315	.246	.239	.188	.184	.146	.143	.115	.113
0.90		.592	.483	.409	.343	.227	.197	.173	.129	.116	.105	.087	.080	.071	.066	.060	.057		
1.00		.375	.276	.217	.177	.144	.126	.112											
	5	(4)	(5)	6	7	8	9	10	11	11	12	13	13	14	14	15	15	16	16
0.60		.881	.867	.861	.861	.865	.870	.877	.884	.848	.859	.869	.838	.851	.820	.835	.806	.821	.794
0.70		.792	.749	.725	.702	.686	.673	.663	.654	.574	.570	.567	.498	.498	.435	.437	.382	.384	.335
0.80		.645	.555	.496	.439	.393	.354	.321	.293	.219	.202	.186	.143	.133	.106	.099	.082	.077	.066
0.90		.467	.352	.273	.219	.180	.152	.130	.114	.096	.087	.080	.072						
	6	(5)	(6)	7	8	9	10	11	12	12	13	14	14	15	16	16	17	17	18
0.60		.915	.889	.874	.867	.863	.863	.864	.867	.827	.834	.841	.805	.814	.823	.791	.801	.770	.782
0.70		.823	.759	.718	.680	.649	.622	.600	.579	.494	.480	.468	.398	.389	.381	.324	.318	.271	.266
0.80		.661	.543	.461	.389	.331	.285	.247	.215	.160	.142	.127	.101	.092	.084	.072	.067	.059	.056
0.90		.464	.333	.247	.194	.159	.135	.117	.103	.092	.084								
	7	(5)	(6)	8	9	10	11	12	12	13	14	15	16	16	17	18	18	19	20
0.60		.884	.850	.886	.873	.864	.858	.855	.808	.809	.812	.815	.819	.781	.787	.793	.758	.765	.772
0.70		.764	.684	.712	.660	.616	.578	.545	.451	.428	.406	.386	.368	.306	.292	.280	.233	.224	.215
0.80		.577	.451	.432	.349	.285	.237	.198	.147	.128	.112	.100	.089	.076	.070	.065	.059	.055	.052
0.90		.401	.285	.231	.181	.150	.128	.113											
	8	(5)	7	8	9	10	11	12	13	14	15	16	17	17	18	19	20	20	21
0.60		.855	.875	.849	.830	.817	.808	.802	.798	.795	.793	.792	.792	.751	.753	.756	.758	.721	.725
0.70		.709	.705	.631	.570	.519	.475	.436	.402	.372	.345	.321	.299	.243	.227	.213	.200	.163	.154
0.80		.512	.452	.346	.271	.218	.178	.149	.127	.111	.097	.087	.079	.070	.065	.061	.057		
0.90		.367	.275	.210	.170	.144	.126												
	9	(6)	7	9	10	11	12	13	14	15	16	17	18	18	19	20	21	21	22
0.60		.886	.843	.864	.841	.823	.809	.798	.789	.782	.776	.772	.768	.724	.722	.721	.720	.681	.681
0.70		.739	.642	.631	.559	.498	.445	.400	.361	.326	.296	.269	.246	.196	.180	.165	.152	.124	.115
0.80		.527	.386	.332	.253	.200	.162	.135	.116	.101	.090	.081	.074	.068					
0.90		.368	.259	.206	.169														
	10	(6)	8	9	10	12	13	14	15	16	17	18	19	19	20	21	22	23	23
0.60		.862	.866	.830	.802	.828	.810	.795	.782	.771	.761	.753	.745	.698	.693	.689	.685	.681	.640
0.70		.692	.657	.562	.485	.479	.419	.369	.326	.288	.256	.228	.204	.162	.146	.132	.120	.109	.090
0.80		.477	.385	.281	.215	.186	.151	.127	.105	.096	.087	.079	.073						
0.90		.351	.258	.202															
	11	(6)	8	10	11	12	13	15	16	17	18	19	19	20	21	22	23	24	24
0.60		.838	.838	.847	.815	.789	.766	.792	.775	.760	.747	.735	.684	.675	.666	.659	.652	.645	.602
0.70		.649	.601	.565	.478	.407	.349	.341	.295	.257	.224	.196	.153	.136	.121	.109	.098	.089	.074
0.80		.439	.340	.274	.207	.165	.137	.121	.105	.094	.085	.078	.072						
0.90		.342	.253	.201															
	12	(7)	8	10	11	13	14	15	16	17	18	19	20	21	22	23	24	25	26
0.60		.868	.811	.816	.780	.796	.770	.747	.727	.709	.693	.679	.665	.653	.641	.630	.620	.611	.601
0.70		.677	.551	.506	.418	.395	.332	.280	.238	.204	.176	.153	.133	.117	.104	.093	.084	.076	.069
0.80		.450	.309	.245	.189	.160	.133	.115	.102	.092	.084								
0.90		.343	.251																
	13	(7)	9	10	12	13	15	16	17	18	19	20	21	22	23	24	25	26	27
0.60		.847	.836	.786	.793	.759	.774	.747	.723	.702	.682	.664	.647	.632	.617	.604	.591	.578	.567
0.70		.638	.567	.455	.415	.340	.316	.263	.220	.186	.158	.136	.118	.104	.092	.082	.074	.067	.062
0.80		.420	.311	.227	.186	.151	.131	.114	.101										
0.90		.338	.251																

TABLE 2

$t^{-1}E_{t,n}(p,v)$

p	n \ t	3	4	5	6	7	8	9	10	11	12	13	14	15	16	17	18	19	20
										P^0=0.900									
	14	(7)	9	11	12	14	15	16	18	19	20	21	22	23	24	25	26	27	28
0.60		.827	.811	.805	.761	.768	.735	.705	.720	.695	.672	.651	.631	.612	.595	.578	.563	.548	.534
0.70		.603	.522	.460	.367	.332	.269	.221	.204	.171	.144	.124	.107	.094	.083	.075	.068	.062	.057
0.80		.397	.290	.225	.177	.149	.128	.112	.101										
0.90		.336																	
	15	(7)	9	11	13	14	16	17	18	19	21	22	23	24	25	26	27	28	29
0.60		.807	.787	.777	.775	.733	.740	.707	.676	.649	.662	.638	.615	.594	.574	.555	.537	.519	.503
0.70		.571	.483	.416	.365	.290	.259	.210	.173	.145	.133	.114	.098	.087	.077	.070	.064	.059	.054
0.80		.380	.276	.215	.175	.146	.127	.112											
0.90		.335																	
	16	(8)	10	12	13	15	16	18	19	20	21	22	24	25	26	27	28	29	29
0.60		.840	.812	.796	.744	.743	.703	.708	.674	.644	.615	.588	.600	.576	.553	.532	.512	.493	.444
0.70		.597	.498	.422	.326	.284	.226	.200	.164	.136	.115	.100	.092	.081	.073	.066	.061	.056	.052
0.80		.386	.277	.214	.171	.145	.126	.112											
0.90		.335																	
	17	(8)	10	12	14	15	17	18	19	21	22	23	24	25	26	27	28	29	30
0.60		.822	.789	.770	.759	.710	.709	.669	.633	.639	.607	.578	.551	.525	.501	.479	.458	.438	.419
0.70		.567	.463	.385	.325	.252	.219	.176	.144	.129	.109	.095	.084	.075	.068	.063	.058	.054	.051
0.80		.373	.268	.208	.171	.144	.126												
0.90		.334																	
	18	(8)	10	12	14	16	17	19	20	21	23	24	25	26	27	28	29	30	31
0.60		.804	.768	.744	.730	.720	.674	.672	.633	.596	.600	.568	.538	.510	.484	.460	.437	.415	.395
0.70		.540	.432	.353	.294	.249	.195	.169	.139	.116	.105	.091	.081	.073	.067	.061	.057	.054	.051
0.80		.362	.261	.205	.169	.144													
0.90		.334																	
	19	(8)	11	13	14	16	18	19	21	22	23	24	26	27	28	29	30	31	32
0.60		.787	.792	.764	.702	.689	.681	.635	.632	.593	.557	.524	.526	.496	.468	.442	.417	.395	.373
0.70		.516	.445	.358	.268	.225	.191	.153	.134	.112	.097	.086	.078	.071	.065	.061	.057		
0.80		.354	.262	.205	.168														
	20	(9)	11	13	15	17	18	20	21	22	24	25	26	27	29	30	31	32	33
0.60		.817	.771	.739	.717	.700	.647	.638	.594	.554	.551	.515	.482	.452	.453	.425	.399	.375	.353
0.70		.537	.417	.331	.269	.222	.174	.149	.123	.105	.094	.084	.076	.069	.064	.060			
0.80		.358	.258	.203	.168														
0.90		.334																	
	25	10	12	15	17	18	20	22	24	25	27	28	29	31	32	33	34	36	37
0.60		.786	.719	.710	.669	.598	.572	.548	.528	.478	.462	.420	.381	.371	.338	.308	.280	.274	.251
0.70		.476	.347	.281	.219	.170	.142	.122	.107	.095	.086	.078	.072						
0.80		.340	.251	.201															
	30	11	13	16	18	20	22	24	26	28	29	31	32	34	35	36	38	39	40
0.60		.758	.674	.647	.593	.548	.509	.474	.444	.417	.367	.346	.305	.289	.256	.227	.216	.192	.171
0.70		.432	.306	.239	.188	.154	.131	.114	.102	.092	.084								
0.80		.335																	
	35	11	15	17	20	22	24	26	28	30	31	33	35	36	38	39	41	42	44
0.60		.697	.673	.593	.561	.505	.456	.413	.376	.344	.294	.270	.248	.214	.198	.172	.159	.139	.130
0.70		.386	.292	.219	.178	.148	.127	.112											
0.80		.334																	
	40	12	16	18	21	23	26	28	30	32	34	35	37	39	40	42	44	45	47
0.60		.679	.637	.546	.503	.440	.411	.363	.322	.287	.256	.215	.193	.174	.149	.135	.123	.106	.098
0.70		.370	.274	.209	.171	.144	.126												
	45	13	16	19	22	25	27	29	32	34	36	38	39	41	43	45	46	48	49
0.60		.663	.573	.506	.453	.410	.351	.303	.279	.242	.212	.186	.155	.138	.123	.110	.095	.086	.075
0.70		.359	.261	.204	.168	.144													
	50	14	17	21	23	26	29	31	33	36	38	40	42	43	45	47	49	50	52
0.60		.648	.546	.498	.410	.362	.322	.271	.230	.208	.179	.155	.135	.113	.100	.090	.081	.071	.065
0.70		.352	.256	.203															
	60	15	19	23	26	29	31	34	37	39	41	43	46	48	50	52	53	55	57
0.60		.594	.501	.435	.363	.306	.247	.213	.185	.154	.131	.112	.101	.089	.079	.071	.064	.059	.055
0.70		.340	.252	.201															
	70	16	21	24	28	31	34	37	39	42	44	47	49	51	54	56	58	60	62
0.60		.550	.465	.367	.310	.253	.209	.176	.144	.125	.107	.095	.084	.075	.069	.063	.059	.055	.051
0.70		.336	.251																
	80	17	22	26	30	33	36	39	42	45	48	50	53	55	57	60	62		
0.60		.514	.416	.333	.271	.217	.177	.148	.126	.109	.096	.085	.077	.070	.065	.060	.057		
0.70		.334																	
	90	18	23	28	31	35	38	42	45	48	50	53	56	58					
0.60		.485	.378	.305	.235	.193	.157	.135	.115	.101	.089	.081	.074	.068					
0.70		.334																	

TABLE 2

$t^{-1}E_{t,n}(p,\nu)$

p	n \ t	3	4	5	6	7	8	9	10	11	12	13	14	15	16	17	18	19	20
										$P^0 = 0.900$									
	100	19	25	29	33	37	41	44	47	50	53	56	59						
0.60		.460	.361	.275	.217	.176	.147	.124	.108	.096	.086	.079	.072						
										$P^0 = 0.950$									
	1	(2)	(3)	(3)	4	4	5	5	5	6	6	6	7	7	7	8	8	8	8
0.60		1.000	1.000	.966	.978	.959	.981	.967	.950	.975	.964	.950	.974	.965	.954	.976	.968	.959	.949
0.70		1.000	1.000	.959	.972	.947	.973	.955	.932	.964	.947	.927	.959	.944	.927	.957	.944	.930	.913
0.80		1.000	1.000	.947	.963	.926	.960	.930	.893	.936	.906	.871	.918	.889	.856	.904	.876	.845	.812
0.90		1.000	1.000	.928	.951	.893	.937	.884	.818	.881	.822	.756	.828	.768	.704	.779	.721	.659	.598
1.00		1.000	1.000	.900	.944	.841	.903	.800	.674	.770	.655	.538	.641	.533	.432	.529	.435	.350	.278
	2	(3)	(4)	(5)	5	6	7	7	8	8	9	9	9	10	10	11	11	11	12
0.60		.965	.970	.978	.942	.958	.971	.952	.967	.951	.966	.952	.936	.955	.941	.958	.946	.934	.952
0.70		.951	.954	.961	.908	.927	.943	.910	.929	.898	.919	.890	.857	.885	.854	.881	.852	.822	.852
0.80		.927	.921	.925	.836	.856	.874	.804	.829	.760	.788	.721	.652	.688	.622	.657	.595	.534	.571
0.90		.889	.863	.853	.704	.717	.730	.600	.622	.504	.528	.424	.335	.359	.284	.306	.242	.190	.207
1.00		.833	.766	.725	.468	.465	.463	.299	.304	.199	.204	.140	.102	.103	.080	.080	.066	.057	.056
	3	(4)	(5)	(6)	7	7	8	9	9	10	10	11	12	12	13	13	14	14	14
0.60		.965	.963	.964	.964	.935	.946	.955	.933	.946	.925	.939	.951	.935	.947	.933	.945	.932	.917
0.70		.940	.930	.925	.922	.866	.875	.884	.834	.848	.799	.816	.832	.787	.804	.761	.779	.737	.693
0.80		.891	.858	.834	.821	.708	.704	.702	.600	.604	.511	.519	.526	.446	.454	.383	.393	.330	.276
0.90		.806	.724	.663	.627	.450	.423	.400	.285	.272	.196	.189	.182	.135	.131	.100	.097	.078	.065
1.00		.667	.508	.403	.328	.195	.168	.147	.111	.100	.086	.079	.074						
	4	(5)	(6)	7	8	9	9	10	11	11	12	13	13	14	15	15	16	16	17
0.60		.973	.963	.955	.955	.956	.928	.935	.941	.917	.926	.935	.914	.925	.934	.916	.926	.909	.920
0.70		.943	.916	.897	.885	.877	.809	.808	.809	.744	.749	.753	.693	.700	.707	.651	.660	.606	.616
0.80		.877	.810	.765	.721	.683	.551	.526	.504	.401	.387	.374	.296	.288	.280	.222	.217	.173	.169
0.90		.753	.622	.533	.448	.380	.250	.217	.190	.136	.122	.110	.088	.081	.075	.065	.061	.055	.052
1.00		.542	.358	.260	.200	.162	.128	.113	.101										
	5	(5)	(6)	8	9	10	10	11	12	13	14	14	15	16	16	17	17	18	19
0.60		.945	.927	.954	.949	.947	.915	.918	.922	.926	.930	.906	.913	.919	.897	.905	.883	.892	.901
0.70		.884	.839	.877	.853	.833	.750	.736	.725	.715	.707	.635	.630	.625	.560	.558	.496	.496	.496
0.80		.763	.665	.701	.632	.571	.432	.392	.357	.327	.300	.227	.209	.194	.149	.139	.109	.102	.096
0.90		.575	.433	.427	.331	.261	.174	.146	.125	.109	.096	.082	.075	.069	.064	.060			
1.00		.354	.258	.216	.172	.145													
	6	(6)	(7)	8	9	10	11	12	13	14	15	16	16	17	18	18	19	20	20
0.60		.958	.937	.922	.913	.908	.905	.905	.905	.906	.908	.911	.884	.888	.893	.867	.873	.879	.855
0.70		.896	.839	.796	.757	.725	.696	.671	.648	.628	.609	.592	.516	.503	.490	.425	.416	.407	.352
0.80		.766	.642	.552	.468	.400	.344	.298	.259	.227	.199	.176	.133	.119	.107	.086	.079	.073	.062
0.90		.560	.396	.290	.220	.174	.144	.122	.107	.095	.086	.078	.072						
1.00		.344	.252	.201															
	7	(6)	(8)	9	10	11	12	13	14	15	16	17	18	18	19	20	21	21	22
0.60		.935	.946	.927	.914	.905	.899	.894	.891	.889	.888	.888	.888	.858	.860	.863	.865	.837	.841
0.70		.843	.840	.785	.734	.688	.648	.612	.579	.550	.522	.497	.474	.401	.383	.366	.350	.295	.283
0.80		.671	.623	.513	.416	.340	.281	.234	.197	.167	.143	.124	.109	.087	.079	.072	.066	.058	.054
0.90		.459	.368	.259	.196	.157	.132	.114	.102	.092	.084								
	8	(7)	8	10	11	12	13	14	15	16	17	18	19	20	21	21	22	23	24
0.60		.949	.919	.932	.916	.904	.893	.885	.879	.874	.870	.866	.864	.862	.860	.829	.828	.829	.829
0.70		.859	.780	.775	.713	.656	.605	.560	.518	.481	.447	.417	.389	.363	.339	.279	.262	.245	.230
0.80		.680	.530	.479	.374	.294	.235	.191	.158	.133	.114	.099	.087	.078	.071	.063	.059	.055	.052
0.90		.455	.301	.239	.182	.149	.128	.112											
	9	(7)	9	10	12	13	14	15	16	17	18	19	20	21	22	23	23	24	25
0.60		.930	.931	.905	.919	.903	.889	.878	.868	.859	.852	.846	.840	.836	.831	.827	.793	.790	.788
0.70		.811	.786	.702	.694	.627	.567	.513	.465	.422	.384	.350	.319	.292	.267	.245	.197	.181	.167
0.80		.603	.520	.386	.339	.260	.203	.163	.135	.114	.099	.087	.078	.071	.065	.061			
0.90		.400	.291	.213	.175	.146	.126												
	10	(7)	9	11	12	14	15	16	17	18	19	20	21	22	23	24	25	25	26
0.60		.909	.908	.913	.890	.903	.886	.871	.858	.847	.836	.827	.818	.810	.803	.796	.790	.752	.747
0.70		.764	.727	.697	.613	.600	.532	.472	.419	.372	.331	.295	.264	.236	.212	.191	.172	.137	.125
0.80		.540	.442	.367	.271	.233	.180	.145	.121	.103	.091	.082	.074	.068	.064				
0.90		.369	.266	.209	.169	.144													
	11	(8)	10	11	13	14	16	17	18	19	20	21	22	23	24	25	26	27	28
0.60		.929	.921	.888	.895	.871	.883	.866	.850	.835	.821	.808	.797	.786	.776	.766	.757	.749	.740
0.70		.787	.736	.630	.600	.518	.501	.435	.378	.329	.287	.251	.220	.194	.171	.152	.135	.120	.108
0.80		.552	.438	.308	.253	.192	.164	.133	.112	.098	.087	.079	.073						
0.90		.369	.263	.203	.168														

TABLE 2

$$t^{-1}E_{t,n}(p,v)$$

p	n \ t	3	4	5	6	7	8	9	10	11	12	13	14	15	16	17	18	19	20
										P^{o}=0.950									
	12	(8)	10	12	13	15	16	17	19	20	21	22	23	24	25	26	27	28	29
0.60		.912	.900	.897	.866	.873	.849	.828	.842	.824	.807	.791	.776	.762	.749	.737	.725	.714	.703
0.70		.744	.681	.629	.529	.499	.422	.358	.343	.293	.251	.216	.186	.162	.141	.124	.109	.097	.087
0.80		.501	.382	.298	.217	.180	.144	.121	.107	.095	.085	.078	.072						
0.90		.352	.255	.202															
	13	(8)	11	12	14	15	17	18	19	21	22	23	24	25	26	27	28	29	30
0.60		.894	.913	.873	.873	.842	.849	.824	.801	.813	.793	.775	.757	.740	.724	.709	.694	.680	.667
0.70		.703	.692	.569	.521	.430	.400	.333	.278	.262	.221	.187	.160	.137	.119	.104	.092	.081	.073
0.80		.461	.380	.262	.209	.163	.139	.117	.103	.093	.084								
0.90		.343	.254	.201															
	14	(9)	11	13	14	16	18	19	20	21	23	24	25	26	27	28	29	30	31
0.60		.914	.894	.884	.845	.846	.849	.821	.795	.770	.781	.759	.738	.718	.700	.682	.664	.648	.632
0.70		.727	.642	.570	.461	.417	.380	.311	.255	.211	.197	.165	.139	.119	.103	.090	.080	.071	.064
0.80		.471	.340	.257	.191	.158	.135	.115	.102	.092	.084								
0.90		.343	.252																
	15	(9)	11	13	15	17	18	20	21	22	23	25	26	27	28	29	30	31	32
0.60		.898	.874	.861	.853	.850	.816	.818	.789	.762	.735	.744	.720	.698	.676	.656	.636	.617	.598
0.70		.689	.595	.517	.455	.404	.323	.291	.236	.193	.160	.147	.124	.106	.092	.080	.072	.065	.059
0.80		.438	.312	.236	.187	.155	.129	.114	.101										
0.90		.339	.251																
	16	(9)	12	14	15	17	19	20	22	23	24	26	27	28	29	30	31	32	33
0.60		.882	.889	.872	.827	.821	.817	.783	.784	.754	.724	.730	.703	.678	.654	.631	.608	.587	.567
0.70		.654	.607	.520	.405	.352	.309	.246	.219	.177	.146	.132	.112	.096	.083	.074	.066	.060	.055
0.80		.412	.312	.233	.178	.149	.128	.112	.101										
0.90		.336																	
	17	(10)	12	14	16	18	19	21	22	24	25	26	28	29	30	31	32	33	34
0.60		.901	.871	.850	.836	.826	.785	.781	.746	.746	.714	.683	.687	.659	.632	.607	.582	.559	.537
0.70		.677	.564	.473	.401	.343	.267	.233	.185	.164	.134	.113	.102	.088	.077	.069	.063	.058	.053
0.80		.420	.292	.220	.176	.148	.126	.112											
0.90		.336																	
	18	(10)	12	14	16	18	20	21	23	24	26	27	28	30	31	32	33	34	35
0.60		.887	.852	.827	.810	.797	.788	.746	.742	.705	.704	.671	.639	.641	.612	.584	.558	.532	.508
0.70		.644	.525	.431	.359	.302	.257	.201	.175	.141	.125	.105	.090	.082	.073	.066	.060	.056	.052
0.80		.399	.278	.212	.172	.145	.126												
0.90		.335																	
	19	(10)	13	15	17	19	20	22	24	25	26	28	29	30	32	33	34	35	36
0.60		.872	.868	.841	.819	.803	.756	.746	.738	.699	.661	.659	.624	.592	.592	.562	.534	.507	.481
0.70		.613	.538	.436	.357	.296	.226	.192	.165	.134	.111	.099	.086	.076	.070	.064	.059	.055	.051
0.80		.383	.279	.212	.171	.145													
0.90		.334																	
	20	(10)	13	15	17	19	21	23	24	26	27	29	30	31	32	34	35	36	37
0.60		.857	.851	.819	.794	.775	.759	.746	.702	.693	.653	.647	.610	.576	.543	.541	.511	.483	.456
0.70		.584	.502	.399	.323	.264	.219	.185	.146	.127	.106	.095	.082	.074	.067	.062	.058	.054	.051
0.80		.370	.270	.207	.169	.144													
0.90		.334																	
	25	12	15	17	19	21	23	25	27	29	30	32	33	35	36	38	39	40	42
0.60		.860	.833	.785	.744	.709	.678	.651	.627	.604	.554	.536	.491	.477	.437	.425	.390	.358	.349
0.70		.553	.436	.326	.250	.198	.161	.135	.115	.101	.089	.080	.073	.068					
0.80		.350	.256	.201															
	30	13	16	19	21	23	26	28	30	31	33	35	37	38	40	41	43	44	46
0.60		.831	.786	.756	.700	.650	.639	.599	.562	.501	.471	.445	.420	.374	.355	.316	.300	.267	.254
0.70		.490	.363	.279	.211	.167	.142	.120	.105	.093	.085	.078	.072						
0.80		.339	.251																
	35	14	17	20	23	25	28	30	32	34	36	38	40	41	43	45	46	48	49
0.60		.805	.744	.698	.661	.599	.575	.525	.480	.440	.404	.371	.341	.296	.273	.252	.219	.203	.177
0.70		.445	.318	.240	.190	.154	.131	.114	.101										
0.80		.335																	
	40	15	18	22	24	27	30	32	34	36	38	40	42	44	46	48	49	51	53
0.60		.782	.705	.677	.596	.555	.519	.462	.411	.367	.328	.294	.264	.237	.213	.193	.164	.149	.136
0.70		.413	.290	.225	.176	.148	.127	.112											
0.80		.334																	
	45	16	20	23	26	29	31	34	36	39	41	43	45	47	49	51	52	54	56
0.60		.761	.701	.628	.567	.515	.445	.408	.354	.327	.285	.250	.219	.193	.170	.151	.127	.114	.102
0.70		.390	.281	.213	.172	.145	.126												
	50	17	21	24	27	30	33	36	38	41	43	45	47	49	51	53	55	57	59
0.60		.742	.668	.583	.513	.455	.405	.362	.307	.277	.237	.203	.175	.152	.133	.116	.103	.091	.082
0.70		.374	.268	.206	.169	.144													
	60	18	23	26	30	33	36	39	42	44	47	49	52	54	56	58	60	62	64
0.60		.679	.610	.507	.448	.379	.322	.276	.238	.196	.171	.143	.127	.109	.094	.083	.074	.066	.060
0.70		.349	.256	.201															

TABLE 2

$$t^{-1}E_{t,n}(p,v)$$

p	n \ t	3	4	5	6	7	8	9	10	11	12	13	14	15	16	17	18	19	20
										P^0=0.950									
	70	20	24	29	32	36	39	42	45	48	51	53	56	58	61	63	65	67	70
0.60		.652	.535	.469	.377	.322	.264	.219	.183	.155	.133	.112	.098	.085	.077	.069	.063	.057	.054
0.70		.341	.252																
	80	21	26	31	35	38	42	45	48	51	54	57	60	62	65	67	70	72	74
0.60		.605	.496	.416	.339	.267	.223	.181	.150	.126	.109	.095	.084	.075	.068	.063	.058	.054	.051
0.70		.336	.251																
	90	22	28	32	37	41	44	48	51	54	58	60	63	66	69	71			
0.60		.564	.463	.359	.295	.237	.187	.156	.130	.110	.098	.085	.077	.070	.065	.060			
0.70		.335																	
	100	23	29	34	39	43	47	51	54	57	61	64	67	70					
0.60		.529	.419	.328	.263	.208	.169	.140	.118	.102	.091	.081	.074	.068					
0.70		.334																	
										P^0=0.975									
	1	(2)	(3)	(4)	4	5	5	6	6	6	7	7	7	8	8	8	9	9	9
0.60		1.000	1.000	1.000	.978	.990	.981	.991	.985	.975	.988	.982	.974	.987	.982	.976	.988	.983	.978
0.70		1.000	1.000	1.000	.972	.987	.973	.987	.977	.964	.981	.971	.959	.977	.968	.957	.975	.967	.957
0.80		1.000	1.000	1.000	.963	.980	.960	.978	.960	.936	.963	.943	.918	.949	.928	.904	.937	.916	.893
0.90		1.000	1.000	1.000	.951	.973	.937	.963	.928	.881	.924	.880	.828	.881	.833	.779	.839	.789	.735
1.00		1.000	1.000	1.000	.944	.968	.903	.942	.870	.770	.843	.748	.641	.730	.630	.529	.621	.526	.436
	2	(4)	(4)	(5)	6	7	7	8	8	9	9	10	10	11	11	12	12	13	13
0.60		1.000	.970	.978	.978	.984	.971	.980	.967	.977	.966	.976	.966	.977	.968	.978	.970	.979	.973
0.70		1.000	.954	.961	.961	.969	.943	.956	.929	.945	.919	.937	.912	.930	.907	.925	.903	.921	.900
0.80		1.000	.921	.925	.924	.931	.874	.890	.829	.850	.788	.813	.752	.779	.720	.748	.690	.720	.663
0.90		1.000	.863	.853	.849	.850	.730	.742	.622	.641	.528	.551	.449	.473	.383	.407	.328	.350	.282
1.00		1.000	.766	.725	.698	.678	.463	.462	.304	.309	.204	.209	.142	.146	.104	.106	.080	.081	.064
	3	(5)	(6)	(7)	8	8	9	10	10	11	12	12	13	13	14	14	15	15	16
0.60		.995	.990	.988	.985	.968	.973	.977	.963	.970	.976	.964	.971	.960	.968	.957	.965	.956	.964
0.70		.989	.977	.970	.962	.924	.927	.931	.893	.901	.909	.873	.883	.846	.858	.820	.834	.796	.812
0.80		.978	.945	.921	.900	.810	.801	.795	.701	.700	.700	.612	.616	.533	.539	.463	.471	.402	.411
0.90		.956	.876	.812	.764	.585	.549	.518	.380	.362	.345	.251	.241	.176	.171	.127	.123	.095	.092
1.00		.917	.742	.600	.489	.272	.229	.196	.130	.116	.104	.084	.077	.069	.064				
	4	(5)	(6)	8	9	10	10	11	12	13	13	14	15	15	16	17	17	18	18
0.60		.973	.963	.979	.977	.977	.959	.962	.965	.968	.953	.958	.963	.949	.955	.961	.949	.955	.943
0.70		.943	.916	.944	.933	.925	.872	.868	.865	.864	.811	.812	.814	.762	.766	.771	.720	.726	.676
0.80		.877	.810	.855	.813	.776	.650	.621	.595	.571	.466	.450	.435	.351	.340	.331	.266	.259	.207
0.90		.753	.622	.667	.571	.489	.325	.280	.243	.213	.148	.132	.119	.091	.083	.077	.065	.060	.054
1.00		.542	.358	.355	.255	.193	.136	.117	.104	.093	.084								
	5	(6)	(7)	9	10	11	12	13	13	14	15	16	16	17	18	19	19	20	20
0.60		.978	.965	.976	.972	.970	.968	.968	.949	.951	.953	.955	.939	.943	.946	.950	.935	.940	.925
0.70		.945	.908	.926	.906	.887	.871	.856	.789	.777	.767	.758	.692	.686	.680	.675	.614	.611	.552
0.80		.867	.773	.794	.726	.663	.606	.555	.430	.394	.362	.333	.254	.235	.217	.201	.155	.145	.113
0.90		.715	.545	.540	.420	.330	.262	.211	.144	.123	.106	.093	.078	.072	.066	.061	.057		
1.00		.458	.291	.249	.185	.150	.128	.112											
	6	(6)	(8)	9	11	12	13	14	15	16	16	17	18	19	20	20	21	22	22
0.60		.958	.968	.954	.969	.964	.960	.958	.956	.955	.935	.935	.937	.938	.939	.921	.924	.927	.909
0.70		.896	.902	.861	.879	.850	.822	.796	.772	.750	.671	.653	.635	.619	.604	.534	.522	.510	.448
0.80		.766	.742	.647	.646	.563	.490	.427	.373	.326	.239	.210	.186	.165	.147	.112	.101	.092	.075
0.90		.560	.487	.354	.319	.238	.183	.147	.122	.104	.088	.080	.073	.068					
1.00		.344	.264	.204	.170	.144													
	7	(7)	(9)	10	11	13	14	15	16	17	18	19	20	20	21	22	23	23	24
0.60		.967	.971	.956	.945	.959	.953	.948	.944	.941	.939	.937	.936	.914	.914	.914	.914	.893	.894
0.70		.906	.898	.847	.799	.814	.775	.738	.703	.671	.640	.612	.585	.506	.484	.463	.444	.379	.364
0.80		.768	.716	.600	.493	.478	.397	.330	.276	.231	.196	.166	.143	.108	.095	.085	.076	.064	.059
0.90		.547	.441	.304	.219	.189	.149	.123	.106	.094	.085	.078	.072						
1.00		.339	.255	.201															
	8	(8)	9	11	12	14	15	16	17	18	19	20	21	22	23	23	24	25	26
0.60		.975	.951	.957	.944	.955	.947	.940	.934	.928	.923	.919	.916	.913	.910	.885	.884	.883	.882
0.70		.916	.844	.835	.776	.780	.730	.682	.638	.596	.558	.522	.489	.458	.429	.359	.337	.316	.297
0.80		.771	.616	.559	.440	.409	.325	.260	.211	.173	.143	.121	.104	.090	.080	.068	.062	.057	.053
0.90		.535	.345	.270	.196	.165	.134	.115	.102	.092	.084								
1.00		.336																	
	9	(8)	10	12	13	14	16	17	18	19	20	21	22	23	24	25	26	27	27
0.60		.961	.957	.959	.945	.931	.942	.932	.923	.915	.908	.902	.896	.891	.886	.881	.877	.873	.845
0.70		.875	.845	.825	.756	.689	.687	.630	.577	.529	.484	.443	.406	.372	.341	.313	.288	.265	.215
0.80		.689	.600	.523	.396	.302	.271	.212	.169	.137	.115	.098	.086	.076	.069	.063	.058	.055	.051
0.90		.451	.326	.247	.183	.148	.129	.112											

TABLE 2

$$t^{-1}E_{t,n}(p,v)$$

p	n \ t	3	4	5	6	7	8	9	10	11	12	13	14	15	16	17	18	19	20
										p^0=0.975									
	10	(8)	11	12	14	15	17	18	19	20	21	22	23	24	25	26	27	28	29
0.60		.946	.962	.941	.945	.929	.937	.925	.914	.903	.894	.885	.876	.868	.861	.854	.847	.841	.835
0.70		.831	.847	.760	.737	.661	.648	.583	.523	.469	.420	.376	.337	.302	.272	.244	.220	.198	.179
0.80		.615	.586	.425	.360	.267	.231	.178	.142	.117	.099	.087	.078	.070	.065	.060			
0.90		.399	.311	.217	.176	.145	.126												
	11	(9)	11	13	14	16	17	19	20	21	22	23	24	26	27	28	28	29	30
0.60		.957	.948	.945	.923	.928	.911	.919	.905	.892	.880	.868	.856	.871	.861	.851	.816	.808	.799
0.70		.848	.796	.754	.661	.634	.556	.539	.474	.416	.365	.320	.281	.278	.245	.216	.170	.151	.134
0.80		.625	.502	.403	.287	.239	.181	.156	.126	.105	.092	.082	.074	.069	.064				
0.90		.398	.277	.212	.169	.144													
	12	(9)	12	13	15	17	18	19	21	22	23	24	26	27	28	29	30	31	32
0.60		.944	.954	.927	.926	.927	.908	.889	.897	.881	.866	.851	.863	.850	.837	.825	.813	.802	.791
0.70		.807	.801	.690	.647	.610	.525	.450	.430	.370	.318	.273	.265	.229	.198	.172	.150	.132	.116
0.80		.563	.494	.337	.268	.218	.165	.132	.116	.099	.087	.079	.073						
0.90		.369	.272	.204	.168														
	13	(10)	12	14	16	17	19	20	22	23	24	25	27	28	29	30	31	32	33
0.60		.955	.940	.932	.928	.904	.906	.884	.889	.871	.853	.835	.845	.829	.814	.799	.784	.770	.756
0.70		.825	.751	.687	.633	.534	.496	.417	.391	.330	.278	.235	.224	.190	.163	.140	.121	.105	.092
0.80		.574	.429	.324	.252	.185	.154	.125	.109	.095	.085	.078	.072						
0.90		.369	.259	.203															
	14	(10)	12	14	16	18	20	21	22	24	25	26	28	29	30	31	32	33	34
0.60		.943	.924	.913	.908	.905	.903	.880	.857	.861	.840	.820	.827	.809	.790	.773	.755	.738	.722
0.70		.786	.701	.628	.566	.515	.470	.388	.320	.296	.245	.204	.191	.160	.136	.116	.100	.087	.077
0.80		.523	.378	.282	.218	.175	.145	.120	.104	.093	.084								
0.90		.353	.254	.201															
	15	(10)	13	15	17	19	20	22	23	25	26	27	29	30	31	32	33	34	35
0.60		.929	.934	.920	.911	.905	.877	.876	.850	.852	.828	.805	.810	.788	.767	.747	.727	.708	.689
0.70		.749	.710	.627	.557	.497	.402	.362	.293	.266	.218	.179	.165	.137	.116	.099	.086	.076	.067
0.80		.481	.376	.275	.210	.168	.135	.117	.102	.092	.084								
0.90		.344	.253	.201															
	16	(11)	13	15	17	19	21	22	24	25	27	28	30	31	32	33	34	35	37
0.60		.943	.918	.902	.890	.882	.876	.846	.844	.816	.816	.790	.793	.769	.745	.722	.699	.677	.684
0.70		.769	.664	.573	.498	.435	.383	.304	.270	.217	.195	.159	.144	.120	.101	.087	.076	.068	.063
0.80		.490	.339	.248	.192	.156	.132	.114	.101										
0.90		.344	.251																
	17	(11)	14	16	18	20	22	23	25	26	28	29	30	32	33	34	35	37	38
0.60		.931	.928	.909	.895	.884	.875	.842	.838	.807	.805	.776	.748	.749	.723	.697	.672	.676	.653
0.70		.733	.673	.574	.491	.422	.365	.286	.249	.198	.175	.143	.118	.107	.091	.079	.069	.065	.058
0.80		.455	.339	.244	.188	.153	.130	.113	.101										
0.90		.339	.251																
	18	(11)	14	16	18	20	22	24	25	27	28	30	31	33	34	35	36	38	39
0.60		.919	.914	.891	.874	.860	.849	.840	.805	.799	.766	.763	.732	.731	.702	.674	.646	.648	.622
0.70		.699	.630	.525	.440	.371	.315	.269	.210	.182	.146	.129	.107	.097	.083	.073	.065	.061	.055
0.80		.427	.312	.228	.179	.148	.127	.112											
0.90		.336																	
	19	(12)	14	17	19	21	23	25	26	28	29	31	32	34	35	36	37	39	40
0.60		.932	.899	.900	.880	.863	.849	.837	.800	.791	.756	.750	.716	.713	.681	.651	.621	.621	.593
0.70		.719	.588	.527	.435	.361	.302	.254	.197	.169	.135	.119	.099	.089	.077	.068	.062	.058	.053
0.80		.435	.293	.226	.177	.147	.127	.112											
0.90		.336																	
	20	(12)	15	17	19	21	23	25	27	28	30	32	33	34	36	37	38	40	41
0.60		.921	.910	.882	.859	.839	.822	.808	.795	.755	.746	.737	.701	.666	.661	.628	.597	.594	.564
0.70		.687	.599	.483	.391	.319	.263	.220	.185	.146	.126	.110	.093	.080	.073	.066	.060	.056	.052
0.80		.412	.293	.216	.173	.145	.126												
0.90		.335																	
	25	14	17	19	22	24	26	28	30	32	34	35	37	38	40	41	43	44	46
0.60		.916	.891	.848	.838	.804	.773	.745	.719	.695	.673	.624	.605	.560	.544	.503	.489	.452	.440
0.70		.643	.515	.384	.316	.244	.193	.156	.130	.111	.097	.084	.076	.069	.064	.060			
0.80		.373	.264	.203	.168														
	30	15	18	21	24	26	29	31	33	35	37	39	40	42	44	45	47	49	50
0.60		.890	.849	.818	.794	.746	.730	.689	.651	.615	.582	.551	.496	.470	.446	.401	.381	.362	.325
0.70		.563	.418	.319	.250	.190	.158	.130	.110	.097	.087	.079	.072						
0.80		.347	.253	.201															
	35	16	20	23	26	28	31	33	36	38	40	42	44	46	47	49	51	52	54
0.60		.865	.836	.791	.753	.692	.664	.611	.590	.545	.503	.464	.429	.397	.347	.321	.297	.260	.241
0.70		.502	.377	.277	.212	.165	.138	.117	.103	.093	.084								
0.80		.338	.251																
	40	17	21	24	27	30	33	36	38	40	42	45	47	49	51	52	54	56	58
0.60		.841	.798	.739	.687	.643	.603	.568	.511	.459	.413	.392	.353	.318	.287	.245	.221	.200	.182
0.70		.456	.329	.240	.187	.153	.130	.113	.101										
0.80		.335																	

TABLE 2

$t^{-1}E_{t,n}(p,v)$

p	n \ t	3	4	5	6	7	8	9	10	11	12	13	14	15	16	17	18	19	20
										$p^0 = 0.975$									
	45	18	22	26	29	32	35	38	40	43	45	47	50	52	54	56	58	59	61
0.60		.820	.762	.718	.654	.598	.549	.505	.442	.409	.358	.315	.292	.257	.227	.200	.177	.149	.133
0.70		.422	.298	.226	.177	.147	.127	.112											
0.80		.334																	
	50	19	23	27	31	34	37	40	42	45	48	50	52	54	57	59	61	63	65
0.60		.799	.727	.670	.623	.558	.500	.449	.384	.346	.313	.269	.231	.199	.182	.158	.138	.121	.106
0.70		.398	.279	.213	.172	.145	.126												
	60	21	26	30	34	37	40	44	47	49	52	55	57	60	62	64	67	69	71
0.60		.763	.693	.612	.545	.465	.398	.359	.309	.254	.220	.192	.160	.141	.120	.103	.092	.081	.072
0.70		.367	.264	.204	.168														
	70	23	28	32	36	40	44	47	50	53	56	59	62	64	67	69	72	74	77
0.60		.731	.637	.539	.458	.392	.338	.278	.231	.193	.163	.139	.120	.101	.089	.078	.070	.063	.058
0.70		.351	.255	.201															
	80	24	30	35	39	43	47	50	54	57	60	63	66	69	72	74	77	79	82
0.60		.678	.589	.498	.407	.335	.278	.223	.189	.155	.130	.110	.095	.084	.075	.067	.061	.056	.053
0.70		.340	.252																
	90	26	32	37	41	46	50	53	57	60	64	67	70	73	76	79	81	84	87
0.60		.655	.546	.445	.350	.291	.235	.185	.154	.127	.110	.095	.083	.074	.068	.062	.058	.054	.051
0.70		.337	.251																
	100	27	33	39	43	48	52	56	60	64	67	71	74	77	80	83			
0.60		.612	.490	.400	.306	.248	.197	.160	.133	.113	.097	.086	.077	.070	.065	.060			
0.70		.335																	
										$p^0 = 0.990$									
	1	(2)	(3)	(4)	5	5	6	6	7	7	7	8	8	9	9	9	10	10	10
0.60		1.000	1.000	1.000	.996	.990	.996	.991	.996	.993	.988	.995	.991	.996	.994	.991	.996	.994	.991
0.70		1.000	1.000	1.000	.995	.987	.994	.987	.994	.989	.981	.990	.985	.992	.988	.982	.990	.986	.981
0.80		1.000	1.000	1.000	.993	.980	.990	.978	.988	.978	.963	.979	.966	.980	.969	.955	.972	.960	.945
0.90		1.000	1.000	1.000	.992	.973	.985	.963	.979	.956	.924	.952	.921	.949	.919	.883	.919	.885	.845
1.00		1.000	1.000	1.000	.995	.968	.982	.942	.965	.917	.843	.895	.821	.875	.802	.715	.786	.703	.614
	2	(4)	(5)	(6)	7	8	8	9	9	10	11	11	12	12	13	13	13	14	14
0.60		1.000	.996	.996	.993	.995	.989	.992	.986	.990	.993	.989	.993	.989	.992	.989	.984	.989	.985
0.70		1.000	.993	.992	.986	.989	.975	.980	.965	.973	.979	.966	.973	.960	.969	.956	.940	.951	.936
0.80		1.000	.987	.981	.971	.972	.938	.945	.903	.915	.925	.884	.898	.854	.871	.825	.774	.797	.747
0.90		1.000	.975	.958	.939	.934	.852	.855	.754	.765	.776	.676	.692	.593	.612	.518	.429	.451	.372
1.00		1.000	.953	.912	.880	.853	.662	.649	.461	.461	.461	.318	.322	.219	.224	.154	.108	.111	.082
	3	(5)	(6)	(7)	9	9	10	11	12	12	13	14	14	15	15	16	16	17	18
0.60		.995	.990	.988	.994	.986	.988	.989	.991	.984	.987	.989	.984	.987	.981	.985	.979	.983	.986
0.70		.989	.977	.970	.983	.961	.961	.962	.963	.939	.943	.947	.922	.928	.901	.908	.879	.889	.897
0.80		.978	.945	.921	.951	.887	.877	.868	.861	.785	.782	.780	.702	.703	.624	.628	.552	.558	.564
0.90		.956	.876	.812	.870	.718	.677	.641	.609	.467	.445	.425	.317	.304	.224	.217	.160	.155	.151
1.00		.917	.742	.600	.678	.402	.335	.281	.238	.147	.129	.115	.085	.078	.067	.062	.057	.054	.051
	4	(6)	(7)	9	10	11	12	13	13	14	15	16	16	17	18	18	19	20	20
0.60		.993	.986	.991	.990	.989	.989	.989	.980	.982	.983	.985	.977	.979	.982	.974	.977	.980	.973
0.70		.983	.964	.973	.964	.957	.951	.946	.909	.906	.904	.902	.862	.862	.863	.821	.823	.825	.782
0.80		.958	.903	.919	.885	.852	.820	.790	.683	.657	.633	.611	.512	.495	.480	.395	.384	.373	.305
0.90		.902	.769	.791	.695	.607	.529	.460	.314	.274	.239	.210	.146	.131	.117	.088	.081	.074	.061
1.00		.792	.522	.509	.356	.256	.192	.151	.110	.097	.087	.079	.072						
	5	(7)	(8)	10	11	12	13	14	15	16	17	18	18	19	20	21	21	22	23
0.60		.993	.985	.989	.986	.983	.982	.981	.981	.981	.981	.981	.972	.973	.975	.976	.967	.969	.971
0.70		.980	.954	.959	.943	.927	.913	.900	.887	.876	.865	.855	.802	.793	.786	.778	.723	.718	.713
0.80		.943	.865	.869	.808	.749	.692	.639	.590	.545	.504	.466	.366	.339	.313	.291	.225	.209	.195
0.90		.855	.675	.660	.527	.417	.331	.264	.213	.174	.145	.122	.092	.081	.073	.067	.059	.055	.052
1.00		.667	.379	.321	.216	.162	.133	.114	.101										
	6	(7)	(9)	11	12	13	14	15	16	17	18	19	20	21	22	23	23	24	25
0.60		.982	.985	.987	.983	.979	.976	.973	.971	.970	.969	.969	.968	.968	.968	.969	.957	.958	.959
0.70		.948	.946	.946	.921	.896	.871	.847	.824	.803	.782	.763	.744	.727	.710	.694	.628	.614	.601
0.80		.860	.832	.818	.731	.647	.570	.502	.441	.387	.340	.300	.264	.234	.207	.184	.139	.124	.112
0.90		.684	.599	.545	.398	.292	.219	.170	.136	.113	.097	.085	.077	.070	.064	.060			
1.00		.406	.305	.240	.177	.146	.126												
	7	(8)	(10)	12	13	14	16	17	18	19	20	21	22	23	23	24	25	26	27
0.60		.986	.986	.986	.980	.974	.981	.977	.974	.972	.969	.967	.966	.964	.950	.949	.949	.948	.948
0.70		.951	.941	.934	.899	.863	.872	.840	.809	.778	.748	.719	.691	.665	.588	.565	.543	.521	.501
0.80		.856	.803	.768	.657	.556	.540	.457	.385	.324	.273	.231	.196	.167	.124	.108	.095	.084	.075
0.90		.659	.536	.452	.309	.220	.191	.146	.118	.100	.088	.079	.073						
1.00		.375	.272	.212	.169														

TABLE 2

$$t^{-1}E_{t,n}(p,v)$$

$p^0 = 0.990$

p	n \ t	3	4	5	6	7	8	9	10	11	12	13	14	15	16	17	18	19	20
	8	(9)	11	12	14	15	17	18	19	20	21	22	23	24	25	26	27	28	29
0.60		.989	.985	.975	.978	.971	.977	.971	.967	.962	.958	.955	.951	.948	.946	.943	.941	.939	.937
0.70		.955	.934	.884	.878	.831	.833	.790	.747	.706	.666	.628	.592	.558	.526	.496	.467	.441	.416
0.80		.853	.783	.641	.591	.477	.446	.359	.290	.235	.191	.158	.132	.112	.096	.083	.074	.066	.060
0.90		.637	.495	.317	.252	.181	.153	.123	.105	.093	.084								
1.00		.357	.259	.201															
	9	(9)	11	13	15	16	18	19	20	21	22	24	25	26	27	28	29	29	30
0.60		.980	.975	.975	.977	.968	.972	.966	.959	.953	.947	.955	.951	.946	.942	.938	.934	.914	.910
0.70		.924	.894	.873	.858	.801	.793	.739	.687	.636	.588	.594	.550	.509	.471	.436	.403	.334	.308
0.80		.774	.682	.600	.532	.411	.369	.286	.223	.177	.143	.132	.110	.093	.080	.071	.064	.057	.053
0.90		.524	.379	.281	.217	.161	.137	.115	.101										
1.00		.335	.251																
	10	(10)	12	14	16	17	19	20	21	22	24	25	26	27	28	29	30	31	32
0.60		.984	.977	.976	.976	.966	.969	.960	.951	.943	.950	.943	.936	.930	.923	.917	.911	.905	.900
0.70		.930	.893	.863	.839	.771	.755	.691	.629	.571	.566	.514	.466	.422	.382	.345	.312	.282	.255
0.80		.776	.664	.563	.481	.357	.308	.232	.179	.141	.127	.105	.089	.078	.069	.063	.058	.054	.051
0.90		.514	.354	.255	.196	.152	.130	.112											
1.00		.334																	
	11	(10)	13	15	16	18	19	21	22	24	25	26	27	28	29	30	31	32	33
0.60		.976	.980	.977	.963	.964	.951	.955	.944	.949	.940	.930	.921	.912	.903	.895	.887	.878	.870
0.70		.898	.892	.854	.773	.743	.666	.645	.575	.560	.499	.443	.393	.348	.308	.273	.241	.214	.190
0.80		.702	.647	.530	.379	.313	.229	.194	.149	.131	.107	.090	.079	.071	.065	.060			
0.90		.442	.335	.238	.176	.147	.126	.112											
	12	(11)	13	15	17	19	20	22	23	25	26	27	28	30	31	32	33	34	35
0.60		.981	.971	.966	.963	.962	.948	.950	.937	.940	.929	.917	.906	.913	.902	.892	.881	.871	.861
0.70		.908	.851	.800	.756	.717	.631	.601	.526	.503	.439	.382	.331	.320	.278	.242	.211	.183	.160
0.80		.708	.561	.439	.347	.278	.201	.168	.131	.114	.096	.083	.075	.069	.064	.060			
0.90		.438	.290	.215	.172	.145													
	13	(11)	14	16	18	19	21	23	24	26	27	28	30	31	32	33	34	35	36
0.60		.973	.974	.967	.963	.946	.945	.945	.930	.932	.918	.904	.909	.896	.883	.870	.857	.844	.831
0.70		.875	.853	.794	.740	.640	.598	.560	.480	.452	.386	.328	.311	.265	.226	.193	.166	.143	.123
0.80		.642	.550	.417	.320	.223	.180	.149	.119	.104	.090	.080	.074	.068					
0.90		.395	.282	.210	.170														
	14	(11)	14	16	18	20	22	24	25	27	28	29	31	32	33	34	36	37	38
0.60		.964	.965	.956	.949	.945	.942	.940	.923	.924	.907	.891	.894	.878	.863	.847	.853	.838	.823
0.70		.841	.810	.738	.674	.617	.567	.523	.439	.406	.339	.284	.264	.221	.186	.157	.147	.126	.108
0.80		.583	.479	.352	.265	.206	.165	.137	.112	.098	.087	.078	.072						
0.90		.368	.264	.204	.168														
	15	(12)	15	17	19	21	23	24	26	28	29	30	32	33	34	36	37	38	39
0.60		.971	.969	.958	.950	.944	.939	.919	.917	.916	.897	.878	.879	.861	.842	.846	.828	.811	.793
0.70		.855	.814	.734	.661	.596	.539	.444	.402	.365	.300	.246	.225	.186	.155	.143	.120	.102	.088
0.80		.592	.472	.338	.250	.192	.154	.123	.107	.095	.085	.078	.072						
0.90		.368	.262	.202															
	16	(12)	15	18	20	22	23	25	27	28	30	31	33	34	35	37	38	39	40
0.60		.963	.959	.961	.952	.944	.920	.915	.911	.889	.886	.865	.864	.843	.822	.824	.803	.783	.763
0.70		.821	.771	.730	.649	.576	.467	.414	.368	.297	.265	.215	.193	.158	.131	.119	.101	.086	.075
0.80		.542	.416	.326	.237	.182	.140	.119	.104	.093	.084								
0.90		.353	.255	.202															
	17	(13)	16	18	20	22	24	26	28	29	31	32	34	35	37	38	39	40	42
0.60		.970	.964	.950	.937	.927	.918	.911	.905	.880	.876	.852	.849	.826	.824	.801	.778	.755	.757
0.70		.836	.776	.677	.588	.511	.444	.387	.339	.268	.236	.189	.168	.137	.123	.102	.087	.075	.069
0.80		.551	.413	.286	.210	.165	.136	.116	.103	.092	.084								
0.90		.353	.254	.201															
	18	(13)	16	19	21	23	25	27	28	30	32	33	35	36	38	39	40	42	43
0.60		.962	.954	.953	.939	.927	.917	.907	.879	.872	.866	.839	.835	.808	.805	.779	.753	.752	.728
0.70		.804	.734	.675	.578	.495	.423	.363	.283	.244	.211	.168	.147	.120	.107	.090	.077	.071	.062
0.80		.508	.370	.278	.204	.160	.133	.115	.101										
0.90		.344	.252																
	19	(13)	17	19	21	24	26	27	29	31	33	34	36	37	39	40	41	43	44
0.60		.953	.960	.941	.924	.928	.915	.884	.874	.865	.856	.827	.820	.791	.786	.757	.729	.726	.698
0.70		.771	.741	.624	.523	.480	.404	.309	.262	.223	.191	.150	.131	.107	.095	.081	.070	.065	.058
0.80		.472	.368	.252	.189	.156	.131	.113	.101										
0.90		.339	.251																
	20	(14)	17	20	22	24	26	28	30	32	33	35	37	38	40	41	43	44	45
0.60		.962	.950	.945	.927	.910	.895	.881	.869	.857	.824	.814	.805	.774	.766	.735	.729	.699	.670
0.70		.788	.700	.624	.516	.425	.351	.291	.243	.204	.159	.136	.118	.097	.087	.074	.068	.061	.055
0.80		.480	.336	.248	.186	.150	.128	.112											
0.90		.339	.251																

TABLE 2

$$t^{-1}E_{t,n}(p,v)$$

$P^0 = 0.990$

p	n \ t	3	4	5	6	7	8	9	10	11	12	13	14	15	16	17	18	19	20
	25	16	19	22	25	27	29	31	34	36	37	39	41	43	44	46	48	49	51
0.60		.954	.934	.918	.907	.879	.851	.825	.823	.800	.754	.732	.712	.693	.649	.632	.616	.575	.561
0.70		.736	.602	.495	.409	.313	.242	.191	.165	.135	.108	.093	.082	.074	.066	.061	.057	.054	.051
0.80		.414	.281	.211	.171	.144													
0.90		.334																	
	30	17	21	24	27	30	32	34	37	39	41	43	45	47	49	50	52	54	55
0.60		.933	.919	.892	.869	.850	.810	.771	.758	.723	.689	.656	.625	.595	.568	.517	.493	.470	.426
0.70		.645	.524	.400	.310	.244	.184	.145	.124	.105	.091	.081	.074	.068					
0.80		.364	.261	.202															
	35	19	22	26	29	32	35	37	40	42	44	46	49	51	52	54	56	58	60
0.60		.930	.885	.867	.832	.800	.771	.720	.696	.649	.605	.563	.547	.510	.453	.421	.391	.364	.339
0.70		.608	.431	.334	.249	.192	.155	.126	.109	.095	.085	.078	.072						
0.80		.349	.252	.201															
	40	20	24	28	31	34	37	40	42	45	47	50	52	54	56	58	60	62	64
0.60		.911	.873	.844	.796	.752	.711	.673	.613	.582	.529	.504	.458	.416	.377	.342	.310	.282	.256
0.70		.542	.389	.289	.213	.166	.137	.117	.102	.092	.084								
0.80		.339	.251																
	45	21	25	29	33	36	39	42	45	48	50	53	55	57	60	62	64	66	68
0.60		.892	.841	.798	.761	.705	.653	.605	.562	.522	.463	.431	.382	.338	.316	.280	.248	.220	.195
0.70		.491	.339	.248	.192	.154	.130	.113	.101										
0.80		.335																	
	50	22	27	31	35	38	41	45	47	50	53	55	58	60	63	65	67	69	72
0.60		.873	.832	.777	.729	.662	.600	.567	.493	.447	.406	.351	.319	.276	.252	.218	.189	.165	.151
0.70		.451	.319	.232	.181	.148	.127	.112											
0.80		.334																	
	60	24	29	34	38	42	45	49	52	55	58	61	64	66	69	71	74	76	78
0.60		.837	.771	.716	.645	.583	.506	.458	.397	.345	.299	.261	.227	.189	.166	.140	.124	.106	.091
0.70		.398	.277	.211	.170	.144													
	70	26	32	37	41	45	49	53	56	59	63	66	69	71	74	77	80	82	85
0.60		.805	.737	.661	.572	.494	.429	.372	.309	.258	.226	.190	.161	.132	.114	.099	.087	.075	.068
0.70		.368	.263	.204															
	80	28	34	39	44	48	52	56	60	64	67	70	73	76	79	82	85	88	90
0.60		.774	.684	.588	.508	.421	.350	.293	.246	.208	.170	.140	.118	.100	.087	.077	.069	.062	.056
0.70		.352	.255	.201															
	90	30	36	42	47	51	56	60	64	67	71	74	78	81	84	87	90	93	96
0.60		.747	.636	.546	.453	.362	.304	.245	.201	.160	.135	.112	.097	.084	.074	.067	.061	.056	.052
0.70		.343	.252																
	100	31	38	44	49	54	59	63	67	71	75	78	82	85	89	92	95	98	101
0.60		.700	.593	.490	.391	.315	.256	.203	.164	.135	.114	.096	.084	.075	.068	.062	.058	.054	.051
0.70		.337	.251																

Selected Tables in Mathematical Statistics
Volume 8, 1985

TABLES OF ADMISSIBLE AND OPTIMAL BALANCED TREATMENT INCOMPLETE BLOCK (BTIB) DESIGNS FOR COMPARING TREATMENTS WITH A CONTROL

Robert E. Bechhofer
Cornell University

and

Ajit C. Tamhane
Northwestern University

ABSTRACT

We consider the problem of comparing simultaneously $p \geq 2$ test treatments with a control treatment in blocks of common size $k < p+1$. Bechhofer and Tamhane (1981) proposed a new class of incomplete block designs called balanced treatment incomplete block (BTIB) designs for this problem of multiple comparisons with a control. In this paper we give tables of admissible BTIB designs and optimal BTIB designs for one-sided and two-sided comparisons for the cases $p = 2(1)6$, $k = 2$ and $p = 3(1)6$, $k = 3$. We restrict consideration to blocks of size 2 and 3 because in our view these cases are the ones of greatest practical interest. The tables are to be used for given (p,k) and specified allowance to design experiments which achieve a specified joint confidence coefficient for one-sided or two-sided comparisons; the optimal designs which we provide accomplish this objective with a minimum total number of blocks.

Received by the editors September 1981 and in revised form August 1982.
AMS(MOS) Subject Classifications (1980): Primary 62Q05; Secondary 62K05, 62K10, 62J15.
This work was supported by U.S. Army Research Office-Durham Contract DAAG29-80-C-0036 and DAAG29-81-K-0168, and Office of Naval Research Contract N00014-75-C-0586 at Cornell University and a subvention grant from the Office of Research and Sponsored Programs, Northwestern University.

0094-8837/85 $1.00 + $.25 per page

TABLE OF CONTENTS

1. INTRODUCTION AND SUMMARY

Suppose that it is desired to compare simultaneously several test treatments with a control treatment (the so-called multiple comparisons with a control problem) using an incomplete block design. For this problem Bechhofer and Tamhane (1981) proposed a new class of incomplete block designs that are balanced with respect to the test treatments (see Definition 3.1 below); they referred to such designs as balanced treatment incomplete block (BTIB) designs. That first paper provided the theory underlying such designs. The main purpose of the present paper is to provide tables of optimal BTIB designs in order to make it possible to implement the theory. More specifically, for one-sided or two-sided comparisons with a control, the tables in the present paper give the optimal BTIB design requiring minimum total number of blocks to achieve a specified joint confidence coefficient for given number of treatments, common block size, and specified allowance on the "width" of the confidence intervals. In this respect, the purpose of these tables is analogous to that of tables that would be employed in the determination of the sample size necessary in a hypothesis testing problem to attain a specified power at a specified alternative. These optimal BTIB designs were obtained from catalogs of admissible BTIB designs; such catalogs are given for all cases considered (with a few exceptions as noted in the sequel). Tables of optimal completely randomized designs for this same multiple comparisons with a control problem are given in Bechhofer and Tamhane (1983).

In these tables we have restricted consideration to blocks of size two and three. Such small block sizes are common in practice. The main reason for this is that in many biological and other applications, naturally occurring blocks are small in size. The block size of two corresponds to a matched pairs design, a standard experimental technique. For example, twins constitute a natural experimental block of size two as do right eye (arm) and left eye (arm), etc. It is also the case that for larger blocks the assumption of homogeneity within a block becomes suspect.

The plan of the present paper is as follows: Section 2 gives the basic assumptions and notation. The class of BTIB designs is described in Section 3. The formulae necessary for computing the joint interval estimates of the differences between the treatment effects and the control effect using a BTIB design are given in Section 4. In Section 5 we give the definitions of an optimal design, an admissible design, and a minimal complete set of generator designs. The minimal complete sets for selected (p,k) are given in Appendix A.1 and are explained in Section 6.2. In Section 6 we also describe the tables of admissible designs and the tables of optimal designs which are provided in Appendices A.2 and A.3, respectively. Computational details are given in Section 7. Illustrations of the uses of the tables are given in Section 8.

2. ASSUMPTIONS, NOTATION, AND STATEMENT OF THE PROBLEM

Suppose that the $p+1$ treatments are indexed by $0,1,\ldots,p$ with 0 denoting the control treatment and $1,2,\ldots,p$ denoting the $p \geq 2$ test treatments. Let $k < p+1$ denote the common size of each block, and let b denote the number of blocks available for experimentation. We assume the usual additive linear model (no treatment × block interaction):

$$Y_{ijh} = \mu + \alpha_i + \beta_j + e_{ijh} \tag{2.1}$$

where Y_{ijh} denotes the random variable if the ith treatment is assigned to the hth plot of the jth block $(0 \leq i \leq p,\ 1 \leq j \leq b,\ 1 \leq h \leq k)$; $\{\alpha_i\}$ and $\{\beta_j\}$ are the unknown treatment and block "effects," respectively, with $\sum_{i=0}^{p} \alpha_i = \sum_{j=1}^{b} \beta_j = 0$. The e_{ijh} are i.i.d. $N(0,\sigma^2)$ random variables with σ^2 being assumed <u>known</u> for <u>design</u> purposes. (See equations (4.6) and (4.7) for Dunnett's (1955) procedure which should be used for <u>analysis</u> purposes when σ^2 is is unknown.)

It is desired to make an exact joint confidence statement employing either (I) <u>one-sided intervals</u> of the form:

$$\{\alpha_0-\alpha_i \leq \hat{\alpha}_0-\hat{\alpha}_i+a\ (1 \leq i \leq p)\}, \tag{2.2}$$

or

(II) two-sided intervals of the form:

$$\{\hat{\alpha}_0-\hat{\alpha}_i-a \le \alpha_0-\alpha_i \le \hat{\alpha}_0-\hat{\alpha}_i+a \quad (1 \le i \le p)\}. \tag{2.3}$$

Here the $\{\hat{\alpha}_i\}$ are the best linear unbiased estimates (BLUE's) of the $\{\alpha_i\}$, and $a > 0$ is a specified allowance associated with these confidence intervals.

3. BALANCED TREATMENT INCOMPLETE BLOCK (BTIB) DESIGNS

Definition 3.1: For given (p,k,b) a design will be referred to as a BTIB design if it has the property that

$$\mathrm{Var}\{\hat{\alpha}_0-\hat{\alpha}_i\} = \tau^2\sigma^2 \quad (1 \le i \le p) \tag{3.1}$$

and

$$\mathrm{Corr}\{\hat{\alpha}_0-\hat{\alpha}_i,\ \hat{\alpha}_0-\hat{\alpha}_j\} = \rho \quad (i \ne j,\ 1 \le i,\ j \le p). \tag{3.2}$$

Here τ^2 and ρ depend, of course, on the particular design employed.

Designs having this same balance property were earlier considered by Pierce (1960).

Let n_{ij} denote the number of replications of the ith treatment in the jth block $(0 \le i \le p,\ 1 \le j \le b)$ and let $N = \sum_{i=0}^{p} \sum_{j=1}^{b} n_{ij}$ denote the total number of observations. Theorem 3.1 of Bechhofer and Tamhane (1981) gives the following combinatorial conditions as both necessary and sufficient for a design to be BTIB:

$$\sum_{j=1}^{b} n_{0j}n_{ij} = \lambda_0 \quad (1 \le i \le p) \tag{3.3}$$

and

$$\sum_{j=1}^{b} n_{i_1j}n_{i_2j} = \lambda_1 \quad (i_1 \ne i_2,\ 1 \le i_1,\ i_2 \le p) \tag{3.4}$$

for some nonnegative integers λ_0 and λ_1. A BTIB design is said to be implementable if $\lambda_0 > 0$.

4. FORMULAE FOR ESTIMATES, THEIR VARIANCES AND CORRELATIONS, AND CONFIDENCE COEFFICIENTS FOR JOINT INTERVAL ESTIMATES

We now give expressions for the BLUE, $\hat{\alpha}_0-\hat{\alpha}_i$, of $\alpha_0-\alpha_i$ $(1 \le i \le p)$, and the variance (3.1) and correlation (3.2) when a BTIB design is used.

(These quantities are derived in the Appendix of Bechhofer and Tamhane (1981).) Let T_i denote the sum of all observations obtained with the ith treatment $(0 \leq i \leq p)$, and let B_j denote the sum of all observations in the jth block $(1 \leq j \leq b)$. Define $B_i^* = \sum_{j=1}^{b} n_{ij}B_j$ and let $Q_i = kT_i - B_i^*$ $(0 \leq i \leq p)$.[1/] Then

$$\hat{\alpha}_0 - \hat{\alpha}_i = \frac{\lambda_1 Q_0 - \lambda_0 Q_i}{\lambda_0(\lambda_0 + p\lambda_1)} \quad (1 \leq i \leq p). \tag{4.1}$$

Also τ^2 of (3.1) is given by

$$\tau^2 = \frac{k(\lambda_0+\lambda_1)}{\lambda_0(\lambda_0+p\lambda_1)}, \tag{4.2}$$

and ρ of (3.2) is given by

$$\rho = \frac{\lambda_1}{\lambda_0 + \lambda_1}. \tag{4.3}$$

For a BTIB design, the confidence coefficient P_1 associated with joint <u>one-sided</u> intervals (2.2) can be written as

$$P_1 = \int_{-\infty}^{\infty} \left[\Phi\left(\frac{x\sqrt{\rho} + a/\tau\sigma}{\sqrt{1-\rho}}\right)\right]^p d\Phi(x), \tag{4.4}$$

and the confidence coefficient P_2 associated with joint <u>two-sided</u> intervals (2.3) can be written as

$$P_2 = \int_{-\infty}^{\infty} \left[\Phi\left(\frac{x\sqrt{\rho} + a/\tau\sigma}{\sqrt{1-\rho}}\right) - \Phi\left(\frac{x\sqrt{\rho} - a/\tau\sigma}{\sqrt{1-\rho}}\right)\right]^p d\Phi(x), \tag{4.5}$$

where $\Phi(\cdot)$ denotes the standard normal cdf, and τ and ρ are given by (4.2) and (4.3), respectively.

[1/] Usually in the literature on BIB designs, Q_i is defined as $T_i - B_i^*/k$ $(1 \leq i \leq k)$ and the Q_i are called "adjusted treatment totals." We use a slightly different definition in order to be consistent with the notation in Bechhofer and Tamhane (1981).

The equations (4.4) and (4.5) are derived under the assumption that σ^2 is known. In practice, the experimenter's knowledge of σ^2 may often be vague. In the absence of good knowledge of σ^2, a conservative upper bound on σ^2 can be specified for design purposes. In either situation one may wish to use the estimate of σ^2 obtained from the data for analysis purposes. Then the error mean square s_ν^2 based on $\nu = N-p-b$ degrees of freedom (df) can be used to estimate σ^2; this can be computed from the following analysis of variance table. In this table G denotes the sum of all observations, and H denotes the sum of squares of all observations.

Table 4.1. Analysis of Variance for a BTIB Design

Source	Sum of Squares (SS)	Degrees of Freedom (df)
Treatments (adjusted for blocks)	$\frac{1}{k(\lambda_0+p\lambda_1)}\left\{\frac{\lambda_1}{\lambda_0} Q_0^2 + \sum_{i=1}^{p} Q_i^2\right\}$	p
Blocks (unadjusted)	$\frac{1}{k}\sum_{j=1}^{b} B_j^2 - \frac{G^2}{N}$	$b-1$
Error	By subtraction	$N-p-b$
Total	$H - \frac{G^2}{N}$	$N-1$

The estimate s_ν^2 should be used with Dunnett's (1955) formulae to make joint confidence statements (analogous to (I) and (II) of (2.2) and (2.3), respectively), i.e.,

(III) a $100(1-\alpha)\%$ joint one-sided confidence statement

$$\{\alpha_0-\alpha_i \leq \hat{\alpha}_0-\hat{\alpha}_i+t_{\nu,p,\rho}^{(\alpha)}\tau s_\nu \quad (1 \leq i \leq p)\}, \tag{4.6}$$

or

(IV) a $100(1-\alpha)\%$ joint two-sided confidence statement

$$\{\alpha_0-\alpha_i \in [\hat{\alpha}_0-\hat{\alpha}_i \pm t'^{(\alpha)}_{\nu,p,\rho}\tau s_\nu] \quad (1 \leq i \leq p)\}. \tag{4.7}$$

In (4.6), $t_{\nu,p,\rho}^{(\alpha)}$ denotes the upper α-point of $\max\{t_i\ (1 \leq i \leq p)\}$ where the vector $(t_1,t_2,\ldots,t_p)$ has a p-variate equicorrelated central t distribution with common correlation ρ, and with df ν; for tables of $t_{\nu,p,\rho}^{(\alpha)}$ see Krishnaiah and Armitage (1966). In (4.7), $t'^{(\alpha)}_{\nu,p,\rho}$ denotes the upper α-point of the distribution of $\max\{|t_i|\ (1 \leq i \leq p)\}$; for tables of $t'^{(\alpha)}_{\nu,p,\rho}$ see Hahn and Hendrickson (1971). Also, Krishnaiah and Armitage (1970) have tabulated $(t'^{(\alpha)}_{\nu,p,\rho})^2$ for selected values of ν, p, ρ and α.

5. OPTIMAL DESIGNS, ADMISSIBLE DESIGNS AND MINIMAL COMPLETE SET OF GENERATOR DESIGNS

To motivate the ideas in this section we give a simple example. Suppose that an experimenter wishes to compare $p = 2$ test treatments with a control treatment using a matched pairs design (i.e., blocks of size $k = 2$) in $b = 13$ blocks. The six possible BTIB designs for this case along with their λ_0, λ_1, τ^2 and ρ-values are listed in Table 5.1. In this table the columns in each design refer to blocks, 0 refers to the control treatment while 1 and 2 refer to the first and second test treatment, respectively. The pairs (0,1), (0,2) and (1,2) within a block can be permuted without changing the design as can the columns in each design.

Specifically, suppose that the experimenter wishes to design an experiment for the purpose of obtaining joint one-sided confidence intervals of the form (2.2) for the differences $(\alpha_0-\alpha_1, \alpha_0-\alpha_2)$ using an allowance $\underline{a} = 0.5$ (say), and suppose further that he is prepared to assume that $\sigma = 1$ (say). The experimenter would like to know which of the (only possible) BTIB designs among the six listed in Table 5.1 will yield the <u>largest</u> confidence coefficient P_1 (4.4) and thus be <u>optimal</u>. The theory and tables given herein will assist the experimenter in answering this question. It will turn out that a design that is optimal for $\underline{a} = 0.5$ may not be optimal for another value of $\underline{a}$; also, a design that is optimal for a particular value of a/σ for <u>one-sided</u> confidence intervals (2.2) may not be optimal for the same value of a/σ for <u>two-sided</u> confidence intervals (2.3).

Table 5.1. Enumeration of BTIB Designs for $p = 2$, $k = 2$ and $b = 13$

D_i	Design	λ_0	λ_1	τ^2	ρ
D_1	$\begin{Bmatrix} 0&0&1&1&1&1&1&1&1&1&1&1&1 \\ 1&2&2&2&2&2&2&2&2&2&2&2&2 \end{Bmatrix}$	1	11	1.043	0.917
D_2	$\begin{Bmatrix} 0&0&0&0&1&1&1&1&1&1&1&1&1 \\ 1&2&1&2&2&2&2&2&2&2&2&2&2 \end{Bmatrix}$	2	9	0.550	0.818
D_3	$\begin{Bmatrix} 0&0&0&0&0&0&1&1&1&1&1&1&1 \\ 1&2&1&2&1&2&2&2&2&2&2&2&2 \end{Bmatrix}$	3	7	0.392	0.700
D_4	$\begin{Bmatrix} 0&0&0&0&0&0&0&0&1&1&1&1&1 \\ 1&2&1&2&1&2&1&2&2&2&2&2&2 \end{Bmatrix}$	4	5	0.321	0.556
D_5	$\begin{Bmatrix} 0&0&0&0&0&0&0&0&0&0&1&1&1 \\ 1&2&1&2&1&2&1&2&1&2&2&2&2 \end{Bmatrix}$	5	3	0.291	0.374
D_6	$\begin{Bmatrix} 0&0&0&0&0&0&0&0&0&0&0&0&1 \\ 1&2&1&2&1&2&1&2&1&2&1&2&2 \end{Bmatrix}$	6	1	0.292	0.143

Alternatively, suppose that the number of blocks (b) is at the disposal of the experimenter and that for a specified value of a/σ one wishes to guarantee a confidence coefficient P_1 (resp., P_2) for joint one-sided (resp., two-sided) confidence intervals of the form (2.2) (resp., (2.3)). Then one would like to know which of the possible BTIB designs accomplishes this objective using the smallest total number of blocks, and thus is optimal. The theory and tables given herein will also assist the experimenter in accomplishing this objective.

For the specific problems described above involving $b = 13$ blocks, it turns out that the experimenter should never use design D_6 in Table 5.1 since that design is inadmissible; this is so because its τ^2 (4.2) is larger than that of D_5 and its ρ (4.3) is smaller than that of D_5 and hence its P_1 (4.4) or P_2 (4.5) are smaller for every a/σ than the corresponding P_1 and P_2 values of D_5; all of the designs $D_1, D_2, \ldots, D_5$ are admissible. These concepts are formalized below in more generality.

We now give the definition of an optimal design.

Definition 5.1: For given (p,k) and specified a/σ and $1-\alpha$, the BTIB design which achieves a joint confidence coefficient P_1 (resp., P_2) $\geq 1-\alpha$ with the smallest b is said to be optimal for that value of $1-\alpha$ for one-sided (resp., two-sided) comparisons. Here P_1 and P_2 are given by (4.4) and (4.5), respectively.

We note that our notion of optimality is different from that of others considered in the literature such as D-, E- and A- optimality of Kiefer (1958). Recently Majumdar and Notz (1983) have studied the performance of BTIB designs using these latter notions of optimality.

In the search for an optimal design we can eliminate inadmissible designs and restrict attention to admissible designs which are defined as follows.

Definition 5.2: Suppose that for given (p,k) we have two BTIB designs D_1 and D_2, with D_i having parameters (b_i,τ_i^2,ρ_i) $(i = 1,2)$ where $b_1 \leq b_2$. If for every a/σ, D_1 yields a joint confidence coefficient (P_1 or P_2) at least as large as (resp., larger than) that yielded by D_2 when $b_1 \leq b_2$ (resp., $b_1 = b_2$), then D_2 is said to be inadmissible with respect to (wrt) D_1. If a design is not inadmissible, then it is said to be admissible. If $b_1 = b_2$, $\tau_1^2 = \tau_2^2$ and $\rho_1 = \rho_2$ then D_1 and D_2 are said to be equivalent.

Theorem 5.1 of Bechhofer and Tamhane (1981) shows that a necessary and sufficient condition that D_2 be inadmissible wrt D_1 is $b_1 \leq b_2$, $\tau_1^2 \leq \tau_2^2$ and $\rho_1 \geq \rho_2$ with at least one inequality strict.

It is clear from the preceding two definitions that for given (p,k) the set of admissible designs is the same whether one considers one-sided joint confidence intervals (2.2) or two-sided joint confidence intervals (2.3). However, as we will see, the optimal designs can be different for the same a/σ in the two cases.

For given (p,k) all BTIB designs can be built up from a set of elementary BTIB designs called generator designs which are defined as follows:

Definition 5.3: For given (p,k) a generator design is a BTIB design no proper subset of whose blocks forms a BTIB design, and no block of which contains only one of the $p+1$ treatments.

Any BTIB design is either a generator design or a union of generator designs. In fact, if for given (p,k) there are $n \geq 2$ generator designs with the ith generator design D_i having the parameters $(b_i, \lambda_0^{(i)}, \lambda_1^{(i)})$ $(1 \leq i \leq n)$, then a BTIB design $D = \bigcup_{i=1}^{n} f_i D_i$ obtained by forming unions of $f_i \geq 0$ replications of D_i $(1 \leq i \leq n)$ has the parameters $\lambda_0 = \sum_{i=1}^{n} f_i \lambda_0^{(i)}$, $\lambda_1 = \sum_{i=1}^{n} f_i \lambda_1^{(i)}$ and $b = \sum_{i=1}^{n} f_i b_i$.

Usually the number of generator designs for given (p,k) is quite large. However, in order to construct all admissible BTIB designs (except equivalent ones), we can restrict attention to a small subset of these designs. This subset is called the minimal complete set of generator designs which is defined as follows:

Definition 5.4: For given (p,k) the smallest set of generator designs $\{D_i \ (1 \leq i \leq n)\}$ from which all admissible designs (except possibly equivalent ones) can be constructed for that (p,k) is called the minimal complete set of generator designs.

To obtain the minimal complete set of generator designs for given (p,k) one begins with the set of all generator designs. All equivalent generator designs and all so-called strongly (S-) inadmissible and combination (C-) inadmissible generator designs then must be deleted from this set. In this way the minimal complete set of generator designs is obtained. The definition of S- and C-inadmissible designs, and the foregoing procedure are given in detail in Section 5 of Bechhofer and Tamhane (1981). Usually it is quite difficult to determine whether one has constructed all of the possible generator designs for a given (p,k) (the only obvious exception being the case $p \geq 2$, $k = 2$). Therefore indirect methods of proof are needed to verify whether or not a conjectured set of generator designs constitutes the minimal complete set. Such proofs are given for the cases $p = 3(1)10$, $k = 3$ in Notz and Tamhane (1983). The results of this latter article are extended to the case $p = k(1)10$, $k = 4$ and 5 in a Berkeley Ph.D. dissertation by Ture (1982).

6. DESCRIPTION OF THE TABLES

6.1 General remarks

In Appendices A.1, A.2, and A.3 we give tables of minimal complete sets of generator designs, admissible designs, and optimal designs, respectively, for $p = k(1)6$ and $k = 2,3$. For given (p,k) the table of the minimal complete set of generator designs is labeled MIN•p•k; the table of the admissible designs is labeled ADM•p•k; and the tables of the optimal designs are labeled OPT1•p•k and OPT2•p•k where OPT1 corresponds to one-sided intervals and OPT2 corresponds to two-sided intervals. An optimal design table gives for selected b and (a/σ)-values, the optimal design and its associated joint confidence coefficient, i.e., P_1 of (4.4) for OPT1 tables or P_2 of (4.5) for OPT2 tables.

6.2 Tables of minimal complete sets of generator designs

For each (p,k) we first give the table of the minimal complete set of generator designs. The designs in each table are listed in the following order: The designs containing controls are listed first in increasing order of their b-values. The designs not containing controls are listed next, also in increasing order of their b-values. The numbers assigned to the generator designs in this table are used to refer to them in the table of admissible designs, and also in the tables of optimal designs.

For each $p \geq 2$, $k = 2$ the minimal complete set of generator designs consists of only two designs, D_1 and D_2, which are given in Table MIN•p•2, Appendix A.1. For each $p \geq 2$, $k = 2$ any BTIB design D can be written as $f_1D_1 \cup f_2D_2$ where $f_1 \geq 1$, $f_2 \geq 0$. For these D we have $b = f_1p + f_2p(p-1)/2$, $\lambda_0 = f_1$ and $\lambda_1 = f_2$.

For $p = 3$, $k = 3$ the minimal complete set of generator designs consists of only two designs, D_1 and D_2, which are given in Table MIN•3•3,

Appendix A.1. For $p = 3$, $k = 3$ any BTIB design D can be written as $f_1D_1 \cup f_2D_2$ where $f_1 \geq 1$, $f_2 \geq 0$. For these D we have $b = 3f_1 + f_2$ $(b = 3,4,\ldots)$, $\lambda_0 = 2f_1$, and $\lambda_1 = f_1 + f_2$.

For each $p \geq 4$, $k = 3$ the minimal complete set of generator designs consists of more than two generator designs; these designs are given in Tables MIN•p•k for each $p = 4(1)6$, $k = 3$, in Appendix A.1. (Proofs of their minimal complete nature are given in Notz and Tamhane (1983) where minimal complete sets also are given for the additional cases $p = 7(1)10$, $k = 3$.)

6.3 Tables of admissible designs

The main purpose of these tables, independent of their theoretical interest, is to provide the experimeter with a list of "good" designs from among which a choice can be made when some practical considerations dictate that the optimal design given in the OPT tables cannot be used. This might happen if, for example, the optimal design has a certain treatment combination in a block which is infeasible to employ in practice. Any design that is not in the list of admissible designs should not be used since it can be uniformly dominated by at least one design in the list while no design in the list can be uniformly dominated by any other design; of course, an equivalent admissible design can be substituted for one not in the table.

For given (p,k), and for each b the admissible designs are listed in the order of increasing τ^2 and increasing ρ-values. It will be seen from the tables of optimal designs that designs with small τ^2-values tend to be better for large $1-\alpha$ and/or for small a/σ, while the designs with large ρ-values tend to be better for small $1-\alpha$ and/or for large a/σ.

It was shown in Bechhofer and Tamhane (1982) that for $k = 2$, the possibility of a BTIB design being inadmissible wrt another BTIB design with smaller b arises only rarely. (In fact, for $p = 2(1)6$, $k = 2$ an exhaustive enumerative search turned up only four such cases, one for $p = 4$, $k = 2$ and three for $p = 6$, $k = 2$.) However, for $k = 2$ and given $b \geq p$ it was shown that a BTIB design $f_1D_1 \cup f_2D_2$ (where D_1 and D_2 are given in Table MIN•p•2) with

$f_1^L \leq f_1 \leq f_1^U$ is inadmissible wrt another BTIB design with the <u>same</u> b; here f_1^U is the upper limit on f_1 and

$$f_1^L = \begin{cases} \text{int}\left[\frac{2b+1}{2} - \sqrt{\frac{b^2}{3} + \frac{1}{4}}\right] & \text{for } p = 2 \\ \text{int}\left[\frac{b+2}{4}\right] & \text{for } p = 3 \\ \text{int}\left[\frac{p(p-3)q-4b}{2p(p-3)} - \sqrt{\frac{4(p-1)^2b^2}{(p+1)p^2(p-3)^2} + \frac{q^2}{4}}\right] & \text{for } p \geq 4 \end{cases} \tag{6.1}$$

is the lower limit on f_1. In (6.1), we have $q = p-1$ if p is even and $q = (p-1)/2$ if p is odd; also int[z] denotes the smallest integer $\geq z$. For $p = 2$, $k = 2$ we have $f_1^U = b/2$ if b is even and $(b-1)/2$ if b is odd. Thus for $2 \leq b \leq 5$ all designs are admissible; for $b = 6$ the design with $(f_1,f_2) = (3,0)$ is inadmissible; for $b = 20$ the designs with (f_1,f_2) (9,2) and (10,0) are inadmissible, etc. Since admissible designs can be characterized in an easy manner for $p \geq 2$, $k = 2$, we have not given tables of such designs for $k = 2$.

It was also shown in Bechhofer and Tamhane (1982) that for $p = 3$, $k = 3$ all BTIB designs that are obtained by forming unions of the generator designs in the minimal complete set (Table MIN•3•3) are admissible. Therefore we have not given a table of admissible designs for this case.

For the cases $p = 4(1)6$, $k = 3$ the admissible designs cannot be characterized in an easy manner. Therefore admissible designs for these cases are listed in special tables labeled ADM•p•k. For given (p,k,b), if there is a set of equivalent admissible generator designs then we list only one design from such a set. The message "NO DESIGNS OF THIS SIZE EXIST" means that for that b-value there are no implementable BTIB designs, i.e., there are no BTIB designs with $\lambda_0 > 0$. For example, for $p = 6$, $k = 3$ (see Table ADM•6•3) there are no implementable BTIB designs for $b = 8,9,10$; thus although the generator design D_5 (see Table MIN•6•3) has $b = 10$, it is not implementable since it contains no controls. The message "ALL DESIGNS OF THIS SIZE ARE INADMISSIBLE" means that all BTIB designs for that b-value are inadmissible wrt some BTIB designs having smaller b-values. For example, for $p = 6$, $k = 3$ (see Table ADM•6•3) all BTIB designs for $b = 12,13,16,19,20$ and 23 are inadmissible wrt some BTIB designs having smaller b-value; for example, for $b = 12$ the design $2D_1$ with $\tau^2 = 0.75$, $\rho = 0$ is inadmissible wrt the design D_3 with $b = 11$, $\tau^2 = 0.4444$, $\rho = 0.25$.

For every admissible design for given b, the table gives its λ_0, λ_1, τ^2, and ρ-values and the frequency vector $(f_1,f_2,\ldots,f_n)$. For example, for $p = 4$, $k = 3$ there are two admissible designs for $b = 25$ (see Table ADM•4•3); the first one has the frequency vector (0,3,1,0,0,0), i.e., it is the design $3D_2 \cup D_3$, while the second one has the frequency vector (0,1,1,0,0,3), i.e., it is the design $D_2 \cup D_3 \cup 3D_6$ where the generator designs D_i $(1 \leq i \leq 6)$ are given in Table MIN•4•3. The first design has $\lambda_0 = 11$, $\lambda_1 = 5$, $\tau^2 = 0.1408$, $\rho = 0.3125$ while the second design has $\lambda_0 = 5$, $\lambda_1 = 9$, $\tau^2 = 0.2049$, $\rho = 0.6429$.

It might be pointed out that for $p = 4$, $k = 3$, the generator design D_1 appears in only one admissible design (by itself for $b = 4$); see Table ADM•4•3. The same is true of the generator design D_1 for $p = 5$, $k = 3$ which is admissible for $b = 5$; see Table ADM•5•3.

6.4 Tables of optimal designs

We now discuss the tables of optimal designs. For given (p,k) in the optimal design tables we list first the implementable design with the smallest b-value, and for selected a/σ we then list optimal designs and their associated joint confidence coefficients for increasing values of b for which admissible designs exist. When computing these tables it sometimes occurs for some a/σ that the joint confidence coefficient decreases as b increases. When this happens the experimenter should, of course, use the smallest value of b that guarantees the specified value of confidence coefficient $1-\alpha$. Designs for which, for given a/σ, the confidence coefficient is no greater than the joint confidence coefficient associated with a design with <u>smaller</u> b are not shown in these tables since according to Definition 5.1 they are not optimal.

Optimal designs with confidence coefficient > 0.9999 are not shown in this table.

For given b and a/σ, the entry in the table gives the frequency vector of the optimal design and the associated joint confidence coefficient for each P_i $(i = 1,2)$ in the following form

$$\begin{bmatrix} f_1 & f_2 & f_3 \\ \cdot & \cdot & \cdot \\ f_{n-1} & f_n & \\ & P_i & \end{bmatrix}.$$

For example, for $p = 6$, $k = 2$ the optimal design for one-sided comparisons for $b = 51$, $a/\sigma = 0.4$ (see Table OPT1•6•2) is $D_1 \cup 3D_2$ with $P_1 = 0.4935$ while the corresponding optimal design for two-sided comparisons is $6D_1 \cup D_2$ with $P_2 = 0.0707$ (see Table OPT2•6•2); the generator designs D_1 and D_2 are given in Table MIN•6•2. Note that in this example, the optimal designs are different for one-sided and two-sided comparisons. Note also that for $b = 54$, $a/\sigma = 0.4$ there is a blank in the OPT1•6•2 table which means that all BTIB designs for $b = 54$ yield P_1 no greater than 0.4935 (= the maximum P_1 obtained for $b = 51$, $a/\sigma = 0.4$), and therefore none of them is optimal for $a/\sigma = 0.4$. However, for $a/\sigma = 0.5$, the design $4D_1 \cup 2D_2$ is optimal for $b = 54$ for one-sided comparisons. Also, for $a/\sigma = 0.4$, $b = 54$ the design $4D_1 \cup 2D_2$ is optimal for two-sided comparisons although, as just noted, there is no optimal design for $a/\sigma = 0.4$, $b = 54$ for one-sided comparisons.

The (a/σ)-scale used on any page of the table lists (a/σ)-values from the lower limit $(a/\sigma)_L$ to the upper limit $(a/\sigma)_U$ in steps of 0.1 where $(a/\sigma)_U-(a/\sigma)_L = 1.1$, i.e., there are twelve (a/σ)-values listed on each page. We begin the first table with $(a/\sigma)_L = 0.5$ and $(a/\sigma)_U = 1.6$. When the table is to be continued on the following page the (a/σ)-scale is "slid" downwards as follows: We find the smallest listed value of $(a/\sigma) \leq (a/\sigma)_U$ for which, for some b-value listed on the current page of the table, the confidence coefficient associated with the corresponding optimal design exceeds 0.995. If for all b and all (a/σ)-values the confidence coefficients are less than or equal to 0.995 then the same (a/σ)-scale is continued on the next page. If, on the other hand, $(a/\sigma)^*$ (say) is the smallest value of (a/σ) for which the confidence coefficient exceeds 0.995, then on the following page we take $(a/\sigma)_U = \max((a/\sigma)^* - 0.1, 1.2)$ and $(a/\sigma)_L = (a/\sigma)_U - 1.1$. In other words, the final (a/σ)-scale is $0.1(0.1)1.2$.

A perusal of the optimal design table shows that for the small b-values commonly employed in practice, confidence coefficients can be embarrassingly small, if the experimenter wishes to specify a small allowance. For example, from Table OPT2·6·2 we find that for $p = 6$, $k = 2$, if the experimenter specifies $a/\sigma = 0.1$ then even with b as large as 96, one can expect to obtain a two-sided joint confidence coefficient of no more than 0.0003 while for $a/\sigma = 1.0$ one can attain the joint confidence coefficient of 0.90 with $b = 57$. This example highlights the type of tradeoffs which are necessary between the experimenter's requirements and the resources available for experimentation. It also emphasizes the need for large experiments if the conclusions are to be highly reliable.

7. DETAILS OF COMPUTATION

A Fortran computer program was written to generate a table of admissible designs and tables of optimal designs for given (p,k). The input to the program consists of the following quantities: (p,k), the smallest b associated with an implementable design, the number of generator designs in the minimal complete set, their (b,λ_0,λ_1)-values, and the initial (a/σ)-scale.

The program proceeds as follows: For each b-value the program first generates all implementable BTIB designs that can be constructed from the given minimal complete set of generator designs. If $\{D_i\ (1 \leq i \leq n)\}$ is the minimal complete set of generator designs with D_i having parameter values $(b_i,\lambda_0^{(i)},\lambda_1^{(i)})$ $(1 \leq i \leq n)$ then the preceding step is equivalent to finding all frequency vectors $\underset{\sim}{f} = (f_1,f_2,\ldots,f_n)$ such that $\sum_{i=1}^n f_i b_i = b$ and $\sum_{i=1}^n f_i\lambda_0^{(i)} > 0$. If at least one such $\underset{\sim}{f}$ can be found, i.e., if at least one implementable BTIB design exists for that b, then the program proceeds to delete all but one of any equivalent BTIB designs, and any inadmissible BTIB designs. If no implementable BTIB designs exist for that b or if all implementable BTIB designs that exist are inadmissible wrt some designs with smaller b-values then the corresponding messages are printed and the subsequent step of finding the "optimal" designs for that b, which is described in the paragraph below, is skipped. If, on the other hand, there is at least one admissible BTIB design for that b then all such designs (i.e., their frequency vectors $\underset{\sim}{f}$) along with their λ_0, λ_1, τ^2 and ρ-values are printed. This generates the admissible design table.

The next task is to generate the optimal design table. For this we begin with the smallest b associated with an implementable design, and for each such b and larger ones we calculate P_1 and P_2 using (4.4) and (4.5), respectively, for every admissible design for that b and for every a/σ in the scale which is chosen as described in Section 6.4. The design with the largest value of P_1 (resp., P_2) for given b and a/σ is printed with its associated P_1 (resp., P_2) value. The optimal designs are found from this table by inspection. Thus if for given a/σ and b the maximum value of P_1 (resp., P_2) is less than or equal to the maximum value of P_1 (resp., P_2) attained for a smaller b then the design associated with the larger b is deleted from the table. The designs that are not deleted using the procedure described above are the optimal designs in the sense of Definition 5.1. For practical reasons, we further delete from this table of optimal designs those designs for which the P_1 (resp., P_2) value exceeds 0.9999.

Romberg quadrature method was used to evaluate the integrals (4.4) and (4.5). The formula (26.2.17) of Abramowitz and Stegun (1964) was used to evaluate $\Phi(\cdot)$; this formula is accurate to within $\pm\ 7.5 \times 10^{-8}$. All calculations were performed on Northwestern University's CDC Cyber 170/730 in single precision which is equivalent to double precision on an IBM machine.

8. USES OF THE TABLES

In this section we give three numerical examples to illustrate the uses of the tables.

Example 1: Suppose that an experimenter wishes to compare two test treatments with a control treatment in blocks of size two, i.e., $p = 2$, $k = 2$. Suppose further that the experimenter wishes to make one-sided comparisons; a common fixed standardized allowance $a/\sigma = 1$ and confidence coefficient $1-\alpha = 0.95$ are specified. From Table OPT1•2•2 one finds that for $a/\sigma = 1.0$ the smallest number of blocks required to guarantee a joint confidence coefficient of 0.95 is 15; the actual achieved joint confidence coefficient is 0.9568. The corresponding optimal design requires 6 replications of $D_1 = \{\begin{smallmatrix} 0 & 0 \\ 1 & 2 \end{smallmatrix}\}$ and 3 replications of $D_2 = \{\begin{smallmatrix} 1 \\ 2 \end{smallmatrix}\}$. Thus out of a total of 30 experimental units, 12 are allocated to the control treatment and 9 are allocated to each of the two test treatments. If two-sided comparisons were desired instead of one-sided comparisons then the smallest number of blocks required would be 19 and the corresponding optimal design would be $8D_1 \cup 3D_2$.

Example 2: Suppose that $p = 6$, $k = 3$ and that the experimenter has $b = 37$ blocks available for experimentation. From Table ADM•6•3 it is seen that the only admissible designs for $b = 37$ are $2D_3 \cup D_4$ and $D_2 \cup 3D_5$ where the generator designs D_i $(1 \leq i \leq 5)$ are given in Table MIN•6•3. Thus designs such as $D_2 \cup 2D_4$ or $D_1 \cup D_3 \cup 2D_5$ which also consist of 37 blocks are inadmissible and therefore should not be used.

If the experimenter specifies $a/\sigma = 0.3$ for one-sided comparisons, then $D_2 \cup 3D_5$ is the optimal design which yields a joint confidence coefficient of 0.4810. On the other hand, if the experimenter specifies $a/\sigma = 0.6$ then $2D_3 \cup D_4$ is the optimal design which yields a joint confidence coefficient of 0.7696. In fact, $D_2 \cup 3D_5$ is the optimal design for $a/\sigma \leq 0.3$ while $2D_3 \cup D_4$ is the optimal design for $a/\sigma \geq 0.6$; for $a/\sigma = 0.4$ and 0.5 optimal designs with less than 37 blocks achieve a larger joint confidence coefficient and hence these latter designs should be used.

The above examples deal with cases in which the specified value of a/σ is tabulated. We now show by means of an example how one can interpolate wrt a/σ if the specified value of a/σ is not tabulated.

Example 3: Suppose that $p = 2$, $k = 2$, $a/\sigma = 0.85$ and the desired one-sided confidence coefficient is 0.99. From Table OPT1•2•2 the experimenter finds that to obtain $1-\alpha \geq 0.99$ one requires $b = 39$ for $a/\sigma = 0.80$ and $b = 31$ for $a/\sigma = 0.90$. Thus for $a/\sigma = 0.85$ the b needed is between 31 and 39. Equations (4.4) and (4.5) show that for fixed ρ, to obtain the same confidence coefficient, $a/\tau\sigma$ should be constant. It is known that for large values of the joint confidence coefficient approaching unity, the integrals (4.4) and (4.5) are approximately independent of ρ for $0 < \rho < 1$. Therefore $a/\tau\sigma$ should be roughly constant to obtain the same joint confidence coefficient. However, τ^2 is approximately proportional to $1/b$, and therefore one should have $b(a/\sigma)^2$ roughly constant. Applying this interpolation rule one obtains for $a/\sigma = 0.85$ and confidence coefficient = 0.99,

$$b \cong \left(\frac{0.80}{0.85}\right)^2 \times 39 = 34.55$$

or, alternatively

$$b \cong \left(\frac{0.90}{0.85}\right)^2 \times 31 = 34.75.$$

Note that both of the values of b obtained by interpolating from either end are very close and thus we choose $b = 35$. The appropriate design may be chosen to be either the optimal design for $b = 35$ for $a/\sigma = 0.80$ or the optimal design for $b = 35$ for $a/\sigma = 0.90$. In the present example they happen to be identical namely $15D_1 \cup 5D_2$ where D_1 and D_2 are given in Table MIN•2•2. The exact confidence coefficient yielded by this design for $a/\sigma = 0.85$ is 0.9909 which is very close to the desired value of 0.99.

9. ACKNOWLEDGMENTS

The writers would like to thank Messrs. Carl Emont and Stephen Mykytyn who in its early stages contributed to the present project in more ways than one.

They wrote the early versions of the computer programs used in the production of the tables of admissible and optimal BTIB designs given herein. They also provided some insights into the construction and admissibility aspects of BTIB designs.

REFERENCES

1. Abramowitz, M. and Stegun, I. (1964). Handbook of Mathematical Functions. National Bureau of Standards Applied Mathematics Ser. 55, Washington, D.C.: U.S. Govt. Printing Office.

2. Bechhofer, R.E. and Tamhane, A.C. (1981). Incomplete block designs for comparing treatments with a control: General theory. Technometrics 23, 45-57. Corrigendum, Technometrics 24, 171.

3. Bechhofer, R.E. and Tamhane, A.C. (1983). Incomplete block designs for comparing treatments with a control (II): Optimal designs for $p = 2(1)6$, $k = 2$ and $p = 3$, $k = 3$. Sankhyā, Ser. B, 45, 193-224.

4. Bechhofer, R.E. and Tamhane, A.C. (1983). Design of experiments for comparing treatments with a control: Tables of optimal allocations of observations. Technometrics 4, 87-95.

5. Dunnett, C.W. (1955). A multiple comparison procedure for comparing several treatments with a control. Journal of the American Statistical Association 50, 1096-1121.

6. Hahn, G.J. and Hendrickson, R.W. (1971). A table of percentage points of the distribution of the largest absolute value of k Student t variates and its applications. Biometrika 58, 323-332.

7. Kiefer, J. (1958). On the nonrandomized optimality and randomized nonoptimality of symmetrical designs. Annals of Mathematical Statistics 29, 675-699.

8. Krishnaiah, P.R. and Armitage, J.V. (1966). Tables for multivariate t distribution. Sankhyā, Ser. B, 28, 31-56.

9. Krishnaiah, P.R. and Armitage, J.V. (1970). On a multivariate F distribution. In Essays in Probability and Statistics (eds. R.C. Bose et al.), Chapel Hill: Univ. of North Carolina Press, 439-468.

10. Majumdar, D. and Notz, W.I. (1983). Optimal incomplete block designs for comparing treatments with a control, Annals of Statistics 11, 258-266

11. Notz, W.I. and Tamhane, A.C. (1983). Incomplete block (BTIB) designs for comparing treatments with a control: Minimal complete sets of generator designs for $k = 3$, $p = 3(1)10$. Communications in Statistics, Ser. A, Theory and Methods 12 (12), 1391-1412.

12. Pierce, S.C. (1960). Supplemented balance. Biometrika 47, 263-271.

13. Ture, T.E. (1982). On the construction and optimality of balanced treatment incomplete block designs. Ph.D. dissertation, Department of Statistics, University of California, Berkeley.

APPENDIX A.1

Tables of Minimal Complete Sets of Generator Designs

Tables MIN•p•k for $p \geq 2$, $k = 2$

and $p = 3(1)6$, $k = 3$

Table MIN•p•2

Minimal Complete Set of Generator Designs for $p \geq 2$, $k = 2$

D_i	Design	b_i	$\lambda_0^{(i)}$	$\lambda_1^{(i)}$
D_1	$\left\{\begin{matrix} 0 & 0 & & 0 \\ 1 & 2 & \cdots & p \end{matrix}\right\}$	p	1	0
D_2	$\left\{\begin{matrix} 1 & 1 & & p-1 \\ 2 & 3 & \cdots & p \end{matrix}\right\}$	$\frac{p(p-1)}{2}$	0	1

Table MIN•3•3

Minimal Complete Set of Generator Designs for $p = 3$, $k = 3$

D_i	Design	b_i	$\lambda_0^{(i)}$	$\lambda_1^{(i)}$
D_1	$\left\{\begin{matrix} 0 & 0 & 0 \\ 1 & 1 & 2 \\ 2 & 3 & 3 \end{matrix}\right\}$	3	2	1
D_2	$\left\{\begin{matrix} 1 \\ 2 \\ 3 \end{matrix}\right\}$	1	0	1

Table MIN•4•3

Minimal Complete Set of Generator Designs for $p = 4,\ k = 3$

D_i	Design	b_i	$\lambda_0^{(i)}$	$\lambda_1^{(i)}$
D_1	$\begin{Bmatrix} 0 & 0 & 0 & 0 \\ 0 & 0 & 0 & 0 \\ 1 & 2 & 3 & 4 \end{Bmatrix}$	4	2	0
D_2	$\begin{Bmatrix} 0 & 0 & 0 & 0 & 0 & 0 \\ 1 & 1 & 1 & 2 & 2 & 3 \\ 2 & 3 & 4 & 3 & 4 & 4 \end{Bmatrix}$	6	3	1
D_3	$\begin{Bmatrix} 0 & 0 & 0 & 0 & 1 & 1 & 2 \\ 0 & 1 & 1 & 2 & 2 & 3 & 3 \\ 3 & 2 & 4 & 4 & 3 & 4 & 4 \end{Bmatrix}$	7	2	2
D_4	$\begin{Bmatrix} 0 & 0 & 1 & 1 & 1 & 1 & 2 & 2 \\ 1 & 3 & 2 & 2 & 3 & 3 & 3 & 4 \\ 2 & 4 & 3 & 4 & 4 & 4 & 3 & 4 \end{Bmatrix}$	8	1	3
D_5	$\begin{Bmatrix} 0 & 0 & 0 & 0 & 0 & 0 & 0 & 0 & 1 & 2 \\ 0 & 0 & 1 & 1 & 1 & 1 & 2 & 2 & 3 & 3 \\ 3 & 4 & 2 & 2 & 3 & 4 & 3 & 4 & 4 & 4 \end{Bmatrix}$	10	4	2
D_6	$\begin{Bmatrix} 1 & 1 & 1 & 2 \\ 2 & 2 & 3 & 3 \\ 3 & 4 & 4 & 4 \end{Bmatrix}$	4	0	2

Table MIN•5•3

Minimal Complete Set of Generator Designs for $p = 5$, $k = 3$

D_i	Design	b_i	$\lambda_0^{(i)}$	$\lambda_1^{(i)}$
D_1	0 0 0 0 0 0 0 0 0 0 1 2 3 4 5	5	2	0
D_2	0 0 0 0 0 1 2 0 1 1 3 4 2 3 2 3 5 4 5 4 5	7	2	1
D_3	0 0 0 0 0 1 1 1 2 2 1 1 2 3 4 2 3 4 3 3 2 3 4 5 5 5 4 5 4 5	10	2	2
D_4	0 0 0 0 0 0 0 0 0 0 1 1 1 1 2 2 2 3 3 4 2 3 4 5 3 4 5 4 5 5	10	4	1
D_5	1 1 1 2 2 3 1 2 2 3 3 4 4 4 3 5 5 4 5 5 4	7	0	2
D_6	1 1 1 1 1 1 2 2 2 3 2 2 2 3 3 4 3 3 4 4 3 4 5 4 5 5 4 5 5 5	10	0	3

Table MIN•6•3

Minimal Complete Set of Generator Designs for p = 6, k = 3

D_i	Design	b_i	$\lambda_0^{(i)}$	$\lambda_1^{(i)}$
D_1	0 0 0 0 0 0 0 0 0 0 0 0 1 2 3 4 5 6	6	2	0
D_2	0 0 0 1 1 2 3 1 2 4 2 5 3 4 3 6 5 4 6 5 6	7	1	1
D_3	0 0 0 0 0 0 0 0 0 1 2 1 1 1 2 2 3 3 4 4 3 5 2 5 6 3 4 5 6 5 6 5 6	11	3	1
D_4	0 0 0 0 0 0 0 0 0 0 0 0 0 0 0 1 1 1 1 1 2 2 2 2 3 3 3 4 4 5 2 3 4 5 6 3 4 5 6 4 5 6 5 6 6	15	5	1
D_5	1 1 1 1 1 2 2 2 3 4 2 2 3 3 4 3 3 4 5 5 5 6 4 6 5 4 5 6 6 6	10	0	2

APPENDIX A.2

Tables of Admissible Designs

Tables ADM•p•k for $p = 4(1)6$, $k = 3$

TABLE ADM.4.3

P=4 K=3 ADMISSIBLE DESIGNS

b	λ_0	λ_1	τ^2	ρ	1	2	3	4	5	6
4	2	0	1.5000	.0000	1	0	0	0	0	0
5	NO DESIGNS OF THIS SIZE EXIST									
6	3	1	.5714	.2500	0	1	0	0	0	0
7	2	2	.6000	.5000	0	0	1	0	0	0
8	1	3	.9231	.7500	0	0	0	1	0	0
9	NO DESIGNS OF THIS SIZE EXIST									
10	4	2	.3750	.3333	0	0	0	0	1	0
	3	3	.4000	.5000	0	1	0	0	0	1
11	2	4	.5000	.6667	0	0	1	0	0	1
12	6	2	.2857	.2500	0	2	0	0	0	0
	1	5	.8571	.8333	0	0	0	1	0	1
13	5	3	.2824	.3750	0	1	1	0	0	0
14	4	4	.3000	.5000	0	0	0	0	1	1
	3	5	.3478	.6250	0	1	0	0	0	2
15	2	6	.4615	.7500	0	0	1	0	0	2
16	7	3	.2256	.3000	0	1	0	0	1	0
	6	4	.2273	.4000	0	2	0	0	0	1
	1	7	.8276	.8750	0	0	0	1	0	2
17	5	5	.2400	.5000	0	1	1	0	0	1
18	9	3	.1905	.2500	0	3	0	0	0	0
	4	6	.2679	.6000	0	0	0	0	1	2
	3	7	.3226	.7000	0	1	0	0	0	3
19	8	4	.1875	.3333	0	2	1	0	0	0
	2	8	.4412	.8000	0	0	1	0	0	3
20	7	5	.1905	.4167	0	2	0	1	0	0
	6	6	.2000	.5000	0	2	0	0	0	2
	1	9	.8108	.9000	0	0	0	1	0	3
21	5	7	.2182	.5833	0	1	1	0	0	2
22	9	5	.1609	.3571	0	3	0	0	0	1
	4	8	.2500	.6667	0	0	0	0	1	3
	3	9	.3077	.7500	0	1	0	0	0	4
23	8	6	.1641	.4286	0	2	1	0	0	1
	2	10	.4286	.8333	0	0	1	0	0	4
24	12	4	.1429	.2500	0	4	0	0	0	0

TABLE ADM.4.3

P=4 K=3 ADMISSIBLE DESIGNS

b	λ_0	λ_1	τ^2	ρ	1	2	3	4	5	6
24	7	7	.1714	.5000	0	2	0	1	0	1
	6	8	.1842	.5714	0	2	0	0	0	3
	1	11	.8000	.9167	0	0	0	1	0	4
25	11	5	.1408	.3125	0	3	1	0	0	0
	5	9	.2049	.6429	0	1	1	0	0	3
26	10	6	.1412	.3750	0	3	0	1	0	0
	9	7	.1441	.4375	0	3	0	0	0	2
	4	10	.2386	.7143	0	0	0	0	1	4
	3	11	.2979	.7857	0	1	0	0	0	5
27	8	8	.1500	.5000	0	2	1	0	0	2
	2	12	.4200	.8571	0	0	1	0	0	5
28	12	6	.1250	.3333	0	4	0	0	0	1
	7	9	.1595	.5625	0	2	0	1	0	2
	6	10	.1739	.6250	0	2	0	0	0	4
	1	13	.7925	.9286	0	0	0	1	0	5
29	11	7	.1259	.3889	0	3	1	0	0	1
	5	11	.1959	.6875	0	1	1	0	0	4
30	15	5	.1143	.2500	0	5	0	0	0	0
	10	8	.1286	.4444	0	3	0	1	0	1
	9	9	.1333	.5000	0	3	0	0	0	3
	4	12	.2308	.7500	0	0	0	0	1	5
	3	13	.2909	.8125	0	1	0	0	0	6
31	14	6	.1128	.3000	0	4	1	0	0	0
	8	10	.1406	.5556	0	2	1	0	0	3
	2	14	.4138	.8750	0	0	1	0	0	6
32	13	7	.1126	.3500	0	4	0	1	0	0
	12	8	.1136	.4000	0	4	0	0	0	2
	7	11	.1513	.6111	0	2	0	1	0	3
	6	12	.1667	.6667	0	2	0	0	0	5
	1	15	.7869	.9375	0	0	0	1	0	6
33	11	9	.1161	.4500	0	3	1	0	0	2
	5	13	.1895	.7222	0	1	1	0	0	5
34	15	7	.1023	.3182	0	5	0	0	0	1
	10	10	.1200	.5000	0	3	0	1	0	2
	9	11	.1258	.5500	0	3	0	0	0	4
	4	14	.2250	.7778	0	0	0	0	1	6
	3	15	.2857	.8333	0	1	0	0	0	7
35	14	8	.1025	.3636	0	4	1	0	0	1
	8	12	.1339	.6000	0	2	1	0	0	4
	2	16	.4091	.8889	0	0	1	0	0	7
36	18	6	.0952	.2500	0	6	0	0	0	0
	13	9	.1036	.4091	0	4	0	1	0	1

TABLE ADM.4.3

P=4 K=3 ADMISSIBLE DESIGNS

b	λ_0	λ_1	τ^2	ρ	1	2	3	4	5	6
36	12	10	.1058	.4545	0	4	0	0	0	3
	7	13	.1453	.6500	0	2	0	1	0	4
	6	14	.1613	.7000	0	2	0	0	0	6
	1	17	.7826	.9444	0	0	0	1	0	7
37	17	7	.0941	.2917	0	5	1	0	0	0
	11	11	.1091	.5000	0	3	1	0	0	3
	5	15	.1846	.7500	0	1	1	0	0	6
38	16	8	.0938	.3333	0	5	0	1	0	0
	15	9	.0941	.3750	0	5	0	0	0	2
	10	12	.1138	.5455	0	3	0	1	0	3
	9	13	.1202	.5909	0	3	0	0	0	5
	4	16	.2206	.8000	0	0	0	0	1	7
	3	17	.2817	.8500	0	1	0	0	0	8
39	14	10	.0952	.4167	0	4	1	0	0	2
	8	14	.1289	.6364	0	2	1	0	0	5
	2	18	.4054	.9000	0	0	1	0	0	8
40	18	8	.0867	.3077	0	6	0	0	0	1
	13	11	.0972	.4583	0	4	0	1	0	2
	12	12	.1000	.5000	0	4	0	0	0	4
	7	15	.1407	.6818	0	2	0	1	0	5
	6	16	.1571	.7273	0	2	0	0	0	7
	1	19	.7792	.9500	0	0	0	1	0	8
41	17	9	.0866	.3462	0	5	1	0	0	1
	11	13	.1039	.5417	0	3	1	0	0	4
	5	17	.1808	.7727	0	1	1	0	0	7
42	21	7	.0816	.2500	0	7	0	0	0	0
	16	10	.0871	.3846	0	5	0	1	0	1
	15	11	.0881	.4231	0	5	0	0	0	3
	10	14	.1091	.5833	0	3	0	1	0	4
	9	15	.1159	.6250	0	3	0	0	0	6
	4	18	.2171	.8182	0	0	0	0	1	8
	3	19	.2785	.8636	0	1	0	0	0	9
43	20	8	.0808	.2857	0	6	1	0	0	0
	14	12	.0899	.4615	0	4	1	0	0	3
	8	16	.1250	.6667	0	2	1	0	0	6
	2	20	.4024	.9091	0	0	1	0	0	9
44	19	9	.0804	.3214	0	6	0	1	0	0
	18	10	.0805	.3571	0	6	0	0	0	2
	13	13	.0923	.5000	0	4	0	1	0	3
	12	14	.0956	.5385	0	4	0	0	0	5
	7	17	.1371	.7083	0	2	0	1	0	6
	6	18	.1538	.7500	0	2	0	0	0	8
	1	21	.7765	.9545	0	0	0	1	0	9
45	17	11	.0810	.3929	0	5	1	0	0	2
	11	15	.0999	.5769	0	3	1	0	0	5

TABLE ACM.4.3

P=4 K=3 ADMISSIBLE DESIGNS

b	λ_0	λ_1	τ^2	ρ	1	2	3	4	5	6
45	5	19	.1778	.7917	0	1	1	0	0	8
46	21	9	.0752	.3000	0	7	0	0	0	1
	16	12	.0820	.4286	0	5	0	1	0	2
	15	13	.0836	.4643	0	5	0	0	0	4
	10	16	.1054	.6154	0	3	0	1	0	5
	9	17	.1126	.6538	0	3	0	0	0	7
	4	20	.2143	.8333	0	0	0	0	1	9
	3	21	.2759	.8750	0	1	0	0	0	10
47	20	10	.0750	.3333	0	6	1	0	0	1
	14	14	.0857	.5000	0	4	1	0	0	4
	8	18	.1219	.6923	0	2	1	0	0	7
	2	22	.4000	.9167	0	0	1	0	0	10
48	24	8	.0714	.2500	0	8	0	0	0	0
	19	11	.0752	.3667	0	6	0	1	0	1
	18	12	.0758	.4000	0	6	0	0	0	3
	13	15	.0885	.5357	0	4	0	1	0	4
	12	16	.0921	.5714	0	4	0	0	0	6
	7	19	.1343	.7308	0	2	0	1	0	7
	6	20	.1512	.7692	0	2	0	0	0	9
	1	23	.7742	.9583	0	0	0	1	0	10
49	23	9	.0707	.2813	0	7	1	0	0	0
	17	13	.0767	.4333	0	5	1	0	0	3
	11	17	.0967	.6071	0	3	1	0	0	6
	5	21	.1753	.8077	0	1	1	0	0	9
50	21	11	.0703	.3438	0	7	0	0	0	2
	16	14	.0781	.4667	0	5	0	1	0	3
	15	15	.0800	.5000	0	5	0	0	0	5
	10	18	.1024	.6429	0	3	0	1	0	6
	9	19	.1098	.6786	0	3	0	0	0	8
	4	22	.2120	.8462	0	0	0	0	1	10
	3	23	.2737	.8846	0	1	0	0	0	11
51	20	12	.0706	.3750	0	6	1	0	0	2
	14	16	.0824	.5333	0	4	1	0	0	5
	8	20	.1193	.7143	0	2	1	0	0	8
	2	24	.3980	.9231	0	0	1	0	0	11
52	24	10	.0664	.2941	0	8	0	0	0	1
	19	13	.0712	.4063	0	6	0	1	0	2
	18	14	.0721	.4375	0	6	0	0	0	4
	13	17	.0855	.5667	0	4	0	1	0	5
	12	18	.0893	.6000	0	4	0	0	0	7
	7	21	.1319	.7500	0	2	0	1	0	8
	6	22	.1489	.7857	0	2	0	0	0	10
	1	25	.7723	.9615	0	0	0	1	0	11

TABLE ADM.5.3

P=5 K=3 ADMISSIBLE DESIGNS

b	λ_0	λ_1	τ^2	ρ	1	2	3	4	5	6
5	2	0	1.5000	.0000	1	0	0	0	0	0
6	NO DESIGNS OF THIS SIZE EXIST									
7	2	1	.6429	.3333	0	1	0	0	0	0
8	NO DESIGNS OF THIS SIZE EXIST									
9	NO DESIGNS OF THIS SIZE EXIST									
10	4	1	.4167	.2000	0	0	0	1	0	0
	2	2	.5000	.5000	0	0	1	0	0	0
11	NO DESIGNS OF THIS SIZE EXIST									
12	ALL DESIGNS OF THIS SIZE ARE INADMISSIBLE									
13	NO DESIGNS OF THIS SIZE EXIST									
14	4	2	.3214	.3333	0	2	0	0	0	0
	2	3	.4412	.6000	0	1	0	0	1	0
15	6	1	.3182	.1429	1	0	0	1	0	0
16	NO DESIGNS OF THIS SIZE EXIST									
17	6	2	.2500	.2500	0	1	0	1	0	0
	4	3	.2763	.4286	0	0	0	1	1	0
	2	4	.4091	.6667	0	0	1	0	1	0
18	NO DESIGNS OF THIS SIZE EXIST									
19	ALL DESIGNS OF THIS SIZE ARE INADMISSIBLE									
20	8	2	.2083	.2000	0	0	0	2	0	0
	6	3	.2143	.3333	0	0	1	1	0	0
	4	4	.2500	.5000	0	0	2	0	0	0
	2	5	.3889	.7143	0	0	1	0	0	1
21	ALL DESIGNS OF THIS SIZE ARE INADMISSIBLE									
22	ALL DESIGNS OF THIS SIZE ARE INADMISSIBLE									
23	NO DESIGNS OF THIS SIZE EXIST									
24	8	3	.1793	.2727	0	2	0	1	0	0
	6	4	.1923	.4000	0	2	1	0	0	0
	4	5	.2328	.5556	0	0	0	1	2	0
	2	6	.3750	.7500	0	0	1	0	2	0
25	ALL DESIGNS OF THIS SIZE ARE INADMISSIBLE									
26	ALL DESIGNS OF THIS SIZE ARE INADMISSIBLE									
27	10	3	.1560	.2308	0	1	0	2	0	0

TABLE ADM.5.3

P=5 K=3 ADMISSIBLE DESIGNS

b	λ_0	λ_1	τ^2	ρ	1	2	3	4	5	6
27	8	4	.1607	.3333	0	0	0	2	1	0
	6	5	.1774	.4545	0	0	1	1	1	0
	4	6	.2206	.6000	0	0	2	0	1	0
	2	7	.3649	.7778	0	1	0	0	0	2
28	ALL DESIGNS OF THIS SIZE ARE INADMISSIBLE									
29	ALL DESIGNS OF THIS SIZE ARE INADMISSIBLE									
30	12	3	.1389	.2000	0	0	0	3	0	0
	10	4	.1400	.2857	0	0	1	2	0	0
	8	5	.1477	.3846	0	0	2	1	0	0
	6	6	.1667	.5000	0	0	3	0	0	0
	4	7	.2115	.6364	0	0	0	1	0	2
	2	8	.3571	.8000	0	0	1	0	0	2
31	ALL DESIGNS OF THIS SIZE ARE INADMISSIBLE									
32	ALL DESIGNS OF THIS SIZE ARE INADMISSIBLE									
33	ALL DESIGNS OF THIS SIZE ARE INADMISSIBLE									
34	12	4	.1250	.2500	0	2	0	2	0	0
	10	5	.1286	.3333	0	2	1	1	0	0
	8	6	.1382	.4286	0	0	0	2	2	0
	6	7	.1585	.5385	0	0	1	1	2	0
	4	8	.2045	.6667	0	0	2	0	2	0
	2	9	.3511	.8182	0	1	0	0	1	2
35	ALL DESIGNS OF THIS SIZE ARE INADMISSIBLE									
36	ALL DESIGNS OF THIS SIZE ARE INADMISSIBLE									
37	14	4	.1134	.2222	0	1	0	3	0	0
	12	5	.1149	.2941	0	0	0	3	1	0
	10	6	.1200	.3750	0	0	1	2	1	0
	8	7	.1308	.4667	0	0	2	1	1	0
	6	8	.1522	.5714	0	0	3	0	1	0
	4	9	.1990	.6923	0	1	1	0	0	2
	2	10	.3462	.8333	0	1	0	0	0	3
38	ALL DESIGNS OF THIS SIZE ARE INADMISSIBLE									
39	ALL DESIGNS OF THIS SIZE ARE INADMISSIBLE									
40	16	4	.1042	.2000	0	0	0	4	0	0
	14	5	.1044	.2632	0	0	1	3	0	0
	12	6	.1071	.3333	0	0	2	2	0	0
	10	7	.1133	.4118	0	0	3	1	0	0
	8	8	.1250	.5000	0	0	4	0	0	0
	6	9	.1471	.6000	0	0	1	1	0	2
	4	10	.1944	.7143	0	0	0	1	0	3
	2	11	.3421	.8462	0	0	1	0	0	3
41	ALL DESIGNS OF THIS SIZE ARE INADMISSIBLE									

TABLE ADM.5.3

P=5 K=3 ADMISSIBLE DESIGNS

b	λ_0	λ_1	τ^2	ρ	1	2	3	4	5	6
42	ALL DESIGNS OF THIS SIZE ARE INADMISSIBLE									
43	ALL DESIGNS OF THIS SIZE ARE INADMISSIBLE									
44	16	5	.0960	.2381	0	2	0	3	0	0
	14	6	.0974	.3000	0	2	1	2	0	0
	12	7	.1011	.3684	0	0	0	3	2	0
	10	8	.1080	.4444	0	0	1	2	2	0
	8	9	.1203	.5294	0	0	2	1	2	0
	6	10	.1429	.6250	0	0	3	0	2	0
	4	11	.1907	.7333	0	2	0	0	0	3
	2	12	.3387	.8571	0	1	0	0	1	3
45	ALL DESIGNS OF THIS SIZE ARE INADMISSIBLE									
46	ALL DESIGNS OF THIS SIZE ARE INADMISSIBLE									
47	18	5	.0891	.2174	0	1	0	4	0	0
	16	6	.0897	.2727	0	0	0	4	1	0
	14	7	.0918	.3333	0	0	1	3	1	0
	12	8	.0962	.4000	0	0	2	2	1	0
	10	9	.1036	.4737	0	0	3	1	1	0
	8	10	.1164	.5556	0	0	4	0	1	0
	6	11	.1393	.6471	0	1	0	1	0	3
	4	12	.1875	.7500	0	1	1	0	0	3
	2	13	.3358	.8667	0	1	0	0	0	4
48	ALL DESIGNS OF THIS SIZE ARE INADMISSIBLE									
49	ALL DESIGNS OF THIS SIZE ARE INADMISSIBLE									
50	18	6	.0833	.2500	0	0	1	4	0	0
	16	7	.0846	.3043	0	0	2	3	0	0
	14	8	.0873	.3636	0	0	3	2	0	0
	12	9	.0921	.4286	0	0	4	1	0	0
	10	10	.1000	.5000	0	0	5	0	0	0
	8	11	.1131	.5789	0	0	0	2	0	3
	6	12	.1364	.6667	0	0	1	1	0	3
	4	13	.1848	.7647	0	0	0	1	0	4
	2	14	.3333	.8750	0	0	1	0	0	4
51	ALL DESIGNS OF THIS SIZE ARE INADMISSIBLE									
52	ALL DESIGNS OF THIS SIZE ARE INADMISSIBLE									
53	ALL DESIGNS OF THIS SIZE ARE INADMISSIBLE									
54	20	6	.0780	.2308	0	2	0	4	0	0
	18	7	.0786	.2800	0	2	1	3	0	0
	16	8	.0804	.3333	0	0	0	4	2	0
	14	9	.0835	.3913	0	0	1	3	2	0
	12	10	.0887	.4545	0	0	2	2	2	0
	10	11	.0969	.5238	0	0	3	1	2	0
	8	12	.1103	.6000	0	0	4	0	2	0

TABLE ADM.5.3

P=5 K=3 ADMISSIBLE DESIGNS

b	λ_0	λ_1	τ^2	ρ	1	2	3	4	5	6
54	6	13	.1338	.6842	0	2	1	0	0	3
	4	14	.1824	.7778	0	2	0	0	0	4
	2	15	.3312	.8824	0	1	0	0	1	4
55	ALL DESIGNS OF THIS SIZE ARE INADMISSIBLE									
56	ALL DESIGNS OF THIS SIZE ARE INADMISSIBLE									
57	22	6	.0734	.2143	0	1	0	5	0	0
	20	7	.0736	.2593	0	0	0	5	1	0
	18	8	.0747	.3077	0	0	1	4	1	0
	16	9	.0768	.3600	0	0	2	3	1	0
	14	10	.0804	.4167	0	0	3	2	1	0
	12	11	.0858	.4783	0	0	4	1	1	0
	10	12	.0943	.5455	0	0	5	0	1	0
	8	13	.1079	.6190	0	1	1	1	0	3
	6	14	.1316	.7000	0	1	0	1	0	4
	4	15	.1804	.7895	0	1	1	0	0	4
	2	16	.3293	.8889	0	1	0	0	0	5
58	ALL DESIGNS OF THIS SIZE ARE INADMISSIBLE									
59	ALL DESIGNS OF THIS SIZE ARE INADMISSIBLE									
60	22	7	.0694	.2414	0	0	1	5	0	0
	20	8	.0700	.2857	0	0	2	4	0	0
	18	9	.0714	.3333	0	0	3	3	0	0
	16	10	.0739	.3846	0	0	4	2	0	0
	14	11	.0776	.4400	0	0	5	1	0	0
	12	12	.0833	.5000	0	0	6	0	0	0
	10	13	.0920	.5652	0	0	1	2	0	3
	8	14	.1058	.6364	0	0	0	2	0	4
	6	15	.1296	.7143	0	0	1	1	0	4
	4	16	.1786	.8000	0	0	0	1	0	5
	2	17	.3276	.8947	0	0	1	0	0	5
61	ALL DESIGNS OF THIS SIZE ARE INADMISSIBLE									
62	ALL DESIGNS OF THIS SIZE ARE INADMISSIBLE									
63	ALL DESIGNS OF THIS SIZE ARE INADMISSIBLE									
64	24	7	.0657	.2258	0	2	0	5	0	0
	22	8	.0660	.2667	0	2	1	4	0	0
	20	9	.0669	.3103	0	0	0	5	2	0
	18	10	.0686	.3571	0	0	1	4	2	0
	16	11	.0713	.4074	0	0	2	3	2	0
	14	12	.0753	.4615	0	0	3	2	2	0
	12	13	.0812	.5200	0	0	4	1	2	0
	10	14	.0900	.5833	0	0	5	0	2	0
	8	15	.1039	.6522	0	2	0	1	0	4
	6	16	.1279	.7273	0	2	1	0	0	4
	4	17	.1770	.8095	0	2	0	0	0	5
	2	18	.3261	.9000	0	1	0	0	1	5

TABLE ADM.5.3

P=5 K=3 ADMISSIBLE DESIGNS

b	λ_0	λ_1	τ^2	ρ	1	2	3	4	5	6
65	ALL DESIGNS OF THIS SIZE ARE INADMISSIBLE									
66	ALL DESIGNS OF THIS SIZE ARE INADMISSIBLE									
67	26	7	.0624	.2121	0	1	0	6	0	0
	24	8	.0625	.2500	0	0	0	6	1	0
	22	9	.0631	.2903	0	0	1	5	1	0
	20	10	.0643	.3333	0	0	2	4	1	0
	18	11	.0662	.3793	0	0	3	3	1	0
	16	12	.0691	.4286	0	0	4	2	1	0
	14	13	.0732	.4815	0	0	5	1	1	0
	12	14	.0793	.5385	0	0	6	0	1	0
	10	15	.0882	.6000	0	1	0	2	0	4
	8	16	.1023	.6667	0	1	1	1	0	4
	6	17	.1264	.7391	0	1	0	1	0	5
	4	18	.1755	.8182	0	1	1	0	0	5
	2	19	.3247	.9048	0	1	0	0	0	6
68	ALL DESIGNS OF THIS SIZE ARE INADMISSIBLE									
69	ALL DESIGNS OF THIS SIZE ARE INADMISSIBLE									
70	26	8	.0594	.2353	0	0	1	6	0	0
	24	9	.0598	.2727	0	0	2	5	0	0
	22	10	.0606	.3125	0	0	3	4	0	0
	20	11	.0620	.3548	0	0	4	3	0	0
	18	12	.0641	.4000	0	0	5	2	0	0
	16	13	.0671	.4483	0	0	6	1	0	0
	14	14	.0714	.5000	0	0	7	0	0	0
	12	15	.0776	.5556	0	0	0	3	0	4
	10	16	.0867	.6154	0	0	1	2	0	4
	8	17	.1008	.6800	0	0	0	2	0	5
	6	18	.1250	.7500	0	0	1	1	0	5
	4	19	.1742	.8261	0	0	0	1	0	6
	2	20	.3235	.9091	0	0	1	0	0	6
71	ALL DESIGNS OF THIS SIZE ARE INADMISSIBLE									
72	ALL DESIGNS OF THIS SIZE ARE INADMISSIBLE									
73	ALL DESIGNS OF THIS SIZE ARE INADMISSIBLE									
74	28	8	.0567	.2222	0	2	0	6	0	0
	26	9	.0569	.2571	0	2	1	5	0	0
	24	10	.0574	.2941	0	0	0	6	2	0
	22	11	.0584	.3333	0	0	1	5	2	0
	20	12	.0600	.3750	0	0	2	4	2	0
	18	13	.0622	.4194	0	0	3	3	2	0
	16	14	.0654	.4667	0	0	4	2	2	0
	14	15	.0698	.5172	0	0	5	1	2	0
	12	16	.0761	.5714	0	0	6	0	2	0
	10	17	.0853	.6296	0	2	1	1	0	4
	8	18	.0995	.6923	0	2	0	1	0	5
	6	19	.1238	.7600	0	2	1	0	0	5

TABLE ADM.5.3

P=5 K=3 ADMISSIBLE DESIGNS

b	λ_0	λ_1	τ^2	ρ	1	2	3	4	5	6
74	4	20	.1731	.8333	0	2	0	0	0	6
	2	21	.3224	.9130	0	1	0	0	1	6
75	ALL DESIGNS OF THIS SIZE ARE INADMISSIBLE									
76	ALL DESIGNS OF THIS SIZE ARE INADMISSIBLE									
77	30	8	.0543	.2105	0	1	0	7	0	0
	28	9	.0543	.2432	0	0	0	7	1	0
	26	10	.0547	.2778	0	0	1	6	1	0
	24	11	.0554	.3143	0	0	2	5	1	0
	22	12	.0565	.3529	0	0	3	4	1	0
	20	13	.0582	.3939	0	0	4	3	1	0
	18	14	.0606	.4375	0	0	5	2	1	0
	16	15	.0639	.4839	0	0	6	1	1	0
	14	16	.0684	.5333	0	0	7	0	1	0
	12	17	.0747	.5862	0	1	1	2	0	4
	10	18	.0840	.6429	0	1	0	2	0	5
	8	19	.0983	.7037	0	1	1	1	0	5
	6	20	.1226	.7692	0	1	0	1	0	6
	4	21	.1720	.8400	0	1	1	0	0	6
	2	22	.3214	.9167	0	1	0	0	0	7
78	ALL DESIGNS OF THIS SIZE ARE INADMISSIBLE									
79	ALL DESIGNS OF THIS SIZE ARE INADMISSIBLE									
80	30	9	.0520	.2308	0	0	1	7	0	0
	28	10	.0522	.2632	0	0	2	6	0	0
	26	11	.0527	.2973	0	0	3	5	0	0
	24	12	.0536	.3333	0	0	4	4	0	0
	22	13	.0549	.3714	0	0	5	3	0	0
	20	14	.0567	.4118	0	0	6	2	0	0
	18	15	.0591	.4545	0	0	7	1	0	0
	16	16	.0625	.5000	0	0	8	0	0	0
	14	17	.0671	.5484	0	0	1	3	0	4
	12	18	.0735	.6000	0	0	0	3	0	5
	10	19	.0829	.6552	0	0	1	2	0	5
	8	20	.0972	.7143	0	0	0	2	0	6
	6	21	.1216	.7778	0	0	1	1	0	6
	4	22	.1711	.8462	0	0	0	1	0	7
	2	23	.3205	.9200	0	0	1	0	0	7
81	ALL DESIGNS OF THIS SIZE ARE INADMISSIBLE									
82	ALL DESIGNS OF THIS SIZE ARE INADMISSIBLE									
83	ALL DESIGNS OF THIS SIZE ARE INADMISSIBLE									
84	32	9	.0499	.2195	0	2	0	7	0	0
	30	10	.0500	.2500	0	2	1	6	0	0
	28	11	.0503	.2821	0	0	0	7	2	0
	26	12	.0510	.3158	0	0	1	6	2	0
	24	13	.0520	.3514	0	0	2	5	2	0

TABLE ADM.5.3

P=5 K=3 ADMISSIBLE DESIGNS

b	λ_0	λ_1	τ^2	ρ	1	2	3	4	5	6
84	22	14	.0534	.3889	0	0	3	4	2	0
	20	15	.0553	.4286	0	0	4	3	2	0
	18	16	.0578	.4706	0	0	5	2	2	0
	16	17	.0613	.5152	0	0	6	1	2	0
	14	18	.0659	.5625	0	0	7	0	2	0
	12	19	.0724	.6129	0	2	0	2	0	5
	10	20	.0818	.6667	0	2	1	1	0	5
	8	21	.0962	.7241	0	2	0	1	0	6
	6	22	.1207	.7857	0	2	1	0	0	6
	4	23	.1702	.8519	0	2	0	0	0	7
	2	24	.3197	.9231	0	1	0	0	1	7
85	ALL DESIGNS OF THIS SIZE ARE INADMISSIBLE									
86	ALL DESIGNS OF THIS SIZE ARE INADMISSIBLE									
87	32	10	.0480	.2381	0	0	0	8	1	0
	30	11	.0482	.2683	0	0	1	7	1	0
	28	12	.0487	.3000	0	0	2	6	1	0
	26	13	.0495	.3333	0	0	3	5	1	0
	24	14	.0505	.3684	0	0	4	4	1	0
	22	15	.0520	.4054	0	0	5	3	1	0
	20	16	.0540	.4444	0	0	6	2	1	0
	18	17	.0566	.4857	0	0	7	1	1	0
	16	18	.0601	.5294	0	0	8	0	1	0
	14	19	.0649	.5758	0	1	0	3	0	5
	12	20	.0714	.6250	0	1	1	2	0	5
	10	21	.0809	.6774	0	1	0	2	0	6
	8	22	.0953	.7333	0	1	1	1	0	6
	6	23	.1198	.7931	0	1	0	1	0	7
	4	24	.1694	.8571	0	1	1	0	0	7
	2	25	.3189	.9259	0	1	0	0	0	8
88	ALL DESIGNS OF THIS SIZE ARE INADMISSIBLE									
89	ALL DESIGNS OF THIS SIZE ARE INADMISSIBLE									
90	34	10	.0462	.2273	0	0	1	8	0	0
	32	11	.0463	.2558	0	0	2	7	0	0
	30	12	.0467	.2857	0	0	3	6	0	0
	28	13	.0472	.3171	0	0	4	5	0	0
	26	14	.0481	.3500	0	0	5	4	0	0
	24	15	.0492	.3846	0	0	6	3	0	0
	22	16	.0508	.4211	0	0	7	2	0	0
	20	17	.0529	.4595	0	0	8	1	0	0
	18	18	.0556	.5000	0	0	9	0	0	0
	16	19	.0591	.5429	0	0	0	4	0	5
	14	20	.0639	.5882	0	0	1	3	C	5
	12	21	.0705	.6364	0	0	0	3	0	6
	10	22	.0800	.6875	0	0	1	2	0	6
	8	23	.0945	.7419	0	0	0	2	0	7
	6	24	.1190	.8000	0	0	1	1	0	7
	4	25	.1686	.8621	0	0	0	1	0	8
	2	26	.3182	.9286	0	0	1	0	0	8

TABLE ADM.5.3

P=5 K=3 ADMISSIBLE DESIGNS

b	λ_0	λ_1	τ^2	ρ	1	2	3	4	5	6
91	ALL DESIGNS OF THIS SIZE ARE INADMISSIBLE									
92	ALL DESIGNS OF THIS SIZE ARE INADMISSIBLE									
93	ALL DESIGNS OF THIS SIZE ARE INADMISSIBLE									
94	36	10	.0446	.2174	0	2	0	8	0	0
	34	11	.0446	.2444	0	2	1	7	0	0
	32	12	.0448	.2727	0	0	0	8	2	0
	30	13	.0453	.3023	0	0	1	7	2	0
	28	14	.0459	.3333	0	0	2	6	2	0
	26	15	.0468	.3659	0	0	3	5	2	0
	24	16	.0481	.4000	0	0	4	4	2	0
	22	17	.0497	.4359	0	0	5	3	2	0
	20	18	.0518	.4737	0	0	6	2	2	0
	18	19	.0546	.5135	0	0	7	1	2	0
	16	20	.0582	.5556	0	0	8	0	2	0
	14	21	.0630	.6000	0	2	1	2	0	5
	12	22	.0697	.6471	0	2	0	2	0	6
	10	23	.0792	.6970	0	2	1	1	0	6
	8	24	.0938	.7500	0	2	0	1	0	7
	6	25	.1183	.8065	0	2	1	0	0	7
	4	26	.1679	.8667	0	2	0	0	0	8
	2	27	.3175	.9310	0	1	0	0	1	8
95	ALL DESIGNS OF THIS SIZE ARE INADMISSIBLE									
96	ALL DESIGNS OF THIS SIZE ARE INADMISSIBLE									
97	36	11	.0430	.2340	0	0	0	9	1	0
	34	12	.0432	.2609	0	0	1	8	1	0
	32	13	.0435	.2889	0	0	2	7	1	0
	30	14	.0440	.3182	0	0	3	6	1	0
	28	15	.0447	.3488	0	0	4	5	1	0
	26	16	.0457	.3810	0	0	5	4	1	0
	24	17	.0470	.4146	0	0	6	3	1	0
	22	18	.0487	.4500	0	0	7	2	1	0
	20	19	.0509	.4872	0	0	8	1	1	0
	18	20	.0537	.5263	0	0	9	0	1	0
	16	21	.0573	.5676	0	1	1	3	0	5
	14	22	.0622	.6111	0	1	0	3	0	6
	12	23	.0689	.6571	0	1	1	2	0	6
	10	24	.0785	.7059	0	1	0	2	0	7
	8	25	.0930	.7576	0	1	1	1	0	7
	6	26	.1176	.8125	0	1	0	1	0	8
	4	27	.1673	.8710	0	1	1	0	0	8
	2	28	.3169	.9333	0	1	0	0	0	9

TABLE ADM.6.3

P=6 K=3 ADMISSIBLE DESIGNS

b	λ_0	λ_1	τ^2	ρ	1	2	3	4	5
6	2	0	1.5000	.0000	1	0	0	0	0
7	1	1	.8571	.5000	0	1	0	0	0
8	NO DESIGNS OF THIS SIZE EXIST								
9	NO DESIGNS OF THIS SIZE EXIST								
10	NO DESIGNS OF THIS SIZE EXIST								
11	3	1	.4444	.2500	0	0	1	0	0
12	ALL DESIGNS OF THIS SIZE ARE INADMISSIBLE								
13	ALL DESIGNS OF THIS SIZE ARE INADMISSIBLE								
14	2	2	.4286	.5000	0	2	0	0	0
15	5	1	.3273	.1667	0	0	0	1	0
16	ALL DESIGNS OF THIS SIZE ARE INADMISSIBLE								
17	1	3	.6316	.7500	0	1	0	0	1
18	4	2	.2813	.3333	0	1	1	0	0
19	ALL DESIGNS OF THIS SIZE ARE INADMISSIBLE								
20	ALL DESIGNS OF THIS SIZE ARE INADMISSIBLE								
21	7	1	.2637	.1250	1	0	0	1	0
	3	3	.2857	.5000	0	0	1	0	1
22	6	2	.2222	.2500	0	0	2	0	0
23	ALL DESIGNS OF THIS SIZE ARE INADMISSIBLE								
24	2	4	.3462	.6667	0	2	0	0	1
25	5	3	.2087	.3750	0	0	0	1	1
26	8	2	.1875	.2000	0	0	1	1	0
27	1	5	.5806	.8333	0	1	0	0	2
28	4	4	.2143	.5000	0	4	0	0	0
29	7	3	.1714	.3000	0	2	0	1	0
30	10	2	.1636	.1667	0	0	0	2	0
31	3	5	.2424	.6250	0	0	1	0	2
32	6	4	.1667	.4000	0	0	2	0	1

TABLE ADM.6.3

P=6 K=3 ADMISSIBLE DESIGNS

b	λ_0	λ_1	τ^2	ρ	1	2	3	4	5
33	9	3	.1481	.2500	0	0	3	0	0
34	2	6	.3158	.7500	0	2	0	0	2
35	5	5	.1714	.5000	0	0	0	1	2
36	8	4	.1406	.3333	0	0	1	1	1
37	11	3	.1317	.2143	0	0	2	1	0
	1	7	.5581	.8750	0	1	0	0	3
38	4	6	.1875	.6000	0	4	0	0	1
39	7	5	.1390	.4167	0	4	1	0	0
40	10	4	.1235	.2857	0	0	0	2	1
41	13	3	.1191	.1875	0	0	1	2	0
	3	7	.2222	.7000	0	0	1	0	3
42	6	6	.1429	.5000	0	0	2	0	2
43	9	5	.1197	.3571	0	0	3	0	1
44	12	4	.1111	.2500	0	0	4	0	0
	2	8	.3000	.8000	0	2	0	0	3
45	15	3	.1091	.1667	0	0	0	3	0
	5	7	.1532	.5833	0	0	0	1	3
46	8	6	.1193	.4286	0	0	1	1	2
47	11	5	.1064	.3125	0	0	2	1	1
	1	9	.5455	.9000	0	1	0	0	4
48	14	4	.1015	.2222	0	0	3	1	0
	4	8	.1731	.6667	0	4	0	0	2
49	7	7	.1224	.5000	0	7	0	0	0
50	10	6	.1043	.3750	0	0	0	2	2
51	13	5	.0966	.2778	0	0	1	2	1
	3	9	.2105	.7500	0	0	1	0	4
52	16	4	.0938	.2000	0	0	2	2	0
	6	8	.1296	.5714	0	0	2	0	3
53	9	7	.1046	.4375	0	0	3	0	2
54	12	6	.0938	.3333	0	0	4	0	1
	2	10	.2903	.8333	0	2	0	0	4
55	15	5	.0889	.2500	0	0	0	3	1

TABLE ADM.6.3

P=6 K=3 ADMISSIBLE DESIGNS

b	λ_0	λ_1	τ^2	ρ	1	2	3	4	5
55	5	9	.1424	.6429	0	0	0	1	4
56	18	4	.0873	.1818	0	0	1	3	0
	8	8	.1071	.5000	0	0	1	1	3
57	11	7	.0926	.3889	0	0	2	1	2
	1	11	.5373	.9167	0	1	0	0	5
58	14	6	.0857	.3000	0	0	3	1	1
	4	10	.1641	.7143	0	4	0	0	3
59	17	5	.0826	.2273	0	0	4	1	0
	7	9	.1124	.5625	0	7	0	0	1
60	20	4	.0818	.1867	0	0	0	4	0
	10	8	.0931	.4444	0	0	0	2	3
61	13	7	.0839	.3500	0	0	1	2	2
	3	11	.2029	.7857	0	0	1	0	5
62	16	6	.0793	.2727	0	0	2	2	1
	6	10	.1212	.6250	0	0	2	0	4
63	19	5	.0773	.2083	0	0	3	2	0
	9	9	.0952	.5000	0	0	3	0	3
64	12	8	.0833	.4000	0	0	4	0	2
	2	12	.2838	.8571	0	2	0	0	5

APPENDIX A.3a

Tables of Optimal Designs for One-Sided Comparisons

Tables OPT1•p•k for p = 2(1)6, k = 2

and p = 3(1)6, k = 3

TABLE OPT1.2.2

P=2 K=2 OPTIMAL DESIGN AND ASSOCIATED CONFIDENCE COEFFICIENT FOR ONE-SIDED INTERVALS

b \ a/σ	.500	.600	.700	.800	.900	1.000	1.100	1.200	1.300	1.400	1.500	1.600
2	1, 0 .4073	1, 0 .4413	1, 0 .4757	1, 0 .5101	1, 0 .5443	1, 0 .5780	1, 0 .6110	1, 0 .6431	1, 0 .6741	1, 0 .7039	1, 0 .7320	1, 0 .7587
3	1, 1 .5172	1, 1 .5547	1, 1 .5917	1, 1 .6278	1, 1 .6629	1, 1 .6965	1, 1 .7285	1, 1 .7586	1, 1 .7869	1, 1 .8130	1, 1 .8371	1, 1 .8590
4	1, 2 .5561	1, 2 .5939	1, 2 .6308	1, 2 .6666	1, 2 .7008	1, 2 .7334	1, 2 .7640	1, 2 .7926	1, 2 .8190	2, 0 .8450	2, 0 .8708	2, 0 .8934
5	1, 3 .5774	1, 3 .6151	2, 1 .6558	2, 1 .7020	2, 1 .7450	2, 1 .7844	2, 1 .8199	2, 1 .8515	2, 1 .8790	2, 1 .9026	2, 1 .9226	2, 1 .9393
6	2, 2 .5943	2, 2 .6450	2, 2 .6932	2, 2 .7381	2, 2 .7794	2, 2 .8166	2, 2 .8495	2, 2 .8782	2, 2 .9028	2, 2 .9235	2, 2 .9406	2, 2 .9546
7	2, 3 .6175	2, 3 .6679	2, 3 .7153	2, 3 .7592	3, 1 .8045	3, 1 .8440	3, 1 .8777	3, 1 .9058	3, 1 .9288	3, 1 .9471	3, 1 .9614	3, 1 .9723
8	2, 4 .6334	3, 2 .6872	3, 2 .7414	3, 2 .7902	3, 2 .8329	3, 2 .8695	3, 2 .9001	3, 2 .9250	3, 2 .9448	3, 2 .9602	3, 2 .9718	3, 2 .9805
9	3, 3 .6511	3, 3 .7092	3, 3 .7623	3, 3 .8094	3, 3 .8502	4, 1 .8860	4, 1 .9159	4, 1 .9394	4, 1 .9573	4, 1 .9706	4, 1 .9802	4, 1 .9870
10	3, 4 .6673	3, 4 .7248	4, 2 .7806	4, 2 .8303	4, 2 .8719	4, 2 .9057	4, 2 .9323	4, 2 .9526	4, 2 .9677	4, 2 .9785	4, 2 .9861	4, 2 .9912
11	4, 3 .6806	4, 3 .7437	4, 3 .7995	4, 3 .8471	4, 3 .8864	4, 3 .9179	4, 3 .9422	5, 1 .9605	5, 1 .9741	5, 1 .9834	5, 1 .9897	5, 1 .9938
12	4, 4 .6965	4, 4 .7586	5, 2 .8131	5, 2 .8619	5, 2 .9010	5, 2 .9311	5, 2 .9536	5, 2 .9696	5, 2 .9807	5, 2 .9882	5, 2 .9929	5, 2 .9959
13	4, 5 .7087	5, 3 .7733	5, 3 .8299	5, 3 .8764	5, 3 .9130	5, 3 .9407	5, 3 .9609	5, 3 .9751	5, 3 .9846	5, 3 .9908	5, 3 .9947	5, 3 .9970
14	5, 4 .7221	5, 4 .7872	5, 4 .8422	6, 2 .8871	6, 2 .9230	6, 2 .9493	6, 2 .9678	6, 2 .9803	6, 2 .9884	6, 2 .9934	6, 2 .9964	6, 2 .9981

TABLE OPT1.2.2

P=2 K=2 OPTIMAL DESIGN AND ASSOCIATED CONFIDENCE COEFFICIENT FOR ONE-SIDED INTERVALS

b \ a/σ	.300	.400	.500	.600	.700	.800	.900	1.000	1.100	1.200	1.300	1.400
15	4, 7 .5866	5, 5 .6609	5, 5 .7340	6, 3 .7988	6, 3 .8551	6, 3 .8995	6, 3 .9328	6, 3 .9568	6, 3 .9733	6, 3 .9841	6, 3 .9909	6, 3 .9950
16	4, 8 .5939	5, 6 .6712	6, 4 .7451	6, 4 .8117	6, 4 .8661	6, 4 .9085	7, 2 .9399	7, 2 .9625	7, 2 .9776	7, 2 .9871	7, 2 .9929	7, 2 .9962
17	4, 9 .6003	5, 7 .6798	6, 5 .7565	6, 5 .8218	7, 3 .8762	7, 3 .9179	7, 3 .9479	7, 3 .9683	7, 3 .9816	7, 3 .9897	7, 3 .9945	7, 3 .9972
18	5, 8 .6078	6, 6 .6898	6, 6 .7658	7, 4 .8329	7, 4 .8860	7, 4 .9256	7, 4 .9536	7, 4 .9724	7, 4 .9843	8, 2 .9915	8, 2 .9956	8, 2 .9978
19	5, 9 .6144	6, 7 .6983	7, 5 .7765	7, 5 .8423	8, 3 .8939	8, 3 .9327	8, 3 .9593	8, 3 .9766	8, 3 .9872	8, 3 .9933	8, 3 .9967	8, 3 .9984
20	6, 8 .6209	7, 6 .7069	7, 6 .7855	8, 4 .8515	8, 4 .9026	8, 4 .9393	8, 4 .9641	8, 4 .9798	8, 4 .9892	8, 4 .9945	8, 4 .9973	8, 4 .9988
21	6, 9 .6276	7, 7 .7153	8, 5 .7946	8, 5 .8601	8, 5 .9095	9, 3 .9447	9, 3 .9682	9, 3 .9827	9, 3 .9910	9, 3 .9956	9, 3 .9980	9, 3 .9991
22	6,10 .6335	8, 6 .7229	8, 6 .8032	9, 4 .8677	9, 4 .9166	9, 4 .9503	9, 4 .9720	9, 4 .9851	9, 4 .9925	9, 4 .9964	9, 4 .9984	9, 4 .9993
23	7, 9 .6399	8, 7 .7310	9, 5 .8110	9, 5 .8757	9, 5 .9227	9, 5 .9547	10, 3 .9750	10, 3 .9871	10, 3 .9937	10, 3 .9971	10, 3 .9987	10, 3 .9995
24	7,10 .6459	8, 8 .7381	9, 6 .8191	9, 6 .8822	10, 4 .9285	10, 4 .9592	10, 4 .9781	10, 4 .9890	10, 4 .9948	10, 4 .9977	10, 4 .9990	10, 4 .9996
25	8, 9 .6517	9, 7 .7456	9, 7 .8260	10, 5 .8893	10, 5 .9339	10, 5 .9629	10, 5 .9805	10, 5 .9904	10, 5 .9956	11, 3 .9981	11, 3 .9992	11, 3 .9997
26	8,10 .6576	9, 8 .7525	10, 6 .8335	10, 6 .8953	11, 4 .9385	11, 4 .9664	11, 4 .9829	11, 4 .9918	11, 4 .9963	11, 4 .9985	11, 4 .9994	11, 4 .9998
27	8,11 .6630	10, 7 .7592	10, 7 .8401	11, 5 .9013	11, 5 .9433	11, 5 .9696	11, 5 .9848	11, 5 .9929	11, 5 .9969	11, 5 .9987	11, 5 .9995	11, 5 .9998

TABLE OPT1.2.2

P=2 K=2 OPTIMAL DESIGN AND ASSOCIATED CONFIDENCE COEFFICIENT FOR ONE-SIDED INTERVALS

b \ a/σ	.100	.200	.300	.400	.500	.600	.700	.800	.900	1.000	1.100	1.200
28	1,26 .4962	6,16 .5676	9,10 .6688	10, 8 .7660	11, 6 .8467	11, 6 .9068	11, 6 .9472	12, 4 .9723	12, 4 .9865	12, 4 .9939	12, 4 .9974	12, 4 .9990
29	1,27 .4970	7,15 .5710	9,11 .6741	11, 7 .7720	11, 7 .8528	12, 5 .9119	12, 5 .9513	12, 5 .9750	12, 5 .9881	12, 5 .9947	12, 5 .9978	12, 5 .9992
30	1,28 .4978	7,16 .5744	10,10 .6794	11, 8 .7785	12, 6 .8586	12, 6 .9169	12, 6 .9548	12, 6 .9772	13, 4 .9894	13, 4 .9954	13, 4 .9982	13, 4 .9993
31	1,29 .4985	7,17 .5776	10,11 .6847	11, 9 .7842	12, 7 .8645	12, 7 .9213	13, 5 .9581	13, 5 .9794	13, 5 .9907	13, 5 .9961	13, 5 .9985	13, 5 .9995
32	1,30 .4992	8,16 .5810	10,12 .6896	12, 8 .7902	13, 6 .8696	13, 6 .9259	13, 6 .9612	13, 6 .9813	13, 6 .9917	13, 6 .9966	13, 6 .9987	14, 4 .9996
33	1,31 .4998	8,17 .5843	11,11 .6948	12, 9 .7958	13, 7 .8750	13, 7 .9298	14, 5 .9639	14, 5 .9830	14, 5 .9926	14, 5 .9971	14, 5 .9989	14, 5 .9996
34	1,32 .5004	9,16 .5874	11,12 .6996	13, 8 .8011	13, 8 .8798	14, 6 .9338	14, 6 .9666	14, 6 .9846	14, 6 .9935	14, 6 .9975	14, 6 .9991	14, 6 .9997
35	1,33 .5010	9,17 .5908	12,11 .7045	13, 9 .8066	14, 7 .8847	14, 7 .9374	14, 7 .9689	15, 5 .9859	15, 5 .9942	15, 5 .9978	15, 5 .9992	15, 5 .9998
36	1,34 .5016	9,18 .5939	12,12 .7092	14, 8 .8114	14, 8 .8892	15, 6 .9408	15, 6 .9713	15, 6 .9873	15, 6 .9949	15, 6 .9981	15, 6 .9994	15, 6 .9998
37	1,35 .5021	10,17 .5970	13,11 .7137	14, 9 .8167	15, 7 .8936	15, 7 .9441	15, 7 .9733	15, 7 .9884	16, 5 .9954	16, 5 .9984	16, 5 .9995	16, 5 .9998
38	2,34 .5027	10,18 .6002	13,12 .7184	14,10 .8214	15, 8 .8978	16, 6 .9470	16, 6 .9752	16, 6 .9895	16, 6 .9960	16, 6 .9986	16, 6 .9996	16, 6 .9999
39	2,35 .5034	10,19 .6032	13,13 .7227	15, 9 .8262	16, 7 .9017	16, 7 .9500	16, 7 .9770	16, 7 .9904	16, 7 .9964	16, 7 .9988	16, 7 .9996	
40	2,36 .5041	11,18 .6063	14,12 .7272	15,10 .8307	16, 8 .9057	16, 8 .9526	17, 6 .9786	17, 6 .9913	17, 6 .9968	17, 6 .9989	17, 6 .9997	

TABLE OPT1.3.2

P=3 K=2 OPTIMAL DESIGN AND ASSOCIATED CONFIDENCE COEFFICIENT FOR ONE-SIDED INTERVALS

b \ a/σ	.500	.600	.700	.800	.900	1.000	1.100	1.200	1.300	1.400	1.500	1.600
3	1, 0 .2599	1, 0 .2932	1, 0 .3281	1, 0 .3643	1, 0 .4015	1, 0 .4394	1, 0 .4776	1, 0 .5157	1, 0 .5534	1, 0 .5904	1, 0 .6263	1, 0 .6609
6	1, 1 .4585	1, 1 .5038	1, 1 .5488	1, 1 .5932	1, 1 .6364	1, 1 .6778	1, 1 .7170	1, 1 .7537	1, 1 .7877	1, 1 .8185	1, 1 .8466	1, 1 .8714
9	1, 2 .5237	1, 2 .5700	1, 2 .6153	2, 1 .6625	2, 1 .7161	2, 1 .7650	2, 1 .8087	2, 1 .8468	2, 1 .8794	2, 1 .9066	2, 1 .9289	2, 1 .9468
12	1, 3 .5581	2, 2 .6144	2, 2 .6737	2, 2 .7288	2, 2 .7787	3, 1 .8270	3, 1 .8685	3, 1 .9023	3, 1 .9291	3, 1 .9497	3, 1 .9651	3, 1 .9763
15	2, 3 .5919	2, 3 .6537	3, 2 .7208	3, 2 .7806	3, 2 .8321	3, 2 .8748	3, 2 .9092	4, 1 .9369	4, 1 .9575	4, 1 .9722	4, 1 .9823	4, 1 .9891
18	3, 3 .6219	3, 3 .6935	4, 2 .7603	4, 2 .8212	4, 2 .8710	4, 2 .9101	4, 2 .9394	4, 2 .9605	4, 2 .9751	4, 2 .9848	4, 2 .9911	4, 2 .9949
21	3, 4 .6500	4, 3 .7273	4, 3 .7952	5, 2 .8537	5, 2 .9002	5, 2 .9347	5, 2 .9589	5, 2 .9751	5, 2 .9855	5, 2 .9919	5, 2 .9957	5, 2 .9977
24	4, 4 .6778	5, 3 .7568	5, 3 .8257	5, 3 .8804	6, 2 .9224	6, 2 .9521	6, 2 .9718	6, 2 .9841	6, 2 .9914	6, 2 .9956	6, 2 .9978	6, 2 .9990
27	5, 4 .7027	6, 3 .7828	6, 3 .8511	6, 3 .9030	6, 3 .9399	7, 2 .9647	7, 2 .9805	7, 2 .9897	7, 2 .9949	7, 2 .9976	7, 2 .9989	7, 2 .9995
30	6, 4 .7253	6, 4 .8067	7, 3 .8725	7, 3 .9210	7, 3 .9537	7, 3 .9744	7, 3 .9866	7, 3 .9934	7, 3 .9969	7, 3 .9986	7, 3 .9994	7, 3 .9998
33	7, 4 .7460	7, 4 .8280	8, 3 .8906	8, 3 .9354	8, 3 .9642	8, 3 .9813	8, 3 .9909	8, 3 .9958	8, 3 .9982	8, 3 .9993	8, 3 .9997	8, 3 .9999
36	7, 5 .7660	8, 4 .8468	8, 4 .9066	9, 3 .9470	9, 3 .9721	9, 3 .9863	9, 3 .9937	9, 3 .9973	9, 3 .9989	9, 3 .9996	9, 3 .9999	
39	8, 5 .7842	9, 4 .8633	9, 4 .9202	9, 4 .9567	10, 3 .9783	10, 3 .9899	10, 3 .9956	10, 3 .9982	10, 3 .9993	10, 3 .9998		

TABLE OPT1.3.2

P=3 K=2 OPTIMAL DESIGN AND ASSOCIATED CONFIDENCE COEFFICIENT FOR ONE-SIDED INTERVALS

b \ a/σ	.100	.200	.300	.400	.500	.600	.700	.800	.900	1.000	1.100	1.200
42	1,13	1,13	6, 8	8, 6	9, 5	10, 4	10, 4	10, 4	10, 4	10, 4	10, 4	10, 4
	.4570	.5053	.5929	.7016	.8008	.8779	.9317	.9647	.9832	.9926	.9970	.9989
45	1,14	2,13	6, 9	8, 7	10, 5	10, 5	11, 4	11, 4	11, 4	11, 4	11, 4	11, 4
	.4602	.5092	.6043	.7163	.8160	.8913	.9414	.9711	.9869	.9946	.9979	.9993
48	1,15	2,14	7, 9	9, 7	10, 6	11, 5	11, 5	12, 4	12, 4	12, 4	12, 4	12, 4
	.4631	.5136	.6154	.7307	.8301	.9032	.9497	.9763	.9898	.9960	.9986	.9995
51	1,16	3,14	8, 9	10, 7	11, 6	12, 5	12, 5	12, 5	13, 4	13, 4	13, 4	13, 4
	.4657	.5184	.6261	.7443	.8433	.9136	.9570	.9806	.9920	.9970	.9990	.9997
54	1,17	4,14	9, 9	11, 7	12, 6	13, 5	13, 5	13, 5	13, 5	13, 5	13, 5	13, 5
	.4681	.5231	.6364	.7570	.8554	.9229	.9632	.9841	.9938	.9978	.9993	.9998
57	1,18	4,15	9,10	12, 7	13, 6	13, 6	14, 5	14, 5	14, 5	14, 5	14, 5	14, 5
	.4703	.5283	.6468	.7690	.8664	.9311	.9684	.9870	.9952	.9984	.9995	.9999
60	1,19	5,15	10,10	12, 8	14, 6	14, 6	15, 5	15, 5	15, 5	15, 5	15, 5	
	.4723	.5334	.6563	.7806	.8766	.9386	.9728	.9893	.9962	.9988	.9997	
63	1,20	5,16	11,10	13, 8	14, 7	15, 6	15, 6	16, 5	16, 5	16, 5	16, 5	
	.4742	.5385	.6664	.7917	.8861	.9452	.9767	.9912	.9971	.9991	.9998	
66	1,21	6,16	11,11	14, 8	15, 7	16, 6	16, 6	16, 6	16, 6	16, 6	16, 6	
	.4760	.5438	.6758	.8021	.8948	.9511	.9800	.9928	.9977	.9993	.9998	
69	1,22	7,16	12,11	15, 8	16, 7	17, 6	17, 6	17, 6	17, 6	17, 6	17, 6	
	.4776	.5490	.6851	.8119	.9029	.9563	.9828	.9941	.9982	.9995	.9999	
72	1,23	7,17	13,11	13, 9	17, 7	17, 7	18, 6	18, 6	18, 6	18, 6		
	.4791	.5542	.6940	.8212	.9103	.9610	.9853	.9951	.9986	.9996		
75	1,24	8,17	14,11	15, 9	18, 7	18, 7	18, 7	19, 6	19, 6	19, 6		
	.4806	.5596	.7027	.8303	.9170	.9652	.9873	.9960	.9989	.9997		
78	1,25	9,17	15,11	17, 9	18, 8	19, 7	19, 7	19, 7	19, 7	19, 7		
	.4819	.5648	.7110	.8388	.9234	.9689	.9891	.9967	.9991	.9998		

TABLE OPT1.3.2

P=3 K=2 OPTIMAL DESIGN AND ASSOCIATED CONFIDENCE COEFFICIENT FOR ONE-SIDED INTERVALS

b \ a/σ	.100	.200	.300	.400	.500	.600	.700	.800	.900	1.000	1.100	1.200
81	1,26 .4832	9,18 .5700	15,12 .7193	18, 9 .8468	19, 8 .9293	20, 7 .9722	20, 7 .9906	20, 7 .9973	20, 7 .9993	20, 7 .9999		
84	1,27 .4844	10,18 .5752	16,12 .7273	19, 9 .8544	20, 8 .9347	20, 8 .9751	21, 7 .9920	21, 7 .9978	21, 7 .9995	21, 7 .9999		
87	1,28 .4855	11,18 .5804	17,12 .7351	19,10 .8616	21, 8 .9396	21, 8 .9778	22, 7 .9931	22, 7 .9982	22, 7 .9996			
90	1,29 .4866	12,18 .5855	18,12 .7426	20,10 .8686	21, 9 .9442	22, 8 .9802	22, 8 .9940	22, 8 .9985	22, 8 .9997			
93	1,30 .4876	12,19 .5906	18,13 .7499	21,10 .8751	22, 9 .9484	23, 8 .9822	23, 8 .9949	23, 8 .9988	23, 8 .9997			
96	1,31 .4886	13,19 .5957	19,13 .7570	22,10 .8813	23, 9 .9523	24, 8 .9841	24, 8 .9956	24, 8 .9990	24, 8 .9998			
99	1,32 .4896	14,19 .6006	20,13 .7640	23,10 .8872	24, 9 .9559	24, 9 .9858	25, 8 .9962	25, 8 .9992	25, 8 .9998			
102	1,33 .4904	14,20 .6056	21,13 .7706	23,11 .8928	25, 9 .9592	25, 9 .9873	25, 9 .9967	25, 9 .9993	25, 9 .9999			
105	1,34 .4913	15,20 .6105	22,13 .7771	24,11 .8981	25,10 .9623	26, 9 .9886	26, 9 .9972	26, 9 .9994				
108	1,35 .4921	16,20 .6154	22,14 .7835	25,11 .9032	26,10 .9652	27, 9 .9898	27, 9 .9976	27, 9 .9995				
111	1,36 .4929	17,20 .6201	23,14 .7897	26,11 .9080	27,10 .9678	27,10 .9909	28, 9 .9979	28, 9 .9996				
114	1,37 .4936	17,21 .6249	24,14 .7957	27,11 .9125	28,10 .9702	28,10 .9919	28,10 .9982	28,10 .9997				
117	1,38 .4944	18,21 .6296	25,14 .8014	27,12 .9169	28,11 .9724	29,10 .9927	29,10 .9984	29,10 .9997				
120	1,39 .4951	19,21 .6343	25,15 .8071	28,12 .9210	29,11 .9745	30,10 .9935	30,10 .9987	30,10 .9998				

TABLE OPT1.4.2

P=4 K=2 OPTIMAL DESIGN AND ASSOCIATED CONFIDENCE COEFFICIENT FOR ONE-SIDED INTERVALS

b \ a/σ	.500	.600	.700	.800	.900	1.000	1.100	1.200	1.300	1.400	1.500	1.600
4	1, 0 .1659	1, 0 .1948	1, 0 .2263	1, 0 .2602	1, 0 .2962	1, 0 .3341	1, 0 .3733	1, 0 .4136	1, 0 .4544	1, 0 .4953	1, 0 .5358	1, 0 .5757
8	2, 0 .2286	2, 0 .2774	2, 0 .3302	2, 0 .3859	2, 0 .4432	2, 0 .5011	2, 0 .5581	2, 0 .6132	2, 0 .6655	2, 0 .7140	2, 0 .7584	2, 0 .7982
10	1, 1 .4255	1, 1 .4769	1, 1 .5286	1, 1 .5797	1, 1 .6294	1, 1 .6769	1, 1 .7216	1, 1 .7630	1, 1 .8006	1, 1 .8343	1, 1 .8640	1, 1 .8898
12										3, 0 .8381	3, 0 .8740	3, 0 .9036
14	2, 1 .4405	2, 1 .5101	2, 1 .5790	2, 1 .6452	2, 1 .7069	2, 1 .7629	2, 1 .8121	2, 1 .8542	2, 1 .8892	2, 1 .9176	2, 1 .9400	2, 1 .9572
16	1, 2 .5094	1, 2 .5628	1, 2 .6150	1, 2 .6651	1, 2 .7124							
18			3, 1 .6287	3, 1 .7025	3, 1 .7683	3, 1 .8245	3, 1 .8708	3, 1 .9076	3, 1 .9357	3, 1 .9565	3, 1 .9714	3, 1 .9817
20	2, 2 .5323	2, 2 .6041	2, 2 .6723	2, 2 .7350	2, 2 .7907	2, 2 .8388	2, 2 .8789	2, 2 .9112	2, 2 .9366			
22	1, 3 .5532	1, 3 .6064	4, 1 .6744	4, 1 .7515	4, 1 .8169	4, 1 .8697	4, 1 .9105	4, 1 .9406	4, 1 .9619	4, 1 .9763	4, 1 .9858	4, 1 .9917
24	3, 2 .5578	3, 2 .6415	3, 2 .7186	3, 2 .7862	3, 2 .8429	3, 2 .8885	3, 2 .9235	3, 2 .9493	3, 2 .9676	3, 2 .9800	3, 2 .9880	3, 2 .9931
26	2, 3 .5845	2, 3 .6556	2, 3 .7214	5, 1 .7929	5, 1 .8553	5, 1 .9030	5, 1 .9376	5, 1 .9614	5, 1 .9770	5, 1 .9868	5, 1 .9927	5, 1 .9961
28		4, 2 .6758	4, 2 .7575	4, 2 .8260	4, 2 .8804	4, 2 .9211	4, 2 .9502	4, 2 .9698	4, 2 .9825	4, 2 .9902	4, 2 .9948	4, 2 .9973
30	3, 3 .6127	3, 3 .6948	3, 3 .7679	3, 3 .8297	6, 1 .8857	6, 1 .9277	6, 1 .9563	6, 1 .9747	6, 1 .9860	6, 1 .9926	6, 1 .9962	6, 1 .9982

TABLE OPT1.4.2

P=4 K=2 OPTIMAL DESIGN AND ASSOCIATED CONFIDENCE COEFFICIENT FOR ONE-SIDED INTERVALS

b \ a/σ	.300	.400	.500	.600	.700	.800	.900	1.000	1.100	1.200	1.300	1.400
32	2, 4 .4688	2, 4 .5446	2, 4 .6186	5, 2 .7071	5, 2 .7908	5, 2 .8578	5, 2 .9080	5, 2 .9434	5, 2 .9668	5, 2 .9815	5, 2 .9902	5, 2 .9950
34	1, 5 .4907	1, 5 .5458	4, 3 .6390	4, 3 .7283	4, 3 .8045	4, 3 .8657	4, 3 .9120	7, 1 .9459	7, 1 .9692	7, 1 .9833	7, 1 .9914	7, 1 .9958
36		3, 4 .5623	3, 4 .6497	6, 2 .7356	6, 2 .8193	6, 2 .8834	6, 2 .9289	6, 2 .9590	6, 2 .9776	6, 2 .9884	6, 2 .9943	6, 2 .9974
38	2, 5 .4932	2, 5 .5694	5, 3 .6636	5, 3 .7574	5, 3 .8343	5, 3 .8929	5, 3 .9345	5, 3 .9622	5, 3 .9793	5, 3 .9893	5, 3 .9948	5, 3 .9976
40	1, 6 .5053	4, 4 .5797	4, 4 .6769	4, 4 .7630	7, 2 .8438	7, 2 .9041	7, 2 .9447	7, 2 .9700	7, 2 .9847	7, 2 .9926	7, 2 .9967	7, 2 .9986
42		3, 5 .5899	6, 3 .6866	6, 3 .7830	6, 3 .8589	6, 3 .9139	6, 3 .9507	6, 3 .9735	6, 3 .9866	6, 3 .9937	6, 3 .9972	6, 3 .9988
44	2, 6 .5121	5, 4 .5968	5, 4 .7014	5, 4 .7911	8, 2 .8650	8, 2 .9210	8, 2 .9568	8, 2 .9779	8, 2 .9894	8, 2 .9953	8, 2 .9980	8, 2 .9992
46	1, 7 .5168	4, 5 .6089	7, 3 .7081	7, 3 .8057	7, 3 .8795	7, 3 .9304	7, 3 .9626	7, 3 .9812	7, 3 .9912	7, 3 .9961	7, 3 .9984	7, 3 .9994
48	3, 6 .5200	6, 4 .6134	6, 4 .7237	6, 4 .8151	6, 4 .8844	9, 2 .9348	9, 2 .9662	9, 2 .9837	9, 2 .9927	9, 2 .9969	9, 2 .9988	9, 2 .9996
50	2, 7 .5272	5, 5 .6267	5, 5 .7293	8, 3 .8259	8, 3 .8969	8, 3 .9435	8, 3 .9713	8, 3 .9865	8, 3 .9941	8, 3 .9976	8, 3 .9991	8, 3 .9997
52	4, 6 .5285	4, 6 .6317	7, 4 .7441	7, 4 .8360	7, 4 .9026	7, 4 .9465	10, 2 .9734	10, 2 .9879	10, 2 .9949	10, 2 .9980	10, 2 .9993	10, 2 .9997
54	3, 7 .5372	6, 5 .6436	6, 5 .7513	9, 3 .8439	9, 3 .9115	9, 3 .9539	9, 3 .9779	9, 3 .9902	9, 3 .9960	9, 3 .9985	9, 3 .9995	9, 3 .9998
56	2, 8 .5396	5, 6 .6503	8, 4 .7629	8, 4 .8542	8, 4 .9176	8, 4 .9572	8, 4 .9796	8, 4 .9910	11, 2 .9964	11, 2 .9987	11, 2 .9995	11, 2 .9999

TABLE OPT1.4.2

P=4 K=2 OPTIMAL DESIGN AND ASSOCIATED CONFIDENCE COEFFICIENT FOR ONE-SIDED INTERVALS

b \ a/σ	.100	.200	.300	.400	.500	.600	.700	.800	.900	1.000	1.100	1.200
58	1, 9 .4231	1, 9 .4783	4, 7 .5472	7, 5 .6597	7, 5 .7711	10, 3 .8599	10, 3 .9240	10, 3 .9623	10, 3 .9829	10, 3 .9929	10, 3 .9973	10, 3 .9990
60			3, 8 .5516	6, 6 .6675	9, 4 .7801	9, 4 .8701	9, 4 .9300	9, 4 .9656	9, 4 .9846	9, 4 .9937	9, 4 .9976	9, 4 .9992
62			5, 7 .5570	8, 5 .6750	8, 5 .7890	8, 5 .8746	11, 3 .9346	11, 3 .9691	11, 3 .9867	11, 3 .9948	11, 3 .9981	11, 3 .9994
64	1,10 .4295	1,10 .4849	4, 8 .5628	7, 6 .6837	10, 4 .7961	10, 4 .8842	10, 4 .9404	10, 4 .9722	10, 4 .9883	10, 4 .9955	10, 4 .9984	10, 4 .9995
66			6, 7 .5668	9, 5 .6896	9, 5 .8053	9, 5 .8892	12, 3 .9437	12, 3 .9746	12, 3 .9896	12, 3 .9961	12, 3 .9987	12, 3 .9996
68			5, 8 .5736	8, 6 .6989	11, 4 .8109	11, 4 .8966	11, 4 .9492	11, 4 .9775	11, 4 .9910	11, 4 .9967	11, 4 .9989	11, 4 .9997
70	1,11 .4351	1,11 .4907	7, 7 .5766	10, 5 .7036	10, 5 .8202	10, 5 .9018	10, 5 .9518	13, 3 .9791	13, 3 .9919	13, 3 .9971	13, 3 .9991	13, 3 .9997
72			6, 8 .5840	9, 6 .7133	9, 6 .8249	12, 4 .9076	12, 4 .9565	12, 4 .9817	12, 4 .9931	12, 4 .9976	12, 4 .9993	12, 4 .9998
74			5, 9 .5877	8, 7 .7186	11, 5 .8338	11, 5 .9129	11, 5 .9592	11, 5 .9829	14, 3 .9936	14, 3 .9979	14, 3 .9994	14, 3 .9998
76	1,12 .4400	1,12 .4957	7, 8 .5942	10, 6 .7269	10, 6 .8390	13, 4 .9173	13, 4 .9627	13, 4 .9850	13, 4 .9946	13, 4 .9983	13, 4 .9995	13, 4 .9999
78			6, 9 .5987	9, 7 .7328	12, 5 .8463	12, 5 .9226	12, 5 .9654	12, 5 .9863	12, 5 .9951	12, 5 .9985	12, 5 .9996	
80		2,12 .4958	8, 8 .6041	11, 6 .7398	11, 6 .8518	14, 4 .9260	14, 4 .9680	14, 4 .9878	14, 4 .9958	14, 4 .9987	14, 4 .9997	
82	1,13 .4444	1,13 .5001	7, 9 .6093	10, 7 .7461	13, 5 .8578	13, 5 .9311	13, 5 .9706	13, 5 .9889	13, 5 .9963	13, 5 .9989	13, 5 .9997	

TABLE OPT1.5.2

P=5 K=2 OPTIMAL DESIGN AND ASSOCIATED CONFIDENCE COEFFICIENT FOR ONE-SIDED INTERVALS

b \ a/σ	.500	.600	.700	.800	.900	1.000	1.100	1.200	1.300	1.400	1.500	1.600
5	1, 0 .1058	1, 0 .1294	1, 0 .1561	1, 0 .1858	1, 0 .2185	1, 0 .2540	1, 0 .2918	1, 0 .3316	1, 0 .3730	1, 0 .4155	1, 0 .4585	1, 0 .5014
10	2, 0 .1581	2, 0 .2013	2, 0 .2503	2, 0 .3041	2, 0 .3617	2, 0 .4216	2, 0 .4824	2, 0 .5427	2, 0 .6011	2, 0 .6564	2, 0 .7077	2, 0 .7544
15	1, 1 .4055	1, 1 .4623	1, 1 .5199	1, 1 .5768	1, 1 .6321	1, 1 .6846	1, 1 .7335	1, 1 .7781	1, 1 .8180	1, 1 .8530	1, 1 .8831	1, 1 .9085
20	2, 1 .4132	2, 1 .4898	2, 1 .5664	2, 1 .6400	2, 1 .7084	2, 1 .7696	2, 1 .8225	2, 1 .8667	2, 1 .9024	2, 1 .9304	2, 1 .9516	2, 1 .9672
25	1, 2 .5040	1, 2 .5636	1, 2 .6216	3, 1 .6962	3, 1 .7688	3, 1 .8298	3, 1 .8789	3, 1 .9167	3, 1 .9446	3, 1 .9643	3, 1 .9777	3, 1 .9866
30	2, 2 .5239	2, 2 .6039	2, 2 .6795	2, 2 .7480	4, 1 .8168	4, 1 .8738	4, 1 .9165	4, 1 .9469	4, 1 .9676	4, 1 .9809	4, 1 .9892	4, 1 .9941
35	1, 3 .5548	3, 2 .6405	3, 2 .7254	3, 2 .7986	3, 2 .8582	5, 1 .9061	5, 1 .9420	5, 1 .9657	5, 1 .9806	5, 1 .9895	5, 1 .9946	5, 1 .9973
40	2, 3 .5859	4, 2 .6740	4, 2 .7638	4, 2 .8373	4, 2 .8935	4, 2 .9338	4, 2 .9609	4, 2 .9780	6, 1 .9883	6, 1 .9941	6, 1 .9972	6, 1 .9987
45	3, 3 .6136	5, 2 .7047	5, 2 .7964	5, 2 .8677	5, 2 .9190	5, 2 .9533	5, 2 .9746	5, 2 .9869	5, 2 .9936	5, 2 .9971	5, 2 .9987	5, 2 .9995
50	4, 3 .6393	4, 3 .7375	6, 2 .8242	6, 2 .8920	6, 2 .9379	6, 2 .9666	6, 2 .9831	6, 2 .9920	6, 2 .9964	6, 2 .9985	6, 2 .9994	6, 2 .9998
55	5, 3 .6633	5, 3 .7662	7, 2 .8482	7, 2 .9115	7, 2 .9520	7, 2 .9758	7, 2 .9886	7, 2 .9950	7, 2 .9979	7, 2 .9992	7, 2 .9997	
60	6, 3 .6859	6, 3 .7913	6, 3 .8716	8, 2 .9273	8, 2 .9628	8, 2 .9824	8, 2 .9922	8, 2 .9968	8, 2 .9988	8, 2 .9996	8, 2 .9999	
65	5, 4 .7090	7, 3 .8135	7, 3 .8910	7, 3 .9415	7, 3 .9711	9, 2 .9870	9, 2 .9947	9, 2 .9980	9, 2 .9993	9, 2 .9998		

TABLE OPT1.5.2

P=5 K=2 OPTIMAL DESIGN AND ASSOCIATED CONFIDENCE COEFFICIENT FOR ONE-SIDED INTERVALS

b \ a/σ	.100	.200	.300	.400	.500	.600	.700	.800	.900	1.000	1.100	1.200
70	2, 6 .3412	2, 6 .4234	2, 6 .5089	4, 5 .6115	6, 4 .7310	8, 3 .8331	8, 3 .9071	8, 3 .9529	8, 3 .9782	8, 3 .9908		
75	1, 7 .3936	1, 7 .4544	1, 7 .5163	5, 5 .6291	7, 4 .7511	9, 3 .8505	9, 3 .9207	9, 3 .9619	9, 3 .9834	9, 3 .9934		
80			2, 7 .5263	6, 5 .6457	8, 4 .7696	8, 4 .8667	10, 3 .9321	10, 3 .9691	10, 3 .9873	10, 3 .9953		
85	1, 8 .4037	1, 8 .4650	3, 7 .5359	7, 5 .6615	9, 4 .7865	9, 4 .8818	11, 3 .9418	11, 3 .9748	11, 3 .9902	11, 3 .9966		
90			4, 7 .5453	8, 5 .6766	10, 4 .8021	10, 4 .8950	10, 4 .9505	12, 3 .9794	12, 3 .9924	12, 3 .9975		
95	1, 9 .4122	1, 9 .4738	5, 7 .5547	9, 5 .6910	9, 5 .8167	11, 4 .9065	11, 4 .9580	11, 4 .9833	11, 4 .9941	11, 4 .9982		
100			6, 7 .5640	9, 6 .7054	10, 5 .8311	12, 4 .9167	12, 4 .9643	12, 4 .9866	12, 4 .9955	12, 4 .9987		
105	1,10 .4195	1,10 .4813	5, 8 .5741	9, 6 .7197	11, 5 .8442	13, 4 .9257	13, 4 .9696	13, 4 .9891	13, 4 .9966	13, 4 .9991		
110			6, 8 .5843	10, 6 .7331	12, 5 .8562	12, 5 .9337	14, 4 .9740	14, 4 .9911	14, 4 .9974	14, 4 .9993		
115	1,11 .4258	1,11 .4878	7, 8 .5942	11, 6 .7458	13, 5 .8672	13, 5 .9412	15, 4 .9778	15, 4 .9928	15, 4 .9980	15, 4 .9995		
120			8, 8 .6039	12, 6 .7578	14, 5 .8773	14, 5 .9479	14, 5 .9811	16, 4 .9941	16, 4 .9984	16, 4 .9996		
125	1,12 .4314	1,12 .4935	9, 8 .6134	13, 6 .7693	15, 5 .8866	15, 5 .9537	15, 5 .9839	15, 5 .9952	15, 5 .9988	15, 5 .9997		
130			10, 8 .6226	12, 7 .7801	14, 6 .8954	16, 5 .9598	16, 5 .9863	16, 5 .9961	16, 5 .9991	16, 5 .9998		

TABLE OPT1.5.2

P=5 K=2 , OPTIMAL DESIGN AND ASSOCIATED CONFIDENCE COEFFICIENT FOR ONE-SIDED INTERVALS

b \ a/σ	.100	.200	.300	.400	.500	.600	.700	.800	.900	1.000	1.100	1.200
135	1,13 .4364	1,13 .4986	9, 9 .6321	13, 7 .7911	15, 6 .9037	17, 5 .9633	17, 5 .9883	17, 5 .9969	17, 5 .9993	17, 5 .9999		
140		2,13 .4995	10, 9 .6415	14, 7 .8014	16, 6 .9112	18, 5 .9673	18, 5 .9900	18, 5 .9974	18, 5 .9994			
145	1,14 .4408	1,14 .5031	11, 9 .6507	15, 7 .8112	17, 6 .9181	17, 6 .9710	19, 5 .9915	19, 5 .9979	19, 5 .9996			
150		2,14 .5057	12, 9 .6596	16, 7 .8205	18, 6 .9245	18, 6 .9742	18, 6 .9927	18, 6 .9983	18, 6 .9997			
155	1,15 .4448	3,14 .5085	13, 9 .6683	17, 7 .8292	19, 6 .9303	19, 6 .9771	19, 6 .9938	19, 6 .9986	19, 6 .9997			
160		4,14 .5115	14, 9 .6767	18, 7 .8375	18, 7 .9357	20, 6 .9796	20, 6 .9947	20, 6 .9989	20, 6 .9998			
165	1,16 .4484	3,15 .5152	13,10 .6851	17, 8 .8456	19, 7 .9408	21, 6 .9819	21, 6 .9955	21, 6 .9991	21, 6 .9998			
170		4,15 .5190	14,10 .6935	18, 8 .8534	20, 7 .9455	22, 6 .9838	22, 6 .9962	22, 6 .9993	22, 6 .9999			
175	1,17 .4518	5,15 .5228	15,10 .7017	19, 8 .8607	21, 7 .9497	21, 7 .9857	21, 7 .9967	23, 6 .9994				
180		6,15 .5267	16,10 .7096	20, 8 .8677	22, 7 .9537	22, 7 .9873	22, 7 .9972	22, 7 .9995				
185	1,18 .4548	7,15 .5306	17,10 .7173	21, 8 .8743	23, 7 .9572	23, 7 .9887	23, 7 .9976	23, 7 .9996				
190		6,16 .5346	18,10 .7248	22, 8 .8805	22, 8 .9605	24, 7 .9899	24, 7 .9980	24, 7 .9997				
195	1,19 .4576	7,16 .5389	19,10 .7321	23, 8 .8864	23, 8 .9637	25, 7 .9910	25, 7 .9983	25, 7 .9997				

TABLE OPT1.6.2

P=6 K=2 OPTIMAL DESIGN AND ASSOCIATED CONFIDENCE COEFFICIENT FOR ONE-SIDED INTERVALS

b \ a/σ	.500	.600	.700	.800	.900	1.000	1.100	1.200	1.300	1.400	1.500	1.600
6	1, 0 .0675	1, 0 .0859	1, 0 .1076	1, 0 .1327	1, 0 .1612	1, 0 .1931	1, 0 .2281	1, 0 .2660	1, 0 .3063	1, 0 .3485	1, 0 .3922	1, 0 .4368
12	2, 0 .1093	2, 0 .1451	2, 0 .1897	2, 0 .2397	2, 0 .2951	2, 0 .3547	2, 0 .4170	2, 0 .4902	2, 0 .5429	2, 0 .6034	2, 0 .6604	2, 0 .7131
18	3, 0 .1512	3, 0 .2065	3, 0 .2708	3, 0 .3424	3, 0 .4184	3, 0 .4959	3, 0 .5718	3, 0 .6436	3, 0 .7091	3, 0 .7672	3, 0 .8171	3, 0 .8590
21	1, 1 .3931	1, 1 .4549	1, 1 .5177	1, 1 .5799	1, 1 .6400	1, 1 .6966	1, 1 .7488	1, 1 .7956	1, 1 .8367	1, 1 .8718	1, 1 .9013	1, 1 .9254
24											4, 0 .9025	4, 0 .9311
27	2, 1 .3955	2, 1 .4785	2, 1 .5619	2, 1 .6421	2, 1 .7158	2, 1 .7810	2, 1 .8361	2, 1 .8810	2, 1 .9162	2, 1 .9427	2, 1 .9620	2, 1 .9755
33	3, 1 .4135	3, 1 .5117	3, 1 .6081	3, 1 .6972	3, 1 .7751	3, 1 .8393	3, 1 .8897	3, 1 .9271	3, 1 .9537	3, 1 .9715	3, 1 .9833	3, 1 .9905
36	1, 2 .5035	1, 2 .5686	1, 2 .6318									
39			4, 1 .6522	4, 1 .7452	4, 1 .8220	4, 1 .8814	4, 1 .9246	4, 1 .9542	4, 1 .9733	4, 1 .9852	4, 1 .9921	4, 1 .9959
42	2, 2 .5222	2, 2 .6094	2, 2 .6911	2, 2 .7640	2, 2 .8261							
45			5, 1 .6930	5, 1 .7862	5, 1 .8591	5, 1 .9120	5, 1 .9479	5, 1 .9707	5, 1 .9843	5, 1 .9920	5, 1 .9961	5, 1 .9982
48	3, 2 .5443	3, 2 .6457	3, 2 .7371	3, 2 .8140	3, 2 .8748	3, 2 .9197	3, 2 .9511	3, 2 .9716				

TABLE OPT1.6.2

P=6 K=2 OPTIMAL DESIGN AND ASSOCIATED CONFIDENCE COEFFICIENT FOR ONE-SIDED INTERVALS

b \ a/σ	.300	.400	.500	.600	.700	.800	.900	1.000	1.100	1.200	1.300	1.400
51	1, 3 .4273	1, 3 .4935	1, 3 .5598			6, 1 .8210	6, 1 .8885	6, 1 .9345	6, 1 .9637	6, 1 .9810	6, 1 .9906	6, 1 .9985
54			4, 2 .5678	4, 2 .6788	4, 2 .7749	4, 2 .8513	4, 2 .9076	4, 2 .9459	4, 2 .9701	4, 2 .9845	4, 2 .9924	4, 2 .9965
57		2, 3 .5016	2, 3 .5919				7, 1 .9117	7, 1 .9512	7, 1 .9746	7, 1 .9876	7, 1 .9943	7, 1 .9975
60				5, 2 .7090	5, 2 .8066	5, 2 .8801	5, 2 .9307	5, 2 .9626	5, 2 .9811	5, 2 .9911	5, 2 .9960	5, 2 .9983
63		3, 3 .5132	3, 3 .6199	3, 3 .7177				8, 1 .9634	8, 1 .9821	8, 1 .9918	8, 1 .9965	8, 1 .9986
66	1, 4 .4615	1, 4 .5285		6, 2 .7366	6, 2 .8335	6, 2 .9027	6, 2 .9474	6, 2 .9736	6, 2 .9877	6, 2 .9947	6, 2 .9978	6, 2 .9992
69			4, 3 .6456	4, 3 .7510	4, 3 .8363							
72		2, 4 .5454		7, 2 .7617	7, 2 .8565	7, 2 .9207	7, 2 .9597	7, 2 .9811	7, 2 .9918	7, 2 .9967	7, 2 .9988	7, 2 .9996
75			5, 3 .6594	5, 3 .7793	5, 3 .8638	5, 3 .9224	5, 3 .9592					
78		3, 4 .5634		8, 2 .7846	8, 2 .8762	8, 2 .9352	8, 2 .9690	8, 2 .9864	8, 2 .9945	8, 2 .9980	8, 2 .9993	8, 2 .9998
81	1, 5 .4857		6, 3 .6916	6, 3 .8038	6, 3 .8859	6, 3 .9394	5, 3 .9706	6, 3 .9869	6, 3 .9946	6, 3 .9980	6, 3 .9993	6, 3 .9998
84		4, 4 .5799	4, 4 .6966	9, 2 .8053	9, 2 .8931	9, 2 .9468	9, 2 .9760	9, 2 .9901	9, 2 .9962	9, 2 .9987	9, 2 .9996	9, 2 .9999
87			7, 3 .7124	7, 3 .8253	7, 3 .9039	7, 3 .9522	7, 3 .9784	7, 3 .9911	7, 3 .9957	7, 3 .9989	7, 3 .9996	

TABLE OPT1.6.2

P=5 K=2 OPTIMAL DESIGN AND ASSOCIATED CONFIDENCE COEFFICIENT FOR ONE-SIDED INTERVALS

b \ a/σ	.100	.200	.300	.400	.500	.600	.700	.800	.900	1.000	1.100	1.200
90	5, 4 .2114	5, 4 .3260		5, 4 .5960	5, 4 .7211		10, 2 .9076	10, 2 .9562	10, 2 .9813	10, 2 .9927	10, 2 .9974	10, 2 .9992
93	3, 5 .2763	3, 5 .3779	3, 5 .4885	3, 5 .5997	8, 3 .7319	8, 3 .8441	8, 3 .9187	8, 3 .9619	8, 3 .9839	8, 3 .9939	8, 3 .9979	8, 3 .9993
96	1, 6 .3708	1, 6 .4365	1, 6 .5040	6, 4 .6118	6, 4 .7431	6, 4 .8455	11, 2 .9200	11, 2 .9639	11, 2 .9854	11, 2 .9946	11, 2 .9982	11, 2 .9995
99				4, 5 .6187	9, 3 .7501	9, 3 .8607	9, 3 .9310	9, 3 .9695	9, 3 .9879	9, 3 .9957	9, 3 .9986	9, 3 .9996
102			2, 6 .5097	7, 4 .6271	7, 4 .7629	7, 4 .8647		12, 2 .9702	12, 2 .9885	12, 2 .9960	12, 2 .9987	12, 2 .9996
105				5, 5 .6367	10, 3 .7671	10, 3 .8754	10, 3 .9412	10, 3 .9754	10, 3 .9908	10, 3 .9969	10, 3 .9991	10, 3 .9998
108			3, 6 .5163	8, 4 .6421	8, 4 .7810	8, 4 .8810	8, 4 .9427	8, 4 .9755	13, 2 .9910	13, 2 .9970	13, 2 .9991	13, 2 .9998
111	1, 7 .3842	1, 7 .4506	1, 7 .5184	6, 5 .6534	11, 3 .7830	11, 3 .8885	11, 3 .9498	11, 3 .9801	11, 3 .9930	11, 3 .9978	11, 3 .9994	11, 3 .9998
114			4, 6 .5237	9, 4 .6565	9, 4 .7975	9, 4 .8951	9, 4 .9523	9, 4 .9809	9, 4 .9932	9, 4 .9979	9, 4 .9994	9, 4 .9999
117			2, 7 .5289	7, 5 .6692	7, 5 .7986	12, 3 .9001	12, 3 .9570	12, 3 .9838	12, 3 .9946	12, 3 .9984	12, 3 .9996	
120			5, 6 .5314	10, 4 .6706	10, 4 .8126	10, 4 .9073	10, 4 .9600	10, 4 .9849	10, 4 .9950	10, 4 .9985	10, 4 .9996	
123			3, 7 .5387	8, 5 .6841	8, 5 .8158	13, 3 .9105	13, 3 .9632	13, 3 .9868	13, 3 .9958	13, 3 .9988	13, 3 .9997	
126	1, 8 .3953	1, 8 .4622	6, 6 .5395	6, 6 .6855	11, 4 .8265	11, 4 .9179	11, 4 .9664	11, 4 .9881	11, 4 .9963	11, 4 .9990	11, 4 .9998	

TABLE OPT1.3.3

P=3 K=3 OPTIMAL DESIGN AND ASSOCIATED CONFIDENCE COEFFICIENT FOR ONE-SIDED INTERVALS

b \ a/σ	.500	.600	.700	.800	.900	1.000	1.100	1.200	1.300	1.400	1.500	1.600
3	1, 0 .4276	1, 0 .4776	1, 0 .5277	1, 0 .5773	1, 0 .6255	1, 0 .6716	1, 0 .7151	1, 0 .7555	1, 0 .7925	1, 0 .8258	1, 0 .8554	1, 0 .8813
4	1, 1 .4935	1, 1 .5456	1, 1 .5969	1, 1 .6464	1, 1 .6935	1, 1 .7374	1, 1 .7778	1, 1 .8144	1, 1 .8469	1, 1 .8753	1, 1 .8997	1, 1 .9204
5	1, 2 .5329	1, 2 .5855	1, 2 .6365	1, 2 .6851	1, 2 .7307	1, 2 .7726	1, 2 .8105	1, 2 .8442	1, 2 .8737	1, 2 .8990	1, 2 .9203	1, 2 .9381
6	1, 3 .5595	1, 3 .6121	2, 0 .6671	2, 0 .7282	2, 0 .7828	2, 0 .8302	2, 0 .8702	2, 0 .9030	2, 0 .9291	2, 0 .9494	2, 0 .9646	2, 0 .9758
7	1, 4 .5790	2, 1 .6411	2, 1 .7061	2, 1 .7648	2, 1 .8162	2, 1 .8598	2, 1 .8956	2, 1 .9242	2, 1 .9463	2, 1 .9629	2, 1 .9750	2, 1 .9836
8	2, 2 .6005	2, 2 .6696	2, 2 .7332	2, 2 .7897	2, 2 .8384	2, 2 .8790	2, 2 .9117	2, 2 .9372	2, 2 .9565	2, 2 .9707	2, 2 .9808	2, 2 .9877
9	2, 3 .6225	2, 3 .6909	3, 0 .7603	3, 0 .8212	3, 0 .8710	3, 0 .9101	3, 0 .9394	3, 0 .9605	3, 0 .9751	3, 0 .9848	3, 0 .9911	3, 0 .9949
10	2, 4 .6400	3, 1 .7162	3, 1 .7852	3, 1 .8430	3, 1 .8893	3, 1 .9247	3, 1 .9506	3, 1 .9687	3, 1 .9809	3, 1 .9888	3, 1 .9936	3, 1 .9965
11	3, 2 .6599	3, 2 .7371	3, 2 .8040	3, 2 .8591	3, 2 .9024	3, 2 .9350	3, 2 .9583	3, 2 .9742	3, 2 .9847	3, 2 .9912	3, 2 .9952	3, 2 .9974
12	3, 3 .6778	4, 0 .7555	4, 0 .8258	4, 0 .8813	4, 0 .9226	4, 0 .9518	4, 0 .9713	4, 0 .9837	4, 0 .9911	4, 0 .9954	4, 0 .9977	4, 0 .9989
13	4, 1 .6930	4, 1 .7750	4, 1 .8426	4, 1 .8949	4, 1 .9330	4, 1 .9593	4, 1 .9765	4, 1 .9870	4, 1 .9932	4, 1 .9966	4, 1 .9984	4, 1 .9992
14	4, 2 .7105	4, 2 .7907	4, 2 .8558	4, 2 .9054	4, 2 .9409	4, 2 .9649	4, 2 .9801	4, 2 .9893	4, 2 .9945	4, 2 .9973	4, 2 .9988	4, 2 .9995
15	4, 3 .7251	5, 0 .8068	5, 0 .8727	5, 0 .9207	5, 0 .9533	5, 0 .9740	5, 0 .9863	5, 0 .9932	5, 0 .9968	5, 0 .9986	5, 0 .9994	5, 0 .9998

TABLE OPT1.3.3

P=3 K=3 OPTIMAL DESIGN AND ASSOCIATED CONFIDENCE COEFFICIENT FOR ONE-SIDED INTERVALS

b \ a/σ	.100	.200	.300	.400	.500	.600	.700	.800	.900	1.000	1.100	1.200
16	1,13 .4334	1,13 .4886	2,10 .5534	4, 4 .6464	5, 1 .7393	5, 1 .8212	5, 1 .8843	5, 1 .9294	5, 1 .9594	5, 1 .9779	5, 1 .9887	5, 1 .9946
17	1,14 .4372	1,14 .4926	3, 8 .5621	4, 5 .6579	5, 2 .7534	5, 2 .8332	5, 2 .8937	5, 2 .9363	5, 2 .9641	5, 2 .9809	5, 2 .9905	5, 2 .9955
18	1,15 .4407	1,15 .4962	3, 9 .5700	5, 3 .6691	5, 3 .7653	6, 0 .8468	6, 0 .9066	6, 0 .9468	6, 0 .9717	6, 0 .9859	6, 0 .9934	6, 0 .9971
19	1,16 .4439	1,16 .4995	4, 7 .5771	5, 4 .6807	6, 1 .7783	6, 1 .8577	6, 1 .9148	6, 1 .9524	6, 1 .9753	6, 1 .9880	6, 1 .9946	6, 1 .9977
20	1,17 .4469	1,17 .5025	4, 8 .5855	5, 5 .6909	6, 2 .7897	6, 2 .8669	6, 2 .9216	6, 2 .9570	6, 2 .9781	6, 2 .9896	6, 2 .9954	6, 2 .9981
21	1,18 .4496	1,18 .5053	4, 9 .5930	6, 3 .7016	7, 0 .8004	7, 0 .8782	7, 0 .9313	7, 0 .9642	7, 0 .9827	7, 0 .9923	7, 0 .9968	7, 0 .9988
22	1,19 .4522	1,19 .5079	5, 7 .6002	6, 4 .7117	7, 1 .8112	7, 1 .8866	7, 1 .9372	7, 1 .9679	7, 1 .9849	7, 1 .9934	7, 1 .9974	7, 1 .9990
23	1,20 .4546	1,20 .5103	5, 8 .6081	7, 2 .7209	7, 2 .8206	7, 2 .8937	7, 2 .9421	7, 2 .9710	7, 2 .9866	7, 2 .9943	7, 2 .9978	7, 2 .9992
24	1,21 .4568	2,18 .5136	5, 9 .6152	7, 3 .7308	8, 0 .8302	8, 0 .9030	8, 0 .9494	8, 0 .9758	8, 0 .9895	8, 0 .9958	8, 0 .9985	8, 0 .9995
25	1,22 .4589	2,19 .5168	6, 7 .6223	7, 4 .7396	8, 1 .8390	8, 1 .9094	8, 1 .9536	8, 1 .9783	8, 1 .9908	8, 1 .9964	8, 1 .9987	8, 1 .9996
26	1,23 .4608	2,20 .5198	6, 8 .6296	8, 2 .7484	8, 2 .8468	8, 2 .9150	8, 2 .9572	8, 2 .9804	8, 2 .9918	8, 2 .9969	8, 2 .9989	8, 2 .9997
27	1,24 .4627	3,18 .5231	6, 9 .6364	8, 3 .7570	9, 0 .8554	9, 0 .9226	9, 0 .9626	9, 0 .9837	9, 0 .9935	9, 0 .9977	9, 0 .9993	9, 0 .9998
28	1,25 .4644	3,19 .5266	7, 7 .6433	8, 4 .7648	9, 1 .8626	9, 1 .9276	9, 1 .9657	9, 1 .9853	9, 1 .9943	9, 1 .9980	9, 1 .9994	9, 1 .9998

TABLE OPT1.3.3

P=3 K=3 OPTIMAL DESIGN AND ASSOCIATED CONFIDENCE COEFFICIENT FOR ONE-SIDED INTERVALS

b \ a/σ	.100	.200	.300	.400	.500	.600	.700	.800	.900	1.000	1.100	1.200
29	1,26 .4661	3,20 .5299	7, 8 .6501	9, 2 .7731	9, 2 .8691	9, 2 .9320	9, 2 .9683	9, 2 .9867	9, 2 .9950	9, 2 .9983	9, 2 .9995	9, 2 .9999
30	1,27 .4676	4,18 .5332	7, 9 .6565	9, 3 .7806	10, 0 .8767	10, 0 .9382	10, 0 .9723	10, 0 .9889	10, 0 .9960	10, 0 .9987	10, 0 .9996	
31	1,28 .4691	4,19 .5368	8, 7 .6632	10, 1 .7879	10, 1 .8827	10, 1 .9421	10, 1 .9746	10, 1 .9900	10, 1 .9965	10, 1 .9989	10, 1 .9997	
32	1,29 .4705	4,20 .5403	8, 8 .6696	10, 2 .7952	10, 2 .8881	10, 2 .9456	10, 2 .9765	10, 2 .9910	10, 2 .9969	10, 2 .9991	10, 2 .9997	
33	1,30 .4719	5,18 .5436	9, 6 .6757	10, 3 .8019	11, 0 .8947	11, 0 .9506	11, 0 .9795	11, 0 .9925	11, 0 .9976	11, 0 .9993	11, 0 .9998	
34	1,31 .4732	5,19 .5473	9, 7 .6821	11, 1 .8087	11, 1 .8997	11, 1 .9537	11, 1 .9811	11, 1 .9932	11, 1 .9979	11, 1 .9994	11, 1 .9999	
35	1,32 .4744	5,20 .5508	9, 8 .6879	11, 2 .8151	11, 2 .9043	11, 2 .9564	11, 2 .9826	11, 2 .9939	11, 2 .9981	11, 2 .9995		
36	1,33 .4756	6,18 .5542	10, 6 .6940	12, 0 .8212	12, 0 .9101	12, 0 .9605	12, 0 .9848	12, 0 .9949	12, 0 .9985	12, 0 .9996		
37	1,34 .4767	6,19 .5578	10, 7 .6998	12, 1 .8274	12, 1 .9143	12, 1 .9629	12, 1 .9860	12, 1 .9954	12, 1 .9987	12, 1 .9997		
38	1,35 .4778	6,20 .5613	11, 5 .7053	12, 2 .8330	12, 2 .9181	12, 2 .9651	12, 2 .9871	12, 2 .9958	12, 2 .9986	12, 2 .9997		
39	1,36 .4788	7,18 .5647	11, 6 .7111	13, 0 .8387	13, 0 .9231	13, 0 .9683	13, 0 .9887	13, 0 .9965	13, 0 .9991	13, 0 .9998		
40	1,37 .4798	7,19 .5683	11, 7 .7166	13, 1 .8441	13, 1 .9266	13, 1 .9703	13, 1 .9896	13, 1 .9969	13, 1 .9992	13, 1 .9998		
41	1,38 .4808	7,20 .5717	12, 5 .7220	13, 2 .8491	13, 2 .9299	13, 2 .9720	13, 2 .9904	13, 2 .9972	13, 2 .9993	13, 2 .9998		

TABLE OPT1.4.3

P=4 K=3 OPTIMAL DESIGN AND ASSOCIATED CONFIDENCE COEFFICIENT FOR ONE-SIDED INTERVALS

b \ a/σ	.500	.600	.700	.800	.900	1.000	1.100	1.200	1.300	1.400	1.500	1.600
4	1, 0, 0 0, 0, 0 .1880	1, 0, 0 0, 0, 0 .2239	1, 0, 0 0, C, 0 .2631	1, 0, 0 0, 0, 0 .3051	1, 0, 0 0, 0, 0 .3493	1, 0, 0 0, 0, 0 .3952	1, 0, 0 0, 0, 0 .4422	1, 0, 0 0, 0, C .4894	1, 0, 0 0, 0, 0 .5363	1, 0, 0 0, 0, 0 .5822	1, 0, 0 0, 0, 0 .6265	1, 0, 0 0, 0, 0 .6687
6	0, 1, 0 0, 0, 0 .3915	0, 1, 0 0, 0, 0 .4582	0, 1, 0 0, 0, 0 .5257	0, 1, 0 0, 0, 0 .5919	C, 1, 0 0, 0, 0 .6552	0, 1, 0 0, 0, 0 .7140	0, 1, C 0, 0, C .7671	0, 1, 0 0, 0, 0 .8140	0, 1, 0 0, 0, 0 .8542	0, 1, 0 0, 0, 0 .8878	0, 1, 0 0, 0, 0 .9153	0, 1, 0 0, 0, 0 .9373
7	0, 0, 1 0, 0, 0 .4652	0, 0, 1 0, 0, 0 .5249	0, 0, 1 0, 0, 0 .5839	0, 0, 1 0, 0, 0 .6409	0, 0, 1 0, 0, 0 .6948	0, 0, 1 0, 0, 0 .7447	0, 0, 1 0, 0, 0 .7898	0, 0, 1 0, 0, 0 .8297	0, 0, 1 0, 0, 0 .8643	0, 0, 1 0, 0, 0 .8937	0, 0, 1 0, 0, 0 .9181	0, 0, 1 0, 0, 0 .9380
8	C, 0, 0 1, 0, 0 .5034	0, 0, 0 1, 0, 0 .5477	0, 0, 0 1, 0, 0 .5914									
10	0, 1, 0 0, 0, 1 .5323	0, 1, 0 0, 0, 1 .6041	0, 1, 0 0, 0, 1 .6723	0, 1, 0 0, 0, 1 .7350	0, 1, 0 0, 0, 1 .7907	0, 1, 0 0, 0, 1 .8388	0, 0, 0 0, 1, 0 .8795	0, 0, 0 0, 1, 0 .9139	0, 0, 0 0, 1, C .9402	0, 0, 0 0, 1, 0 .9596	0, 0, 0 0, 1, 0 .9735	0, 0, 0 0, 1, 0 .9831
11	0, 0, 1 0, 0, 1 .5508	0, 0, 1 0, 0, 1 .6113										
12		0, 2, 0 0, 0, 0 .6231	0, 2, 0 0, 0, 0 .7083	0, 2, 0 0, 0, 0 .7825	0, 2, 0 C, C, 0 .8439	0, 2, 0 0, 0, 0 .8921	0, 2, 0 0, 0, 0 .9282	0, 2, 0 0, 0, 0 .9539	0, 2, 0 0, 0, 0 .9715	0, 2, 0 0, 0, 0 .9830	0, 2, 0 0, 0, 0 .9902	0, 2, 0 0, 0, 0 .9946
13	0, 1, 1 0, 0, 0 .5665	0, 1, 1 0, 0, 0 .6533	0, 1, 1 0, 0, 0 .7323	0, 1, 1 0, 0, 0 .8005	0, 1, 1 0, 0, 0 .8567	0, 1, 1 0, 0, 0 .9009	0, 1, 1 0, 0, C .9339	0, 1, 1 0, 0, 0 .9575	0, 1, 1 0, 0, 0 .9737	0, 1, 1 0, 0, 0 .9843	0, 1, 1 0, 0, 0 .9910	0, 1, 1 0, 0, 0 .9950
14	0, 1, 0 0, 0, 2 .5973	C, 1, 0 0, 0, 2 .6680	0, 0, 0 0, 1, 1 .7399	0, 0, 0 0, 1, 1 .8029								
15												

TABLE OPT1.4.3

P=4 K=3 OPTIMAL DESIGN AND ASSOCIATED CONFIDENCE COEFFICIENT FOR ONE-SIDED INTERVALS

b \ a/σ	.400	.500	.600	.700	.800	.900	1.000	1.100	1.200	1.300	1.400	1.500
16						0, 2, 0 0, 0, 1 .9034	0, 2, 0 0, 0, 1 .9387	0, 2, 0 0, 0, 1 .9628	0, 2, 0 0, 0, 1 .9785	0, 1, 0 0, 1, 0 .9881	0, 1, 0 0, 1, 0 .9938	0, 1, 0 0, 1, 0 .9969
17												
18				0, 3, 0 0, 0, 0 .8193	0, 3, 0 0, 0, 0 .8834	0, 3, 0 0, 0, 0 .9289	0, 3, 0 0, 0, 0 .9590	0, 3, 0 0, 0, 0 .9776	0, 3, 0 0, 0, 0 .9884	0, 3, 0 0, 0, 0 .9943	0, 3, 0 0, 0, 0 .9974	0, 3, 0 0, 0, 0 .9988
19			0, 2, 1 0, 0, 0 .7521	0, 2, 1 0, 0, 0 .8311	0, 2, 1 0, 0, 0 .8912	0, 2, 1 0, 0, 0 .9338	0, 2, 1 0, 0, 0 .9619	0, 2, 1 0, 0, 0 .9793	0, 2, 1 0, 0, 0 .9893	0, 2, 1 0, 0, 0 .9948	0, 2, 1 0, 0, 0 .9976	0, 2, 1 0, 0, 0 .9990
20	0, 2, 0 0, 0, 2 .5797	0, 2, 0 0, 0, 2 .6769	0, 2, 0 0, 0, 2 .7630	0, 2, 0 1, 0, 0 .8358	0, 2, 0 1, 0, 0 .8932	0, 2, 0 1, 0, 0 .9342						
21	0, 1, 1 0, 0, 2 .5878	0, 1, 1 0, 0, 2 .6781										
22	0, 0, 0 0, 1, 3 .5908	0, 3, 0 0, 0, 1 .7003	0, 3, 0 0, 0, 1 .7950	0, 3, 0 0, 0, 1 .8686	0, 3, 0 0, 0, 1 .9211	0, 3, 0 0, 0, 1 .9556	0, 3, 0 0, 0, 1 .9766	0, 3, 0 0, 0, 1 .9885	0, 3, 0 0, 0, 1 .9947	0, 3, 0 0, 0, 1 .9977	0, 3, 0 0, 0, 1 .9990	0, 3, 0 0, 0, 1 .9996
23	0, 2, 1 0, 0, 1 .6024	0, 2, 1 0, 0, 1 .7091	0, 2, 1 0, 0, 1 .7995	0, 2, 1 0, 0, 1 .8701								
24	0, 2, 0 0, 0, 3 .6171	0, 4, 0 0, 0, 0 .7140	0, 4, 0 0, 0, 0 .8140	0, 4, 0 0, 0, 0 .8878	0, 4, 0 0, 0, 0 .9373	0, 4, 0 0, 0, 0 .9674	0, 4, 0 0, 0, 0 .9843	0, 4, 0 0, 0, 0 .9929	0, 4, 0 0, 0, 0 .9970	0, 4, 0 0, 0, 0 .9988	0, 4, 0 0, 0, 0 .9996	0, 4, 0 0, 0, 0 .9999
25	0, 1, 1 0, 0, 3 .6183	0, 3, 1 0, 0, 0 .7280	0, 3, 1 0, 0, 0 .8237	0, 3, 1 0, 0, 0 .8941	0, 3, 1 0, 0, 0 .9410	0, 3, 1 0, 0, 0 .9695	0, 3, 1 0, 0, 0 .9854	0, 3, 1 0, 0, 0 .9935	0, 3, 1 0, 0, 0 .9973	0, 3, 1 0, 0, 0 .9990		

TABLE OPT1.4.3

P=4 K=3 OPTIMAL DESIGN AND ASSOCIATED CONFIDENCE COEFFICIENT FOR ONE-SIDED INTERVALS

b \ a/σ	.100	.200	.300	.400	.500	.600	.700	.800	.900	1.000	1.100	1.200
26	0, 1, 0 0, 0, 5 .3786	0, 1, 0 0, 0, 5 .4543	0, 1, 0 0, 0, 5 .5316	0, 3, 0 0, 0, 2 .6343	0, 3, 0 0, 0, 2 .7429	0, 3, 0 0, 0, 2 .8313	0, 3, 0 1, 0, 0 .8969	0, 3, 0 1, 0, C .9422	0, 3, 0 1, 0, 0 .9699	0, 3, 0 1, 0, 0 .9855		
27	0, 0, 1 0, 0, 5 .4040	0, 0, 1 0, 0, 5 .4671		0, 2, 1 0, 0, 2 .6409	C, 2, 1 0, 0, 2 .7447							
28	0, 0, 0 1, 0, 5 .4343	0, 0, 0 1, 0, 5 .4796	0, 2, 0 0, 0, 4 .5413	0, 4, 0 0, 0, 1 .6452	0, 4, 0 0, 0, 1 .7629	0, 4, 0 0, 0, 1 .8542	0, 4, 0 0, 0, 1 .9176	0, 4, 0 0, 0, 1 .9572	0, 4, 0 0, 0, 1 .9796	0, 4, 0 0, 0, 1 .9910	0, 4, 0 0, 0, 1 .9964	0, 4, 0 0, 0, 1 .9986
29			0, 1, 1 0, 0, 4 .5458	0, 3, 1 0, 0, 1 .6556	0, 3, 1 C, 0, 1 .7691	0, 3, 1 0, 0, 1 .8572	0, 3, 1 0, 0, 1 .9188	0, 3, 1 0, 0, 1 .9575				
30			0, 0, C 0, 1, 5 .5477	0, 3, 0 0, 0, 3 .6675	0, 5, 0 0, 0, C .7761	0, 5, 0 0, 0, 0 .8689	0, 5, 0 0, 0, C .9302	0, 5, 0 0, 0, 0 .9661	0, 5, 0 0, 0, 0 .9850	0, 5, 0 0, C, 0 .9939	0, 5, 0 0, 0, 0 .9977	0, 5, 0 0, 0, 0 .9992
31			0, 2, 1 0, 0, 3 .5541	0, 2, 1 0, C, 3 .6694	0, 4, 1 0, 0, 0 .7856	0, 4, 1 0, 0, 0 .8750	0, 4, 1 0, 0, 0 .9337	0, 4, 1 0, 0, 0 .9680	0, 4, 1 0, 0, 0 .9859	0, 4, 1 0, 0, 0 .9943	0, 4, 1 0, C, 0 .9979	0, 4, 1 0, 0, 0 .9993
32	0, 0, 0 1, 0, 6 .4417	0, 0, C 1, 0, 6 .4872	0, 2, 0 0, 0, 5 .5628	0, 4, 0 0, 0, 2 .6835	0, 4, 0 C, 0, 2 .7963	0, 4, 0 0, 0, 2 .8802	0, 4, 0 0, 0, 2 .9357	0, 4, 0 1, 0, 0 .9688	0, 4, 0 1, 0, 0 .9862	0, 4, 0 1, 0, 0 .9945	0, 4, 0 1, 0, 0 .9980	
33			0, 1, 1 0, C, 5 .5640	0, 3, 1 0, C, 2 .6889	0, 3, 1 0, 0, 2 .7980							
34			0, 3, 0 0, 0, 4 .5729	0, 5, 0 0, 0, 1 .6948	0, 5, 0 0, 0, 1 .8128	0, 5, 0 0, 0, 1 .8965	0, 5, 0 0, 0, 1 .9484	0, 5, 0 0, 0, 1 .9768	0, 5, 0 0, 0, 1 .9906	0, 5, 0 0, 0, 1 .9965	0, 5, 0 0, C, 1 .9988	0, 5, 0 0, 0, 1 .9997
35		0, 0, 1 0, 0, 7 .4877	0, 2, 1 0, 0, 4 .5771	0, 4, 1 0, 0, 1 .7028	0, 4, 1 0, 0, 1 .8174	0, 4, 1 0, 0, 1 .8986	0, 4, 1 0, 0, 1 .9492	0, 4, 1 0, 0, 1 .9771				

TABLE OPT1.5.3

P=5 K=3 OPTIMAL DESIGN AND ASSOCIATED CONFIDENCE COEFFICIENT FOR ONE-SIDED INTERVALS

b \ a/σ	.500	.600	.700	.800	.900	1.000	1.100	1.200	1.300	1.400	1.500	1.600
5	1, 0, 0 0, 0, 0 .1238	1, 0, 0 0, 0, 0 .1540	1, 0, 0 0, 0, 0 .1884	1, 0, 0 0, 0, 0 .2267	1, 0, 0 0, 0, 0 .2685	1, 0, 0 0, 0, 0 .3134	1, 0, 0 0, 0, 0 .3606	1, 0, 0 0, 0, 0 .4093	1, 0, 0 0, 0, 0 .4589	1, 0, 0 0, 0, 0 .5085	1, 0, 0 0, 0, 0 .5574	1, 0, 0 0, 0, 0 .6047
7	0, 1, 0 0, 0, 0 .3453	0, 1, 0 0, 0, 0 .4056	0, 1, 0 0, 0, 0 .4679	0, 1, 0 0, 0, 0 .5307	0, 1, 0 0, 0, 0 .5925	0, 1, 0 0, 0, 0 .6517	0, 1, 0 0, 0, 0 .7072	0, 1, 0 0, 0, 0 .7578	0, 1, 0 0, 0, 0 .8031	0, 1, 0 0, 0, 0 .8426	0, 1, 0 0, 0, 0 .8763	0, 1, 0 0, 0, 0 .9044
10	0, 0, 1 0, 0, 0 .4494	0, 0, 1 0, 0, 0 .5157	0, 0, 1 0, 0, 0 .5815	0, 0, 1 0, 0, 0 .6448	0, 0, 1 0, 0, 0 .7042	0, 0, 1 0, 0, 0 .7585	0, 0, 0 1, 0, 0 .8156	0, 0, 0 1, 0, 0 .8635	0, 0, 0 1, 0, 0 .9016	0, 0, 0 1, 0, 0 .9310	0, 0, 0 1, 0, 0 .9529	0, 0, 0 1, 0, 0 .9686
14	0, 1, 0 0, 1, 0 .5100	0, 1, 0 0, 1, 0 .5776	0, 2, 0 0, 0, 0 .6459	0, 2, 0 0, 0, 0 .7236	0, 2, 0 0, 0, 0 .7913	0, 2, 0 0, 0, 0 .8477	0, 2, 0 0, 0, 0 .8926	0, 2, 0 0, 0, 0 .9268	0, 2, 0 0, 0, 0 .9518	0, 2, 0 0, 0, 0 .9693	0, 2, 0 0, 0, 0 .9811	0, 2, 0 0, 0, 0 .9887
15												1, 0, 0 1, 0, 0 .9888
17	0, 0, 1 0, 1, 0 .5502	0, 0, 0 1, 1, 0 .6286	0, 0, 0 1, 1, 0 .7110	0, 1, 0 1, 0, 0 .7843	0, 1, 0 1, 0, 0 .8505	0, 1, 0 1, 0, 0 .9009	0, 1, 0 1, 0, 0 .9372	0, 1, 0 1, 0, 0 .9619	0, 1, 0 1, 0, 0 .9779	0, 1, 0 1, 0, 0 .9877	0, 1, 0 1, 0, 0 .9934	0, 1, 0 1, 0, 0 .9966
20	0, 0, 2 0, 0, 0 .5861	0, 0, 2 0, 0, 0 .6742	0, 0, 1 1, 0, 0 .7638	0, 0, 1 1, 0, 0 .8373	0, 0, 1 1, 0, 0 .8935	0, 0, 0 2, 0, 0 .9345	0, 0, 0 2, 0, 0 .9623	0, 0, 0 2, 0, 0 .9794	0, 0, 0 2, 0, 0 .9893	0, 0, 0 2, 0, 0 .9947	0, 0, 0 2, 0, 0 .9975	0, 0, 0 2, 0, 0 .9989
24	0, 0, 0 1, 2, 0 .6206	0, 2, 1 0, 0, 0 .7197	0, 2, 0 1, 0, 0 .8061	0, 2, 0 1, 0, 0 .8764	0, 2, 0 1, 0, 0 .9259	0, 2, 0 1, 0, 0 .9583	0, 2, 0 1, 0, 0 .9779	0, 2, 0 1, 0, 0 .9889	0, 2, 0 1, 0, 0 .9948	0, 2, 0 1, 0, 0 .9977	0, 2, 0 1, 0, 0 .9990	0, 2, 0 1, 0, 0 .9996
27	0, 0, 1 1, 1, 0 .6562	0, 0, 0 2, 1, 0 .7578	0, 0, 0 2, 1, 0 .8426	0, 1, 0 2, 0, 0 .9054	0, 1, 0 2, 0, 0 .9478	0, 1, 0 2, 0, 0 .9731	0, 1, 0 2, 0, 0 .9870	0, 1, 0 2, 0, 0 .9942	0, 1, 0 2, 0, 0 .9975	0, 1, 0 2, 0, 0 .9990	0, 1, 0 2, 0, 0 .9996	0, 1, 0 2, 0, 0 .9999
30	0, 0, 2 1, 0, 0 .6880	0, 0, 1 2, 0, 0 .7904	0, 0, 1 2, 0, 0 .8727	0, 0, 1 2, 0, 0 .9286	0, 0, 1 2, 0, 0 .9629	0, 0, 0 3, 0, 0 .9824	0, 0, 0 3, 0, 0 .9922	0, 0, 0 3, 0, 0 .9968	0, 0, 0 3, 0, 0 .9988	0, 0, 0 3, 0, 0 .9996	0, 0, 0 3, 0, 0 .9999	

TABLE OPT1.5.3

P=5 K=3 OPTIMAL DESIGN AND ASSOCIATED CONFIDENCE COEFFICIENT FOR ONE-SIDED INTERVALS

b \ a/σ	.100	.200	.300	.400	.500	.600	.700	.800	.900	1.000	1.100	1.200
34	0, 1, 0 0, 1, 2 .3655	0, 1, 0 0, 1, 2 .4342	0, 1, 0 0, 1, 2 .5049	0, 0, 1 1, 2, 0 .6009	0, 2, 1 1, 0, 0 .7167	0, 2, 1 1, 0, 0 .8202	0, 2, 0 2, 0, 0 .8965	0, 2, 0 2, 0, 0 .9461	0, 2, 0 2, 0, 0 .9742	0, 2, 0 2, 0, 0 .9887	0, 2, 0 2, 0, 0 .9954	0, 2, 0 2, 0, 0 .9983
37	0, 1, 0 0, 0, 3 .3749	0, 1, 0 0, 0, 3 .4442	0, 1, 0 0, 0, 3 .5152	0, 0, 2 1, 1, 0 .6237	0, 0, 1 2, 1, 0 .7446	0, 0, 0 3, 1, 0 .8458	0, 0, 0 3, 1, 0 .9161	0, 1, 0 3, 0, 0 .9590	0, 1, 0 3, 0, 0 .9819	0, 1, 0 3, 0, 0 .9927	0, 1, 0 3, 0, 0 .9973	0, 1, 0 3, 0, 0 .9991
40	0, 0, 1 0, 0, 3 .3831	0, 0, 1 0, 0, 3 .4529	0, 0, 0 1, 0, 3 .5267	0, 0, 3 1, 0, 0 .6451	0, 0, 2 2, 0, 0 .7696	0, 0, 1 3, 0, 0 .8674	0, 0, 1 3, 0, 0 .9324	0, 0, 1 3, 0, 0 .9690	0, 0, 1 3, 0, 0 .9872	0, 0, 0 4, 0, 0 .9952	0, 0, 0 4, 0, 0 .9984	0, 0, 0 4, 0, 0 .9995
44	0, 1, 0 0, 1, 3 .3904	0, 1, 0 0, 1, 3 .4606	0, 0, 3 0, 2, 0 .5384	0, 0, 1 2, 2, 0 .6667	0, 2, 1 2, 0, 0 .7918	0, 2, 1 2, 0, 0 .8864	0, 2, 0 3, 0, 0 .9453	0, 2, 0 3, 0, 0 .9766	0, 2, 0 3, 0, 0 .9910	0, 2, 0 3, 0, 0 .9969	0, 2, 0 3, 0, 0 .9990	0, 2, 0 3, 0, 0 .9997
47	0, 1, 0 0, 0, 4 .3970	0, 1, 0 0, 0, 4 .4674	0, 1, 0 1, 0, 3 .5517	0, 0, 2 2, 1, 0 .6869	0, 0, 1 3, 1, 0 .8126	0, 0, 0 4, 1, 0 .9028	0, 0, 0 4, 1, 0 .9557	0, 1, 0 4, 0, 0 .9822	0, 1, 0 4, 0, 0 .9937	0, 1, 0 4, 0, 0 .9980	0, 1, 0 4, 0, 0 .9994	0, 1, 0 4, 0, 0 .9999
50	0, 0, 1 0, 0, 4 .4029	0, 0, 1 0, 0, 4 .4735	0, 0, 0 2, 0, 3 .5648	0, 0, 3 2, 0, 0 .7059	0, 0, 2 3, 0, 0 .8315	0, 0, 1 4, 0, 0 .9167	0, 0, 1 4, 0, 0 .9643	0, 0, 1 4, 0, 0 .9866	0, 0, 1 4, 0, 0 .9955	0, 0, 1 4, 0, 0 .9987	0, 0, 1 4, 0, 0 .9997	
54	0, 1, 0 0, 1, 4 .4082	0, 1, 0 0, 1, 4 .4791	0, 0, 4 0, 2, 0 .5776	0, 0, 1 3, 2, 0 .7242	0, 2, 1 3, 0, 0 .8483	0, 2, 1 3, 0, 0 .9287	0, 2, 0 4, 0, 0 .9711	0, 2, 0 4, 0, 0 .9898	0, 2, 0 4, 0, 0 .9969	0, 2, 0 4, 0, 0 .9991	0, 2, 0 4, 0, 0 .9998	
57	0, 1, 0 0, 0, 5 .4131	0, 1, 0 0, 0, 5 .4841	0, 0, 5 0, 1, 0 .5910	0, 0, 2 3, 1, 0 .7416	0, 0, 1 4, 1, 0 .8635	0, 0, 0 5, 1, 0 .9391	0, 0, 0 5, 1, 0 .9766	0, 1, 0 5, 0, 0 .9923	0, 1, 0 5, 0, 0 .9978	0, 1, 0 5, 0, 0 .9994	0, 1, 0 5, 0, 0 .9999	
60	0, 0, 1 0, 0, 5 .4175	0, 0, 1 0, 0, 5 .4888	0, 0, 6 0, 0, 0 .6039	0, 0, 3 3, 0, 0 .7578	0, 0, 2 4, 0, 0 .8775	0, 0, 1 5, 0, 0 .9479	0, 0, 1 5, 0, 0 .9812	0, 0, 1 5, 0, 0 .9942	0, 0, 1 5, 0, 0 .9984	0, 0, 1 5, 0, 0 .9996		
64	0, 1, 0 0, 1, 5 .4216	0, 1, 0 0, 1, 5 .4930	0, 0, 4 1, 2, 0 .6165	0, 0, 1 4, 2, 0 .7732	0, 2, 1 4, 0, 0 .8899	0, 2, 1 4, 0, 0 .9554	0, 2, 0 5, 0, 0 .9848	0, 2, 0 5, 0, 0 .9956	0, 2, 0 5, 0, 0 .9989			

TABLE OPT1.5.3

P=5 K=3 OPTIMAL DESIGN AND ASSOCIATED CONFIDENCE COEFFICIENT FOR ONE-SIDED INTERVALS

b \ a/σ	.100	.200	.300	.400	.500	.600	.700	.800	.900	1.000	1.100	1.200
67	0, 1, 0 0, 0, 6 .4254	0, 1, 0 0, 0, 6 .4969	0, 0, 5 1, 1, 0 .6293	0, 0, 2 4, 1, 0 .7879	0, 0, 1 5, 1, 0 .9010	0, 0, 0 6, 1, 0 .9619	0, 0, 0 6, 1, 0 .9877	0, 1, 0 6, 0, 0 .9966	0, 1, 0 6, 0, 0 .9992			
70	0, 0, 1 0, 0, 6 .4290	0, 0, 1 0, 0, 6 .5006	0, 0, 6 1, 0, 0 .6416	0, 0, 3 4, 0, 0 .8015	0, 0, 2 5, 0, 0 .9112	0, 0, 1 6, 0, 0 .9674	0, 0, 1 6, 0, 0 .9901	0, 0, 1 6, 0, 0 .9974	0, 0, 1 6, 0, 0 .9994	0, 0, 1 6, 0, 0 .9999		
74	0, 1, 0 0, 1, 6 .4323	0, 1, 0 0, 1, 6 .5040	0, 0, 3 3, 2, 0 .6535	0, 0, 1 5, 2, 0 .8142	0, 2, 1 5, 0, 0 .9203	0, 2, 1 5, 0, 0 .9721	0, 2, 0 6, 0, 0 .9920	0, 2, 0 6, 0, 0 .9981	0, 2, 0 6, 0, 0 .9996			
77	0, 1, 0 0, 0, 7 .4354	0, 1, 1 0, 0, 6 .5076	0, 0, 5 2, 1, 0 .6655	0, 0, 2 5, 1, 0 .8264	0, 0, 0 7, 1, 0 .9284	0, 0, 0 7, 1, 0 .9762	0, 0, 0 7, 1, 0 .9935	0, 0, 0 7, 1, 0 .9985	0, 1, 0 7, 0, 0 .9997			
80	0, 0, 1 0, 0, 7 .4382	0, 0, 0 1, 0, 7 .5119	0, 0, 6 2, 0, 0 .6770	0, 0, 3 5, 0, 0 .8378	0, 0, 2 6, 0, 0 .9358	0, 0, 1 7, 0, 0 .9796	0, 0, 1 7, 0, 0 .9947	0, 0, 1 7, 0, 0 .9989	0, 0, 1 7, 0, 0 .9998			
84	0, 1, 0 0, 1, 7 .4410	0, 2, 1 0, 0, 6 .5166	0, 0, 3 4, 2, 0 .6881	0, 0, 0 7, 2, 0 .8483	0, 2, 1 6, 0, 0 .9424	0, 2, 1 6, 0, 0 .9826	0, 2, 0 7, 0, 0 .9957	0, 2, 0 7, 0, 0 .9991	0, 2, 0 7, 0, 0 .9999			
87	0, 1, 0 0, 0, 8 .4435	0, 1, 1 1, 0, 6 .5214	0, 0, 5 3, 1, 0 .6990	0, 0, 2 6, 1, 0 .8583	0, 0, 0 8, 1, 0 .9483	0, 0, 0 8, 1, 0 .9851	0, 0, 0 8, 1, 0 .9965	0, 0, 0 8, 1, 0 .9993				
90	0, 0, 1 0, 0, 8 .4459	0, 0, 0 2, 0, 7 .5267	0, 0, 6 3, 0, 0 .7096	0, 0, 3 6, 0, 0 .8677	0, 0, 2 7, 0, 0 .9537	0, 0, 1 8, 0, 0 .9873	0, 0, 1 8, 0, 0 .9972	0, 0, 1 8, 0, 0 .9995				
94	0, 1, 0 0, 1, 8 .4482	0, 2, 1 1, 0, 6 .5321	0, 0, 3 5, 2, 0 .7198	0, 0, 0 8, 2, 0 .8764	0, 2, 1 7, 0, 0 .9585	0, 2, 1 7, 0, 0 .9891	0, 2, 0 8, 0, 0 .9977	0, 2, 0 8, 0, 0 .9996				

TABLE OPT1.6.3

P=6 K=3 OPTIMAL DESIGN AND ASSOCIATED CONFIDENCE COEFFICIENT FOR ONE-SIDED INTERVALS

b \ a/σ	.500	.600	.700	.800	.900	1.000	1.100	1.200	1.300	1.400	1.500	1.600
6	1, 0, 0 0, 0 .0815	1, 0, 0 0, 0 .1060	1, 0, 0 0, 0 .1349	1, 0, 0 0, 0 .1685	1, 0, 0 0, 0 .2065	1, 0, 0 0, 0 .2485	1, 0, 0 0, 0 .2940	1, 0, 0 0, 0 .3424	1, 0, 0 0, 0 .3927	1, 0, 0 0, 0 .4442	1, 0, 0 0, 0 .4959	1, 0, 0 0, 0 .5468
7	0, 1, 0 0, 0 .3384	0, 1, 0 0, 0 .3869	0, 1, 0 0, 0 .4372	0, 1, 0 0, 0 .4883	0, 1, 0 0, 0 .5395	0, 1, 0 0, 0 .5900	0, 1, 0 0, 0 .6389	0, 1, 0 0, 0 .6855	0, 1, 0 0, 0 .7293	0, 1, 0 0, 0 .7696	0, 1, 0 0, 0 .8063	0, 1, 0 0, 0 .8391
11		0, 0, 1 0, 0 .4037	0, 0, 1 0, 0 .4837	0, 0, 1 0, 0 .5636	0, 0, 1 0, 0 .6402	0, 0, 1 0, 0 .7109	0, 0, 1 0, 0 .7737	0, 0, 1 0, 0 .8275	0, 0, 1 0, 0 .8719	0, 0, 1 0, 0 .9073	0, 0, 1 0, 0 .9346	0, 0, 1 0, 0 .9551
14	0, 2, 0 0, 0 .4408	0, 2, 0 0, 0 .5132	0, 2, 0 0, 0 .5850	0, 2, 0 0, 0 .6538	0, 2, 0 0, 0 .7177	0, 2, 0 0, 0 .7751	0, 2, 0 0, 0 .8250	0, 2, 0 0, 0 .8671	0, 2, 0 0, 0 .9016	0, 2, 0 0, 0 .9289	0, 2, 0 0, 0 .9499	0, 2, 0 0, 0 .9656
15					0, 0, 0 1, 0 .7276	0, 0, 0 1, 0 .7996	0, 0, 0 1, 0 .8579	0, 0, 0 1, 0 .9028	0, 0, 0 1, 0 .9358	0, 0, 0 1, 0 .9591	0, 0, 0 1, 0 .9747	0, 0, 0 1, 0 .9849
17	0, 1, 0 0, 1 .4990	0, 1, 0 0, 1 .5532	0, 1, 0 0, 1 .6063	0, 1, 0 0, 1 .6576								
18		0, 1, 1 0, 0 .5560	0, 1, 1 0, 0 .6485	0, 1, 1 0, 0 .7322	0, 1, 1 0, 0 .8038	0, 1, 1 0, 0 .8620	0, 1, 1 0, 0 .9067	0, 1, 1 0, 0 .9394	0, 1, 1 0, 0 .9622	0, 1, 1 0, 0 .9773	0, 1, 1 0, 0 .9869	0, 1, 1 0, 0 .9927
21	0, 0, 1 0, 1 .5222	0, 0, 1 0, 1 .6094	0, 0, 1 0, 1 .6911	0, 0, 1 0, 1 .7640	0, 0, 1 0, 1 .8261	0, 0, 1 0, 1 .8764	0, 0, 1 0, 1 .9153	1, 0, 0 1, 0 .9447	1, 0, 0 1, 0 .9671	1, 0, 0 1, 0 .9812	1, 0, 0 1, 0 .9897	1, 0, 0 1, 0 .9945
22			0, 0, 2 0, 0 .7042	0, 0, 2 0, 0 .7916	0, 0, 2 0, 0 .8607	0, 0, 2 0, 0 .9116	0, 0, 2 0, 0 .9468	0, 0, 2 0, 0 .9695	0, 0, 2 0, 0 .9834	0, 0, 2 0, 0 .9913	0, 0, 2 0, 0 .9957	0, 0, 2 0, 0 .9980
24	0, 2, 0 0, 1 .5540	0, 2, 0 0, 1 .6274										

TABLE OPT1.6.3

P=6 K=3 OPTIMAL DESIGN AND ASSOCIATED CONFIDENCE COEFFICIENT FOR ONE-SIDED INTERVALS

b \ a/σ	.300	.400	.500	.600	.700	.800	.900	1.000	1.100	1.200	1.300	1.400
25	0, 0, 0 1, 1 .3362	0, 0, 0 1, 1 .4424		0, 0, 0 1, 1 .6571	0, 0, 0 1, 1 .7504	0, 0, 0 1, 1 .8276	0, 0, 0 1, 1 .8871	0, 0, 0 1, 1 .9299	0, 0, 0 1, 1 .9587	0, 0, 0 1, 1 .9769	0, 0, 0 1, 1 .9878	0, 0, 0 1, 1 .9938
26					0, 0, 1 1, 0 .7522	0, 0, 1 1, 0 .8376	0, 0, 1 1, 0 .9001	0, 0, 1 1, 0 .9422	0, 0, 1 1, 0 .9685	0, 0, 1 1, 0 .9838	0, 0, 1 1, 0 .9921	0, 0, 1 1, 0 .9964
27	0, 1, 0 0, 2 .4487	0, 1, 0 0, 2 .5037	0, 1, 0 0, 2 .5586									
28			0, 4, 0 0, 0 .5900	0, 4, 0 0, 0 .6855	0, 4, 0 0, 0 .7696	0, 4, 0 0, 0 .8391						
29				0, 2, 0 1, 0 .6993	0, 2, 0 1, 0 .7966	0, 2, 0 1, 0 .8713	0, 2, 0 1, 0 .9238	0, 2, 0 1, 0 .9578	0, 2, 0 1, 0 .9781	0, 2, 0 1, 0 .9893	0, 2, 0 1, 0 .9951	0, 2, 0 1, 0 .9979
30						0, 0, 0 2, 0 .8736	0, 0, 0 2, 0 .9279	0, 0, 0 2, 0 .9617	0, 0, 0 2, 0 .9810	0, 0, 0 2, 0 .9911	0, 0, 0 2, 0 .9961	0, 0, 0 2, 0 .9984
31		0, 0, 1 0, 2 .5184	0, 0, 1 0, 2 .6087									
32			0, 0, 2 0, 1 .6235	0, 0, 2 0, 1 .7310	0, 0, 2 0, 1 .8197	0, 0, 2 0, 1 .8868	0, 0, 2 0, 1 .9335	0, 0, 2 0, 1 .9635	0, 0, 2 0, 1 .9812			
33				0, 0, 3 0, 0 .7366	0, 0, 3 0, 0 .8335	0, 0, 3 0, 0 .9027	0, 0, 3 0, 0 .9474	0, 0, 3 0, 0 .9736	0, 0, 3 0, 0 .9877	0, 0, 3 0, 0 .9947	0, 0, 3 0, 0 .9978	0, 0, 3 0, 0 .9992
34	0, 2, 0 0, 2 .4579	0, 2, 0 0, 2 .5347										

TABLE OPT1.6.3

P=6 K=3 OPTIMAL DESIGN AND ASSOCIATED CONFIDENCE COEFFICIENT FOR ONE-SIDED INTERVALS

b \ a/σ	.100	.200	.300	.400	.500	.600	.700	.800	.900	1.000	1.100	1.200
35	0, 0, 0 1, 2 .2189	0, 0, 0 1, 2 .3138		0, 0, 0 1, 2 .5367	0, 0, 0 1, 2 .6475	0, 0, 0 1, 2 .7465						
36					0, 0, 1 1, 1 .6547	0, 0, 1 1, 1 .7686	0, 0, 1 1, 1 .8567	0, 0, 1 1, 1 .9182	0, 0, 1 1, 1 .9568	0, 0, 1 1, 1 .9790	0, 0, 1 1, 1 .9905	0, 0, 1 1, 1 .9960
37	0, 1, 0 0, 3 .3723	0, 1, 0 0, 3 .4260	0, 1, 0 0, 3 .4810			0, 0, 2 1, 0 .7696	0, 0, 2 1, 0 .8634	0, 0, 2 1, 0 .9259	0, 0, 2 1, 0 .9631	0, 0, 2 1, 0 .9831	0, 0, 2 1, 0 .9929	0, 0, 2 1, 0 .9972
38				0, 4, 0 0, 1 .5578	0, 4, 0 0, 1 .6586							
39					0, 4, 1 0, 0 .6811	0, 4, 1 0, 0 .7872	0, 4, 1 0, 0 .8685					
40					0, 0, 0 2, 1 .6839	0, 0, 0 2, 1 .8004	0, 0, 0 2, 1 .8651	0, 0, 0 2, 1 .9397	0, 0, 0 2, 1 .9711	0, 0, 0 2, 1 .9873	0, 0, 0 2, 1 .9949	0, 0, 0 2, 1 .9981
41				0, 0, 1 0, 3 .5688			0, 0, 1 2, 0 .8877	0, 0, 1 2, 0 .9432	0, 0, 1 2, 0 .9738	0, 0, 1 2, 0 .9890	0, 0, 1 2, 0 .9957	0, 0, 1 2, 0 .9985
42				0, 0, 2 0, 2 .5799	0, 0, 2 0, 2 .6966							
43					0, 0, 3 0, 1 .7110	0, 0, 3 0, 1 .8198	0, 0, 3 0, 1 .8977	0, 0, 3 0, 1 .9471	0, 0, 3 0, 1 .9751	0, 0, 3 0, 1 .9893	0, 0, 3 0, 1 .9958	
44			0, 2, 0 0, 3 .4926			0, 0, 4 0, 0 .8275	0, 0, 4 0, 0 .9073	0, 0, 4 0, 0 .9551	0, 0, 4 0, 0 .9803	0, 0, 4 0, 0 .9921	0, 0, 4 0, 0 .9971	0, 0, 4 0, 0 .9991

APPENDIX A.3b

Tables of Optimal Designs for Two-Sided Comparisons

Tables OPT2•p•k for $p = 2(1)6$, $k = 2$

and $p = 3(1)6$, $k = 3$

TABLE OPT2.2.2

P=2 K=2 OPTIMAL DESIGN AND ASSOCIATED CONFIDENCE COEFFICIENT FOR TWO-SIDED INTERVALS

b \ a/σ	.500	.600	.700	.800	.900	1.000	1.100	1.200	1.300	1.400	1.500	1.600
2	1, 0 .0764	1, 0 .1080	1, 0 .1439	1, 0 .1835	1, 0 .2261	1, 0 .2709	1, 0 .3173	1, 0 .3646	1, 0 .4122	1, 0 .4594	1, 0 .5057	1, 0 .5507
3	1, 1 .1271	1, 1 .1758	1, 1 .2313	1, 1 .2890	1, 1 .3483	1, 1 .4079	1, 1 .4666	1, 1 .5234	1, 1 .5775	1, 1 .6263	1, 1 .6755	1, 1 .7188
4	1, 2 .1580	1, 2 .2167	1, 2 .2792	1, 2 .3434	1, 2 .4074	1, 2 .4698	2, 0 .5310	2, 0 .5927	2, 0 .6503	2, 0 .7031	2, 0 .7506	2, 0 .7928
5	2, 1 .1994	2, 1 .2729	2, 1 .3503	2, 1 .4287	2, 1 .5053	2, 1 .5779	2, 1 .6451	2, 1 .7059	2, 1 .7595	2, 1 .8061	2, 1 .8457	2, 1 .8788
6	2, 2 .2353	2, 2 .3177	2, 2 .4020	2, 2 .4847	2, 2 .5631	2, 2 .6353	2, 2 .7001	2, 2 .7569	2, 2 .8058	2, 2 .8471	2, 2 .8813	2, 2 .9092
7	3, 1 .2625	3, 1 .3536	3, 1 .4460	3, 1 .5354	3, 1 .6185	3, 1 .6932	3, 1 .7581	3, 1 .8130	3, 1 .8582	3, 1 .8944	3, 1 .9229	3, 1 .9447
8	3, 2 .2995	3, 2 .3981	3, 2 .4953	3, 2 .5865	3, 2 .6688	3, 2 .7403	3, 2 .8007	3, 2 .8502	3, 2 .8896	3, 2 .9203	3, 2 .9437	3, 2 .9610
9	3, 3 .3281	3, 3 .4311	3, 3 .5303	3, 3 .6212	4, 1 .7037	4, 1 .7748	4, 1 .8331	4, 1 .8793	4, 1 .9148	4, 1 .9413	4, 1 .9604	4, 1 .9740
10	4, 2 .3559	4, 2 .4662	4, 2 .5709	4, 2 .6649	4, 2 .7456	4, 2 .8121	4, 2 .8649	4, 2 .9054	4, 2 .9354	4, 2 .9571	4, 2 .9722	4, 2 .9824
11	4, 3 .3845	4, 3 .4983	4, 3 .6036	4, 3 .6960	4, 3 .7735	4, 3 .8360	4, 3 .8844	5, 1 .9213	5, 1 .9482	5, 1 .9668	5, 1 .9794	5, 1 .9875
12	4, 4 .4079	5, 2 .5251	5, 2 .6335	5, 2 .7269	5, 2 .8031	5, 2 .8627	5, 2 .9073	5, 2 .9393	5, 2 .9615	5, 2 .9763	5, 2 .9859	5, 2 .9918
13	5, 3 .4345	5, 3 .5555	5, 3 .6634	5, 3 .7540	5, 3 .8264	5, 3 .8816	5, 3 .9219	5, 3 .9501	5, 3 .9692	5, 3 .9816	5, 3 .9894	5, 3 .9941
14	5, 4 .4577	5, 4 .5796	5, 4 .6862	6, 2 .7764	6, 2 .8468	6, 2 .8989	6, 2 .9358	6, 2 .9606	6, 2 .9767	6, 2 .9867	6, 2 .9927	6, 2 .9961

TABLE OPT2.2.2

P=2 K=2 OPTIMAL DESIGN AND ASSOCIATED CONFIDENCE COEFFICIENT FOR TWO-SIDED INTERVALS

b \ a/σ	.400	.500	.600	.700	.800	.900	1.000	1.100	1.200	1.300	1.400	1.500
15	5, 5 .3450	6, 3 .4796	6, 3 .6050	6, 3 .7129	6, 3 .7998	6, 3 .8659	6, 3 .9137	6, 3 .9466	6, 3 .9682	6, 3 .9818	6, 3 .9899	6, 3 .9947
16	6, 4 .3642	6, 4 .5020	6, 4 .6277	6, 4 .7337	6, 4 .8174	7, 2 .8803	7, 2 .9252	7, 2 .9552	7, 2 .9742	7, 2 .9858	7, 2 .9925	7, 2 .9962
17	6, 5 .3811	6, 5 .5209	7, 3 .6483	7, 3 .7544	7, 3 .8364	7, 3 .8959	7, 3 .9367	7, 3 .9631	7, 3 .9794	7, 3 .9890	7, 3 .9944	7, 3 .9973
18	7, 4 .3975	7, 4 .5419	7, 4 .6694	7, 4 .7731	7, 4 .8515	7, 4 .9073	7, 4 .9448	8, 2 .9686	8, 2 .9830	8, 2 .9913	8, 2 .9957	8, 2 .9980
19	7, 5 .4143	7, 5 .5602	7, 5 .6868	8, 3 .7895	8, 3 .8659	8, 3 .9188	8, 3 .9532	8, 3 .9744	8, 3 .9866	8, 3 .9933	8, 3 .9968	8, 3 .9986
20	7, 6 .4292	8, 4 .5779	8, 4 .7059	8, 4 .8061	8, 4 .8788	8, 4 .9282	8, 4 .9596	8, 4 .9784	8, 4 .9890	8, 4 .9947	8, 4 .9976	8, 4 .9989
21	8, 5 .4452	8, 5 .5955	8, 5 .7220	8, 5 .8194	9, 3 .8897	9, 3 .9364	9, 3 .9653	9, 3 .9821	9, 3 .9912	9, 3 .9959	9, 3 .9982	9, 3 .9993
22	8, 6 .4599	9, 4 .6107	9, 4 .7378	9, 4 .8339	9, 4 .9008	9, 4 .9441	9, 4 .9702	9, 4 .9850	9, 4 .9929	9, 4 .9968	9, 4 .9986	9, 4 .9995
23	9, 5 .4741	9, 5 .6275	9, 5 .7528	9, 5 .8458	9, 5 .9095	10, 3 .9501	10, 3 .9742	10, 3 .9874	10, 3 .9942	10, 3 .9975	10, 3 .9990	10, 3 .9996
24	9, 6 .4885	9, 6 .6420	10, 4 .7660	10, 4 .8574	10, 4 .9185	10, 4 .9563	10, 4 .9780	10, 4 .9896	10, 4 .9953	10, 4 .9980	10, 4 .9992	10, 4 .9997
25	9, 7 .5014	10, 5 .6556	10, 5 .7798	10, 5 .8680	10, 5 .9259	10, 5 .9610	10, 5 .9808	11, 3 .9911	11, 3 .9962	11, 3 .9984	11, 3 .9994	11, 3 .9998
26	10, 6 .5152	10, 6 .6704	10, 6 .7914	11, 4 .8774	11, 4 .9329	11, 4 .9657	11, 4 .9836	11, 4 .9927	11, 4 .9969	11, 4 .9988	11, 4 .9995	11, 4 .9998
27	10, 7 .5278	11, 5 .6831	11, 5 .8036	11, 5 .8868	11, 5 .9392	11, 5 .9696	11, 5 .9858	11, 5 .9938	11, 5 .9975	11, 5 .9990	11, 5 .9997	11, 5 .9999

TABLE OPT2.2.2

P=2 K=2 OPTIMAL DESIGN AND ASSOCIATED CONFIDENCE COEFFICIENT FOR TWO-SIDED INTERVALS

b \ a/σ	.100	.200	.300	.400	.500	.600	.700	.800	.900	1.000	1.100	1.200
28	9,10 .0498	10, 8 .1828	10, 8 .3597	11, 6 .5402	11, 6 .6962	11, 6 .8143	11, 6 .8946	12, 4 .9447	12, 4 .9731	12, 4 .9878	12, 4 .9948	12, 4 .9980
29	10, 9 .0516	10, 9 .1886	11, 7 .3695	11, 7 .5525	11, 7 .7077	12, 5 .8246	12, 5 .9027	12, 5 .9500	12, 5 .9762	12, 5 .9895	12, 5 .9957	12, 5 .9984
30	10,10 .0533	11, 8 .1942	11, 8 .3793	11, 8 .5637	12, 6 .7198	12, 6 .8344	12, 6 .9096	12, 6 .9544	13, 4 .9788	13, 4 .9909	13, 4 .9964	13, 4 .9987
31	11, 9 .0550	11, 9 .2000	11, 9 .3885	12, 7 .5757	12, 7 .7307	13, 5 .8432	13, 5 .9163	13, 5 .9588	13, 5 .9813	13, 5 .9922	13, 5 .9970	13, 5 .9989
32	11,10 .0568	11,10 .2056	12, 8 .3981	12, 8 .5865	13, 6 .7413	13, 6 .8522	13, 6 .9224	13, 6 .9625	13, 6 .9833	13, 6 .9932	13, 6 .9974	14, 4 .9991
33	11,11 .0585	12, 9 .2112	12, 9 .4072	13, 7 .5974	13, 7 .7517	13, 7 .8599	14, 5 .9279	14, 5 .9660	14, 5 .9853	14, 5 .9942	14, 5 .9979	14, 5 .9993
34	12,10 .0602	12,10 .2168	13, 8 .4161	13, 8 .6080	14, 6 .7611	14, 6 .8679	14, 6 .9333	14, 6 .9692	14, 6 .9869	14, 6 .9949	14, 6 .9982	14, 6 .9994
35	12,11 .0619	12,11 .2222	13, 9 .4252	14, 7 .6179	14, 7 .7708	14, 7 .8750	14, 7 .9378	15, 5 .9719	15, 5 .9884	15, 5 .9956	15, 5 .9985	15, 5 .9995
36	12,12 .0636	13,10 .2277	13,10 .4337	14, 8 .6281	14, 8 .7795	15, 6 .8818	15, 6 .9425	15, 6 .9746	15, 6 .9897	15, 6 .9962	15, 6 .9987	15, 6 .9996
37	13,11 .0653	13,11 .2332	14, 9 .4425	14, 9 .6375	15, 7 .7883	15, 7 .8883	15, 7 .9466	15, 7 .9768	16, 5 .9908	16, 5 .9967	16, 5 .9989	16, 5 .9997
38	13,12 .0670	14,10 .2384	14,10 .4509	15, 8 .6471	15, 8 .7965	16, 6 .8942	16, 6 .9505	16, 6 .9790	16, 6 .9919	16, 6 .9972	16, 6 .9991	16, 6 .9997
39	13,13 .0687	14,11 .2439	15, 9 .4590	15, 9 .6562	16, 7 .8044	16, 7 .9002	16, 7 .9540	16, 7 .9808	16, 7 .9928	16, 7 .9975	16, 7 .9992	16, 7 .9998
40	14,12 .0703	14,12 .2491	15,10 .4674	16, 8 .6649	16, 8 .8121	16, 8 .9054	17, 6 .9572	17, 6 .9826	17, 6 .9936	17, 6 .9979	17, 6 .9994	17, 6 .9998

TABLE OPT2.3.2

P=3 K=2 OPTIMAL DESIGN AND ASSOCIATED CONFIDENCE COEFFICIENT FOR TWO-SIDED INTERVALS

b \ a/σ	.500	.600	.700	.800	.900	1.000	1.100	1.200	1.300	1.400	1.500	1.600
3	1, 0 .0211	1, 0 .0355	1, 0 .0546	1, 0 .0766	1, 0 .1075	1, 0 .1410	1, 0 .1788	1, 0 .2202	1, 0 .2646	1, 0 .3114	1, 0 .3597	1, 0 .4087
6	1, 1 .0750	1, 1 .1202	1, 1 .1751	1, 1 .2378	1, 1 .3055	1, 1 .3757	1, 1 .4459	1, 1 .5142	1, 1 .5790	1, 1 .6392	1, 1 .6941	1, 1 .7434
9	2, 1 .1249	2, 1 .1955	2, 1 .2775	2, 1 .3659	2, 1 .4559	2, 1 .5431	2, 1 .6243	2, 1 .6971	2, 1 .7604	2, 1 .8140	2, 1 .8581	2, 1 .8937
12	2, 2 .1794	2, 2 .2702	2, 2 .3686	3, 1 .4709	3, 1 .5709	3, 1 .6618	3, 1 .7406	3, 1 .8062	3, 1 .8588	3, 1 .8996	3, 1 .9303	3, 1 .9527
15	3, 2 .2336	3, 2 .3444	3, 2 .4592	3, 2 .5689	3, 2 .6672	3, 2 .7508	4, 1 .8196	4, 1 .8745	4, 1 .9153	4, 1 .9445	4, 1 .9647	4, 1 .9781
18	4, 2 .2836	4, 2 .4097	4, 2 .5346	4, 2 .6481	4, 2 .7441	4, 2 .8208	4, 2 .8789	4, 2 .9210	4, 2 .9502	4, 2 .9696	4, 2 .9821	4, 2 .9898
21	4, 3 .3344	4, 3 .4690	5, 2 .5985	5, 2 .7114	5, 2 .8018	5, 2 .8697	5, 2 .9179	5, 2 .9503	5, 2 .9711	5, 2 .9838	5, 2 .9913	5, 2 .9955
24	5, 3 .3818	5, 3 .5259	5, 3 .6555	6, 2 .7626	6, 2 .8457	6, 2 .9045	6, 2 .9436	6, 2 .9682	6, 2 .9829	6, 2 .9912	6, 2 .9956	6, 2 .9979
27	6, 3 .4254	6, 3 .5758	6, 3 .7055	6, 3 .8069	6, 3 .8800	7, 2 .9296	7, 2 .9610	7, 2 .9795	7, 2 .9897	7, 2 .9951	7, 2 .9978	7, 2 .9990
30	6, 4 .4673	7, 3 .6200	7, 3 .7475	7, 3 .8426	7, 3 .9076	7, 3 .9488	7, 3 .9732	7, 3 .9868	7, 3 .9938	7, 3 .9973	7, 3 .9988	7, 3 .9995
33	7, 4 .5069	7, 4 .6603	8, 3 .7831	8, 3 .8713	8, 3 .9265	8, 3 .9627	8, 3 .9817	8, 3 .9916	8, 3 .9963	8, 3 .9985	8, 3 .9994	8, 3 .9998
36	8, 4 .5431	8, 4 .6971	8, 4 .8140	9, 3 .8944	9, 3 .9443	9, 3 .9726	9, 3 .9874	9, 3 .9946	9, 3 .9978	9, 3 .9992	9, 3 .9997	
39	9, 4 .5765	9, 4 .7295	9, 4 .8410	9, 4 .9136	10, 3 .9565	10, 3 .9798	10, 3 .9913	10, 3 .9965	10, 3 .9987	10, 3 .9995	10, 3 .9998	

TABLE OPT2.3.2

P=3 K=2 OPTIMAL DESIGN AND ASSOCIATED CONFIDENCE COEFFICIENT FOR TWO-SIDED INTERVALS

b \ a/σ	.100	.200	.300	.400	.500	.600	.700	.800	.900	1.000	1.100	1.200
42	7, 7 .0126	8, 6 .0872	8, 6 .2358	9, 5 .4244	9, 5 .6085	10, 4 .7582	10, 4 .8638	10, 4 .9295	10, 4 .9664	10, 4 .9852	10, 4 .9939	10, 4 .9977
45	8, 7 .0140	8, 7 .0955	9, 6 .2545	9, 6 .4511	10, 5 .6381	10, 5 .7839	11, 4 .8832	11, 4 .9423	11, 4 .9739	11, 4 .9891	11, 4 .9958	11, 4 .9985
48	8, 8 .0153	9, 7 .1038	9, 7 .2728	10, 6 .4774	11, 5 .6651	11, 5 .8074	12, 4 .8996	12, 4 .9527	12, 4 .9797	12, 4 .9920	12, 4 .9971	12, 4 .9990
51	9, 8 .0167	9, 8 .1121	10, 7 .2914	11, 6 .5023	12, 5 .6900	12, 5 .8281	12, 5 .9141	12, 5 .9611	13, 4 .9841	13, 4 .9941	13, 4 .9980	13, 4 .9994
54	9, 9 .0181	10, 8 .1206	11, 7 .3094	12, 6 .5259	12, 6 .7136	13, 5 .8464	13, 5 .9264	13, 5 .9682	13, 5 .9876	13, 5 .9956	13, 5 .9986	13, 5 .9996
57	10, 9 .0196	11, 8 .1290	11, 8 .3268	12, 7 .5487	13, 6 .7354	14, 5 .8627	14, 5 .9368	14, 5 .9740	14, 5 .9904	14, 5 .9968	14, 5 .9990	14, 5 .9997
60	10,10 .0211	11, 9 .1376	12, 8 .3444	13, 7 .5706	14, 6 .7554	14, 6 .8776	15, 5 .9457	15, 5 .9786	15, 5 .9925	15, 5 .9976	15, 5 .9993	15, 5 .9998
63	11,10 .0226	12, 9 .1462	13, 8 .3615	14, 7 .5913	15, 6 .7737	15, 6 .8907	15, 6 .9534	16, 5 .9824	16, 5 .9941	16, 5 .9982	16, 5 .9995	16, 5 .9999
66	11,11 .0242	12,10 .1547	14, 8 .3780	15, 7 .6110	15, 7 .7909	16, 6 .9024	16, 6 .9601	16, 6 .9856	16, 6 .9954	16, 6 .9987	16, 6 .9997	
69	12,11 .0258	13,10 .1634	14, 9 .3945	15, 8 .6300	16, 7 .8068	17, 6 .9127	17, 6 .9657	17, 6 .9882	17, 6 .9964	17, 6 .9990	17, 6 .9998	
72	12,12 .0273	14,10 .1719	15, 9 .4106	16, 8 .6481	17, 7 .8215	17, 7 .9221	18, 6 .9705	18, 6 .9903	18, 6 .9972	18, 6 .9993	18, 6 .9998	
75	13,12 .0290	14,11 .1865	16, 9 .4261	17, 8 .6652	18, 7 .8349	18, 7 .9304	18, 7 .9746	19, 6 .9920	19, 6 .9978	19, 6 .9995	19, 6 .9999	
78	14,12 .0306	15,11 .1891	16,10 .4414	18, 8 .6814	18, 8 .8474	19, 7 .9378	19, 7 .9782	19, 7 .9934	19, 7 .9983	19, 7 .9996		

TABLE OPT2.3.2

P=3 K=2 OPTIMAL DESIGN AND ASSOCIATED CONFIDENCE COEFFICIENT FOR TWO-SIDED INTERVALS

b \ a/σ	.100	.200	.300	.400	.500	.600	.700	.800	.900	1.000	1.100	1.200
81	14,13 .0323	15,12 .1976	17,10 .4565	18, 9 .6971	19, 8 .8590	20, 7 .9444	20, 7 .9813	20, 7 .9945	20, 7 .9986	20, 7 .9997		
84	15,13 .0340	16,12 .2062	18,10 .4711	19, 9 .7120	20, 8 .8697	20, 8 .9503	21, 7 .9639	21, 7 .9955	21, 7 .9989	21, 7 .9998		
87	15,14 .0357	17,12 .2147	19,10 .4852	20, 9 .7261	21, 8 .8795	21, 8 .9556	22, 7 .9861	22, 7 .9963	22, 7 .9992	22, 7 .9998		
90	16,14 .0374	17,13 .2231	19,11 .4993	21, 9 .7394	22, 8 .8886	22, 8 .9603	22, 8 .9881	22, 8 .9970	22, 8 .9993	22, 8 .9999		
93	16,15 .0392	18,13 .2316	20,11 .5129	21,10 .7522	22, 9 .8970	23, 8 .9645	23, 8 .9898	23, 8 .9975	23, 8 .9995			
96	17,15 .0409	19,13 .2400	21,11 .5261	22,10 .7645	23, 9 .9048	24, 8 .9682	24, 8 .9912	24, 8 .9979	24, 8 .9996			
99	17,16 .0427	19,14 .2483	21,12 .5390	23,10 .7760	24, 9 .9120	24, 9 .9716	25, 8 .9924	25, 8 .9983	25, 8 .9997			
102	18,16 .0445	20,14 .2567	22,12 .5516	24,10 .7870	25, 9 .9186	25, 9 .9746	25, 9 .9935	25, 9 .9986	25, 9 .9998			
105	18,17 .0463	21,14 .2649	23,12 .5639	24,11 .7974	25,10 .9247	26, 9 .9773	26, 9 .9944	26, 9 .9988	26, 9 .9998			
108	19,17 .0482	21,15 .2731	24,12 .5758	25,11 .8074	26,10 .9304	27, 9 .9797	27, 9 .9951	27, 9 .9990	27, 9 .9998			
111	20,17 .0500	22,15 .2813	24,13 .5875	26,11 .8169	27,10 .9357	27,10 .9818	28, 9 .9958	28, 9 .9992	28, 9 .9999			
114	20,18 .0519	23,15 .2894	25,13 .5989	27,11 .8259	28,10 .9405	28,10 .9837	28,10 .9964	28,10 .9994				
117	21,18 .0537	23,16 .2974	26,13 .6100	28,11 .8344	28,11 .9449	29,10 .9854	29,10 .9969	29,10 .9995				
120	21,19 .0556	24,16 .3054	27,13 .6207	28,12 .8426	29,11 .9491	30,10 .9869	30,10 .9973	30,10 .9996				

TABLE OPT2.4.2

P=4 K=2 OPTIMAL DESIGN AND ASSOCIATED CONFIDENCE COEFFICIENT FOR TWO-SIDED INTERVALS

b \ a/σ	.500	.600	.700	.800	.900	1.000	1.100	1.200	1.300	1.400	1.500	1.600
4	1, 0 .0058	1, 0 .0117	1, 0 .0207	1, 0 .0337	1, 0 .0511	1, 0 .0734	1, 0 .1007	1, 0 .1330	1, 0 .1699	1, 0 .2111	1, 0 .2558	1, 0 .3033
8	2, 0 .0215	2, 0 .0416	2, 0 .0709	2, 0 .1103	2, 0 .1594	2, 0 .2172	2, 0 .2819	2, 0 .3513	2, 0 .4229	2, 0 .4943	2, 0 .5634	2, 0 .6286
10	1, 1 .0515	1, 1 .0938	1, 1 .1500	1, 1 .2178	1, 1 .2937	1, 1 .3736	1, 1 .4539	1, 1 .5314	1, 1 .6039	1, 1 .6699	1, 1 .7286	1, 1 .7799
12							3, 0 .4567	3, 0 .5428	3, 0 .6236	3, 0 .6966	3, 0 .7604	3, 0 .8144
14	2, 1 .0887	2, 1 .1565	2, 1 .2420	2, 1 .3390	2, 1 .4402	2, 1 .5391	2, 1 .6306	2, 1 .7112	2, 1 .7797	2, 1 .8357	2, 1 .8802	2, 1 .9145
16	1, 2 .1004	1, 2 .1697	1, 2 .2517	1, 2 .3400								
18	3, 1 .1268	3, 1 .2174	3, 1 .3256	3, 1 .4413	3, 1 .5544	3, 1 .6573	3, 1 .7452	3, 1 .8166	3, 1 .8720	3, 1 .9132	3, 1 .9428	3, 1 .9634
20	2, 2 .1545	2, 2 .2539	2, 2 .3655	2, 2 .4786	2, 2 .5847	2, 2 .6787	2, 2 .7580	2, 2 .8226	2, 2 .8732			
22	4, 1 .1660	4, 1 .2768	4, 1 .4024	4, 1 .5292	4, 1 .6455	4, 1 .7443	4, 1 .8228	4, 1 .8818	4, 1 .9239	4, 1 .9527	4, 1 .9716	4, 1 .9835
24	3, 2 .2030	3, 2 .3249	3, 2 .4549	3, 2 .5791	3, 2 .6882	3, 2 .7777	3, 2 .8473	3, 2 .8987	3, 2 .9352	3, 2 .9599	3, 2 .9761	3, 2 .9862
26	2, 3 .2110	5, 1 .3340	5, 1 .4723	5, 1 .6043	5, 1 .7182	5, 1 .8089	5, 1 .8761	5, 1 .9231	5, 1 .9541	5, 1 .9737	5, 1 .9855	5, 1 .9923
28	4, 2 .2486	4, 2 .3880	4, 2 .5295	4, 2 .6571	4, 2 .7623	4, 2 .8427	4, 2 .9005	4, 2 .9397	4, 2 .9649	4, 2 .9805	4, 2 .9895	4, 2 .9946
30	3, 3 .2673	3, 3 .4053	3, 3 .5407	6, 1 .6681	6, 1 .7762	6, 1 .8569	6, 1 .9130	6, 1 .9496	6, 1 .9720	6, 1 .9852	6, 1 .9924	6, 1 .9963

TABLE OPT2.4.2

P=4 K=2 OPTIMAL DESIGN AND ASSOCIATED CONFIDENCE COEFFICIENT FOR TWO-SIDED INTERVALS

b \ a/σ	.400	.500	.600	.700	.800	.900	1.000	1.100	1.200	1.300	1.400	1.500
32	5, 2 .1575	5, 2 .2921	5, 2 .4451	5, 2 .5930	5, 2 .7193	5, 2 .8171	5, 2 .8871	5, 2 .9337	5, 2 .9630	5, 2 .9803	5, 2 .9900	5, 2 .9952
34	4, 3 .1767	4, 3 .3174	4, 3 .4703	4, 3 .6130	4, 3 .7324	4, 3 .8242	7, 1 .8928	7, 1 .9387	7, 1 .9667	7, 1 .9828	7, 1 .9915	7, 1 .9960
36	6, 2 .1839	6, 2 .3337	6, 2 .4971	6, 2 .6475	6, 2 .7694	6, 2 .8584	6, 2 .9181	6, 2 .9552	6, 2 .9768	6, 2 .9886	6, 2 .9947	6, 2 .9977
38	5, 3 .2065	5, 3 .3633	5, 3 .5266	5, 3 .6718	5, 3 .7865	5, 3 .8692	5, 3 .9244	5, 3 .9586	5, 3 .9786	5, 3 .9895	5, 3 .9951	8, 1 .9979
40	4, 4 .2178	4, 4 .3736	7, 2 .5444	7, 2 .6946	7, 2 .8101	7, 2 .8898	7, 2 .9401	7, 2 .9694	7, 2 .9853	7, 2 .9933	7, 2 .9971	7, 2 .9988
42	6, 3 .2354	6, 3 .4058	6, 3 .5761	6, 3 .7204	6, 3 .8284	6, 3 .9015	6, 3 .9470	6, 3 .9733	6, 3 .9873	6, 3 .9943	6, 3 .9976	6, 3 .9991
44	5, 4 .2504	5, 4 .4211	8, 2 .5875	8, 2 .7353	8, 2 .8433	8, 2 .9139	8, 2 .9559	8, 2 .9789	8, 2 .9905	8, 2 .9960	8, 2 .9984	8, 2 .9994
46	7, 3 .2635	7, 3 .4453	7, 3 .6199	7, 3 .7611	7, 3 .8612	7, 3 .9252	7, 3 .9624	7, 3 .9824	7, 3 .9923	7, 3 .9968	7, 3 .9988	7, 3 .9996
48	6, 4 .2813	6, 4 .4641	6, 4 .6343	9, 2 .7704	9, 2 .8705	9, 2 .9325	9, 2 .9674	9, 2 .9853	9, 2 .9938	9, 2 .9976	9, 2 .9991	9, 2 .9997
50	8, 3 .2909	8, 3 .4823	8, 3 .6589	8, 3 .7953	8, 3 .8873	8, 3 .9427	8, 3 .9730	8, 3 .9882	8, 3 .9952	8, 3 .9982	8, 3 .9994	8, 3 .9998
52	7, 4 .3108	7, 4 .5032	7, 4 .6754	7, 4 .8059	7, 4 .8930	10, 2 .9470	10, 2 .9758	10, 2 .9897	10, 2 .9960	10, 2 .9985	10, 2 .9995	10, 2 .9998
54	6, 5 .3218	9, 3 .5168	9, 3 .6937	9, 3 .8243	9, 3 .9081	9, 3 .9559	9, 3 .9805	9, 3 .9920	9, 3 .9970	9, 3 .9989	9, 3 .9997	9, 3 .9999
56	8, 4 .3390	8, 4 .5391	8, 4 .7112	8, 4 .8357	8, 4 .9145	8, 4 .9592	8, 4 .9821	11, 2 .9928	11, 2 .9973	11, 2 .9991	11, 2 .9997	

TABLE OPT2.4.2

P=4 K=2 OPTIMAL DESIGN AND ASSOCIATED CONFIDENCE COEFFICIENT FOR TWO-SIDED INTERVALS

b \ a/σ	.100	.200	.300	.400	.500	.600	.700	.800	.900	1.000	1.100	1.200
58	7, 5 .0035	7, 5 .0454	7, 5 .1666	7, 5 .3523	7, 5 .5500	10, 3 .7248	10, 3 .8489	10, 3 .9248	10, 3 .9658	10, 3 .9858	10, 3 .9946	
60	6, 6 .0038	6, 6 .0481	6, 6 .1725	9, 4 .3661	9, 4 .5722	9, 4 .7427	9, 4 .8605	9, 4 .9313	9, 4 .9692	9, 4 .9873	9, 4 .9952	
62	5, 7 .0039	8, 5 .0506	8, 5 .1832	8, 5 .3811	8, 5 .5850	11, 3 .7526	11, 3 .8699	11, 3 .9363	11, 3 .9734	11, 3 .9896	11, 3 .9962	
64	7, 6 .0043	7, 6 .0539	7, 6 .1907	10, 4 .3921	10, 4 .6028	10, 4 .7704	10, 4 .8812	10, 4 .9445	10, 4 .9765	10, 4 .9910	10, 4 .9968	
66	6, 7 .0045	6, 7 .0560	9, 5 .1995	9, 5 .4085	9, 5 .6169	9, 5 .7793	12, 3 .8879	12, 3 .9493	12, 3 .9792	12, 3 .9923	12, 3 .9974	
68	8, 6 .0048	8, 6 .0596	8, 6 .2083	8, 6 .4180	11, 4 .6311	11, 4 .7949	11, 4 .8986	11, 4 .9550	11, 4 .9820	11, 4 .9935	11, 4 .9979	
70	7, 7 .0051	7, 7 .0624	10, 5 .2156	10, 5 .4346	10, 5 .6460	10, 5 .8045	10, 5 .9037	13, 3 .9582	13, 3 .9837	13, 3 .9943	13, 3 .9982	
72	9, 6 .0053	9, 6 .0654	9, 6 .2255	9, 6 .4456	12, 4 .6573	12, 4 .8166	12, 4 .9132	12, 4 .9634	12, 4 .9861	12, 4 .9953	12, 4 .9985	
74	8, 7 .0056	8, 7 .0687	8, 7 .2319	11, 5 .4595	11, 5 .6726	11, 5 .8264	11, 5 .9185	11, 5 .9659	14, 3 .9872	14, 3 .9958	14, 3 .9987	
76	7, 8 .0059	10, 6 .0712	10, 6 .2423	10, 6 .4716	13, 4 .6815	13, 4 .8358	13, 4 .9256	13, 4 .9701	13, 4 .9893	13, 4 .9966	13, 4 .9990	
78	9, 7 .0062	9, 7 .0750	9, 7 .2498	12, 5 .4832	12, 5 .6971	12, 5 .8457	12, 5 .9308	12, 5 .9725	12, 5 .9903	12, 5 .9969	12, 5 .9991	
80	8, 8 .0065	8, 8 .0777	11, 6 .2587	11, 6 .4963	11, 6 .7063	14, 4 .8530	14, 4 .9361	14, 4 .9755	14, 4 .9917	14, 4 .9975	14, 4 .9993	
82	10, 7 .0068	10, 7 .0812	10, 7 .2672	13, 5 .5059	13, 5 .7195	13, 5 .8626	13, 5 .9412	13, 5 .9778	13, 5 .9926	13, 5 .9978	13, 5 .9994	

TABLE OPT2.5.2

P=5 K=2 OPTIMAL DESIGN AND ASSOCIATED CONFIDENCE COEFFICIENT FOR TWO-SIDED INTERVALS

b \ a/σ	.500	.600	.700	.800	.900	1.000	1.100	1.200	1.300	1.400	1.500	1.600
5	1, 0 .0016	1, 0 .0038	1, 0 .0079	1, 0 .0144	1, 0 .0243	1, 0 .0382	1, 0 .0567	1, 0 .0803	1, 0 .1091	1, 0 .1431	1, 0 .1619	1, 0 .2251
10	2, 0 .0082	2, 0 .0188	2, 0 .0366	2, 0 .0636	2, 0 .1007	2, 0 .1483	2, 0 .2054	2, 0 .2704	2, 0 .3410	2, 0 .4145	2, 0 .4882	2, 0 .5597
15	1, 1 .0393	1, 1 .0802	1, 1 .1388	1, 1 .2126	1, 1 .2970	1, 1 .3863	1, 1 .4753	1, 1 .5600	1, 1 .6377	1, 1 .7067	1, 1 .7664	1, 1 .8170
20	2, 1 .0688	2, 1 .1353	2, 1 .2250	2, 1 .3305	2, 1 .4422	2, 1 .5510	2, 1 .6501	2, 1 .7354	2, 1 .8056	2, 1 .8610	2, 1 .9033	2, 1 .9344
25	3, 1 .1003	3, 1 .1903	3, 1 .3048	3, 1 .4309	3, 1 .5551	3, 1 .6671	3, 1 .7607	3, 1 .8345	3, 1 .8895	3, 1 .9287	3, 1 .9555	3, 1 .9731
30	2, 2 .1436	2, 2 .2526	4, 1 .3793	4, 1 .5179	4, 1 .6454	4, 1 .7520	4, 1 .8346	4, 1 .8944	4, 1 .9353	4, 1 .9619	4, 1 .9784	4, 1 .9882
35	3, 2 .1890	3, 2 .3224	3, 2 .4665	3, 2 .6023	3, 2 .7180	5, 1 .8148	5, 1 .8848	5, 1 .9317	5, 1 .9613	5, 1 .9791	5, 1 .9891	5, 1 .9946
40	4, 2 .2320	4, 2 .3845	4, 2 .5404	4, 2 .6785	4, 2 .7880	4, 2 .8678	4, 2 .9218	4, 2 .9560	6, 1 .9766	6, 1 .9883	6, 1 .9944	6, 1 .9974
45	5, 2 .2735	5, 2 .4408	5, 2 .6030	5, 2 .7383	5, 2 .8387	5, 2 .9067	5, 2 .9492	5, 2 .9738	5, 2 .9873	5, 2 .9941	5, 2 .9974	5, 2 .9989
50	4, 3 .3170	6, 2 .4922	6, 2 .6566	6, 2 .7860	6, 2 .8762	6, 2 .9332	6, 2 .9663	6, 2 .9840	6, 2 .9929	6, 2 .9970	6, 2 .9988	6, 2 .9996
55	5, 3 .3622	5, 3 .5426	7, 2 .7026	7, 2 .8244	7, 2 .9044	7, 2 .9517	7, 2 .9773	7, 2 .9900	7, 2 .9959	7, 2 .9984	7, 2 .9994	7, 2 .9998
60	6, 3 .4041	6, 3 .5913	6, 3 .7450	8, 2 .8556	8, 2 .9258	8, 2 .9648	8, 2 .9845	8, 2 .9937	8, 2 .9976	8, 2 .9991	8, 2 .9997	
65	7, 3 .4432	7, 3 .6343	7, 3 .7834	7, 3 .8832	7, 3 .9423	9, 2 .9741	9, 2 .9893	9, 2 .9959	9, 2 .9986	9, 2 .9995	9, 2 .9999	

TABLE OPT2.5.2

P=5 K=2 OPTIMAL DESIGN AND ASSOCIATED CONFIDENCE COEFFICIENT FOR TWO-SIDED INTERVALS

b \ a/σ	.100	.200	.300	.400	.500	.600	.700	.800	.900	1.000	1.100	1.200
70	4, 5 .0009	6, 4 .0213	6, 4 .1079	6, 4 .2743	8, 3 .4797	8, 3 .6723	8, 3 .8154	8, 3 .9060	8, 3 .9564	8, 3 .9815	8, 3 .9928	
75	5, 5 .0010	5, 5 .0247	7, 4 .1216	7, 4 .3030	7, 4 .5155	9, 3 .7062	9, 3 .8423	9, 3 .9239	9, 3 .9668	9, 3 .9868	9, 3 .9952	
80	6, 5 .0012	6, 5 .0285	6, 5 .1358	8, 4 .3305	8, 4 .5510	10, 3 .7363	10, 3 .8649	10, 3 .9382	10, 3 .9745	10, 3 .9905	10, 3 .9968	
85	5, 6 .0014	7, 5 .0323	7, 5 .1514	9, 4 .3570	9, 4 .5836	9, 4 .7653	11, 3 .8841	11, 3 .9496	11, 3 .9804	11, 3 .9931	11, 3 .9978	
90	6, 6 .0016	6, 6 .0362	8, 5 .1667	8, 5 .3831	10, 4 .6137	10, 4 .7914	10, 4 .9011	12, 3 .9588	12, 3 .9848	12, 3 .9950	12, 3 .9985	
95	7, 6 .0018	7, 6 .0407	9, 5 .1818	9, 5 .4102	11, 4 .6414	11, 4 .8143	11, 4 .9162	11, 4 .9667	11, 4 .9882	11, 4 .9963	13, 3 .9990	
100	6, 7 .0020	8, 6 .0451	8, 6 .1969	10, 5 .4360	12, 4 .6671	12, 4 .8345	12, 4 .9287	12, 4 .9731	12, 4 .9911	12, 4 .9974	12, 4 .9993	
105	7, 7 .0023	9, 6 .0497	9, 6 .2133	11, 5 .4606	11, 5 .6924	13, 4 .8523	13, 4 .9392	13, 4 .9782	13, 4 .9932	13, 4 .9981	13, 4 .9995	
110	8, 7 .0026	8, 7 .0545	10, 6 .2293	12, 5 .4840	12, 5 .7159	14, 4 .8680	14, 4 .9481	14, 4 .9823	14, 4 .9947	14, 4 .9986	14, 4 .9997	
115	9, 7 .0029	9, 7 .0597	11, 6 .2450	11, 6 .5065	13, 5 .7375	13, 5 .8828	15, 4 .9556	15, 4 .9856	15, 4 .9959	15, 4 .9990	15, 4 .9998	
120	8, 8 .0032	10, 7 .0648	12, 6 .2604	12, 6 .5296	14, 5 .7573	14, 5 .8960	14, 5 .9621	16, 4 .9882	16, 4 .9968	16, 4 .9993	16, 4 .9999	
125	9, 8 .0035	11, 7 .0699	11, 7 .2764	13, 6 .5515	15, 5 .7755	15, 5 .9076	15, 5 .9679	15, 5 .9905	15, 5 .9976	15, 5 .9995		
130	10, 8 .0038	10, 8 .0755	12, 7 .2925	14, 6 .5723	16, 5 .7922	16, 5 .9177	16, 5 .9727	16, 5 .9923	16, 5 .9981	16, 5 .9996		

TABLE OPT2.5.2

P=5 K=2 OPTIMAL DESIGN AND ASSOCIATED CONFIDENCE COEFFICIENT FOR TWO-SIDED INTERVALS

b \ a/σ	.100	.200	.300	.400	.500	.600	.700	.800	.900	1.000	1.100	1.200
135	9, 9 .0042	11, 8 .0812	13, 7 .3082	15, 6 .5921	15, 6 .8085	17, 5 .9267	17, 5 .9767	17, 5 .9937	17, 5 .9986	17, 5 .9997		
140	10, 9 .0045	12, 8 .0869	14, 7 .3236	16, 6 .6109	16, 6 .8235	18, 5 .9346	18, 5 .9801	18, 5 .9949	18, 5 .9989	18, 5 .9998		
145	11, 9 .0049	11, 9 .0926	15, 7 .3386	15, 7 .6293	17, 6 .8372	17, 6 .9420	19, 5 .9829	19, 5 .9958	19, 5 .9991	19, 5 .9999		
150	10,10 .0053	12, 9 .0987	14, 8 .3543	16, 7 .6473	18, 6 .8497	18, 6 .9485	18, 6 .9855	18, 6 .9966	18, 6 .9993			
155	11,10 .0057	13, 9 .1048	15, 8 .3696	17, 7 .6644	19, 6 .8612	19, 6 .9543	19, 6 .9877	19, 6 .9972	19, 6 .9995			
160	12,10 .0062	14, 9 .1110	16, 8 .3845	18, 7 .6805	20, 6 .8718	20, 6 .9593	20, 6 .9895	20, 6 .9978	20, 6 .9996			
165	11,11 .0066	13,10 .1171	17, 8 .3991	19, 7 .6958	19, 7 .8819	21, 6 .9638	21, 6 .9910	21, 6 .9982	21, 6 .9997			
170	12,11 .0071	14,10 .1236	18, 8 .4133	20, 7 .7103	20, 7 .8912	22, 6 .9677	22, 6 .9923	22, 6 .9985	22, 6 .9998			
175	13,11 .0076	15,10 .1301	17, 9 .4279	19, 8 .7246	21, 7 .8998	21, 7 .9713	21, 7 .9934	23, 6 .9988	23, 6 .9998			
180	12,12 .0080	16,10 .1365	18, 9 .4422	20, 8 .7383	22, 7 .9075	22, 7 .9745	22, 7 .9944	22, 7 .9990	22, 7 .9999			
185	13,12 .0086	15,11 .1430	19, 9 .4561	21, 8 .7512	23, 7 .9147	23, 7 .9774	23, 7 .9952	23, 7 .9992				
190	14,12 .0091	16,11 .1498	20, 9 .4696	22, 8 .7634	24, 7 .9212	24, 7 .9799	24, 7 .9959	24, 7 .9993				
195	15,12 .0096	17,11 .1566	21, 9 .4828	23, 8 .7750	23, 8 .9275	25, 7 .9821	25, 7 .9965	25, 7 .9995				

TABLE OPT2.6.2

P=6 K=2 OPTIMAL DESIGN AND ASSOCIATED CONFIDENCE COEFFICIENT FOR TWO-SIDED INTERVALS

b \ a/σ	.500	.600	.700	.800	.900	1.000	1.100	1.200	1.300	1.400	1.500	1.600
6	1, 0 .0004	1, 0 .0013	1, 0 .0030	1, 0 .0062	1, 0 .0116	1, 0 .0199	1, 0 .0320	1, 0 .0495	1, 0 .0700	1, 0 .0970	1, 0 .1294	1, 0 .1670
12	2, 0 .0032	2, 0 .0085	2, 0 .0169	2, 0 .0366	2, 0 .0637	2, 0 .1012	2, 0 .1497	2, 0 .2082	2, 0 .2750	2, 0 .3475	2, 0 .4229	2, 0 .4983
18	3, 0 .0094	3, 0 .0241	3, 0 .0509	3, 0 .0928	3, 0 .1509	3, 0 .2240	3, 0 .3087	3, 0 .3999	3, 0 .4925	3, 0 .5814	3, 0 .6630	3, 0 .7349
21	1, 1 .0324	1, 1 .0731	1, 1 .1351	1, 1 .2158	1, 1 .3089	1, 1 .4071	1, 1 .5036	1, 1 .5936	1, 1 .6742	1, 1 .7440	1, 1 .8027	1, 1 .8508
24										4, 0 .7458	4, 0 .8131	4, 0 .8662
27	2, 1 .0572	2, 1 .1236	2, 1 .2184	2, 1 .3329	2, 1 .4545	2, 1 .5717	2, 1 .6761	2, 1 .7635	2, 1 .8329	2, 1 .8856	2, 1 .9240	2, 1 .9510
33	3, 1 .0841	3, 1 .1748	3, 1 .2960	3, 1 .4322	3, 1 .5661	3, 1 .6848	3, 1 .7815	3, 1 .8549	3, 1 .9075	3, 1 .9433	3, 1 .9665	3, 1 .9809
36	1, 2 .0915	1, 2 .1757										
39	4, 1 .1134	4, 1 .2267	4, 1 .3689	4, 1 .5182	4, 1 .6547	4, 1 .7665	4, 1 .8503	4, 1 .9086	4, 1 .9468	4, 1 .9703	4, 1 .9841	4, 1 .9919
42	2, 2 .1403	2, 2 .2599	2, 2 .3972	2, 2 .5327								
45	5, 1 .1446	5, 1 .2786	5, 1 .4369	5, 1 .5926	5, 1 .7253	5, 1 .8263	5, 1 .8964	5, 1 .9415	5, 1 .9686	5, 1 .9839	5, 1 .9921	5, 1 .9963
48	3, 2 .1843	3, 2 .3299	3, 2 .4871	3, 2 .6317	3, 2 .7504	3, 2 .8397	3, 2 .9022	3, 2 .9432				

TABLE OPT2.6.2

P=6 K=2 OPTIMAL DESIGN AND ASSOCIATED CONFIDENCE COEFFICIENT FOR TWO-SIDED INTERVALS

b \ a/σ	.400	.500	.600	.700	.800	.900	1.000	1.100	1.200	1.300	1.400	1.500
51	6, 1 .0707			6, 1 .4997	6, 1 .6565	6, 1 .7816	6, 1 .8704	6, 1 .9278	6, 1 .9620	6, 1 .9811	6, 1 .9911	6, 1 .9960
54	4, 2 .0974	4, 2 .2250	4, 2 .3918	4, 2 .5604	4, 2 .7054	4, 2 .8157	4, 2 .8919	4, 2 .9403	4, 2 .9689	4, 2 .9847	4, 2 .9929	4, 2 .9969
57	2, 3 .1005				7, 1 .7110	7, 1 .8263	7, 1 .9031	7, 1 .9494	7, 1 .9752	7, 1 .9885	7, 1 .9950	7, 1 .9979
60	5, 2 .1183	5, 2 .2663	5, 2 .4476	5, 2 .6218	5, 2 .7623	5, 2 .8617	5, 2 .9252	5, 2 .9622	5, 2 .9821	5, 2 .9921	5, 2 .9967	5, 2 .9987
63	3, 3 .1300	3, 3 .2754				8, 1 .8620	8, 1 .9273	8, 1 .9544	8, 1 .9836	8, 1 .9929	8, 1 .9971	8, 1 .9989
66	6, 2 .1397	6, 2 .3053	6, 2 .4984	6, 2 .6738	6, 2 .8070	6, 2 .8951	6, 2 .9473	6, 2 .9754	6, 2 .9893	6, 2 .9957	6, 2 .9984	6, 2 .9994
69	4, 3 .1578	4, 3 .3262	4, 3 .5110	4, 3 .6743								
72	7, 2 .1615	7, 2 .3431	7, 2 .5447	7, 2 .7183	7, 2 .8425	7, 2 .9197	7, 2 .9623	7, 2 .9837	7, 2 .9935	7, 2 .9976	7, 2 .9992	7, 2 .9997
75	5, 3 .1846	5, 3 .3714	5, 3 .5665	5, 3 .7292	5, 3 .8451							
78	8, 2 .1836	8, 2 .3796	8, 2 .5870	8, 2 .7565	8, 2 .8711	8, 2 .9381	8, 2 .9728	8, 2 .9890	8, 2 .9959	8, 2 .9986	8, 2 .9996	8, 2 .9999
81	6, 3 .2108	6, 3 .4132	6, 3 .6145	6, 3 .7731	6, 3 .8790	6, 3 .9411	6, 3 .9738	6, 3 .9893	6, 3 .9960	6, 3 .9986	6, 3 .9996	
84	4, 4 .2158	9, 2 .4147	9, 2 .6255	9, 2 .7893	9, 2 .8942	9, 2 .9520	9, 2 .9802	9, 2 .9925	9, 2 .9974	9, 2 .9992	9, 2 .9998	
87	7, 3 .2365	7, 3 .4519	7, 3 .6543	7, 3 .8088	7, 3 .9044	7, 3 .9568	7, 3 .9822	7, 3 .9933	7, 3 .9977	7, 3 .9993	7, 3 .9998	

TABLE OPT2.6.2

P=6 K=2 OPTIMAL DESIGN AND ASSOCIATED CONFIDENCE COEFFICIENT FOR TWO-SIDED INTERVALS

b \ a/σ	.100	.200	.300	.400	.500	.600	.700	.800	.900	1.000	1.100	1.200
90	5, 4 .0003	5, 4 .0130	5, 4 .0858	5, 4 .2474	5, 4 .4552	10, 2 .6607	10, 2 .8176	10, 2 .9129	10, 2 .9626	10, 2 .9854	10, 2 .9948	
93	3, 5 .0003	3, 5 .0131	8, 3 .0884	8, 3 .2617	8, 3 .4880	8, 3 .6931	8, 3 .8382	8, 3 .9239	8, 3 .9679	8, 3 .9877	8, 3 .9958	
96	6, 4 .0003	6, 4 .0152	6, 4 .0984	6, 4 .2772	6, 4 .4980		11, 2 .8420	11, 2 .9282	11, 2 .9707	11, 2 .9892	11, 2 .9964	
99	4, 5 .0004	4, 5 .0151	4, 5 .0993	9, 3 .2865	9, 3 .5218	9, 3 .7256	9, 3 .8625	9, 3 .9390	9, 3 .9758	9, 3 .9914	9, 3 .9972	
102	7, 4 .0004	7, 4 .0174	7, 4 .1110	7, 4 .3056	7, 4 .5366	7, 4 .7310	12, 2 .8630	12, 2 .9406	12, 2 .9770	12, 2 .9920	12, 2 .9975	
105	5, 5 .0005	5, 5 .0189	5, 5 .1149	10, 3 .3109	10, 3 .5534	10, 3 .7543	10, 3 .8829	10, 3 .9509	10, 3 .9817	10, 3 .9939	10, 3 .9982	
108	8, 4 .0005	8, 4 .0198	8, 4 .1236	8, 4 .3329	8, 4 .5717	8, 4 .7635	8, 4 .8856	8, 4 .9510	13, 2 .9819	13, 2 .9941	13, 2 .9982	
111	6, 5 .0005	6, 5 .0219	6, 5 .1300	6, 5 .3351	11, 3 .5830	11, 3 .7799	11, 3 .9000	11, 3 .9602	11, 3 .9860	11, 3 .9956	11, 3 .9988	
114	4, 6 .0006	9, 4 .0223	9, 4 .1363	9, 4 .3591	9, 4 .6037	9, 4 .7914	9, 4 .9047	9, 4 .9618	9, 4 .9865	9, 4 .9957	9, 4 .9988	
117	7, 5 .0006	7, 5 .0248	7, 5 .1449	7, 5 .3665	12, 3 .6107	12, 3 .8026	12, 3 .9144	12, 3 .9676	12, 3 .9892	12, 3 .9968	12, 3 .9992	
120	5, 6 .0007	5, 6 .0258	10, 4 .1491	10, 4 .3843	10, 4 .6330	10, 4 .8156	10, 4 .9202	10, 4 .9699	10, 4 .9900	10, 4 .9971	10, 4 .9992	
123	8, 5 .0007	8, 5 .0278	8, 5 .1595	8, 5 .3951	13, 3 .6366	13, 3 .8229	13, 3 .9265	13, 3 .9736	13, 3 .9917	13, 3 .9977	13, 3 .9994	
126	6, 6 .0008	6, 6 .0294	11, 4 .1619	11, 4 .4087	11, 4 .6600	11, 4 .8366	11, 4 .9329	11, 4 .9761	11, 4 .9926	11, 4 .9980	11, 4 .9995	

TABLE OPT2.3.3

P=3 K=3 OPTIMAL DESIGN AND ASSOCIATED CONFIDENCE COEFFICIENT FOR TWO-SIDED INTERVALS

b \ a/σ	.500	.600	.700	.800	.900	1.000	1.100	1.200	1.300	1.400	1.500	1.600
3	1, 0 .0735	1, 0 .1186	1, 0 .1741	1, 0 .2380	1, 0 .3078	1, 0 .3808	1, 0 .4542	1, 0 .5259	1, 0 .5939	1, 0 .6567	1, 0 .7136	1, 0 .7641
4	1, 1 .1090	1, 1 .1709	1, 1 .2432	1, 1 .3220	1, 1 .4033	1, 1 .4836	1, 1 .5602	1, 1 .6309	1, 1 .6947	1, 1 .7510	1, 1 .7996	1, 1 .8409
5	1, 2 .1395	1, 2 .2132	1, 2 .2956	1, 2 .3815	1, 2 .4666	1, 2 .5475	1, 2 .6220	1, 2 .6888	1, 2 .7475	1, 2 .7980	1, 2 .8407	1, 2 .8761
6	2, 0 .1784	2, 0 .2713	2, 0 .3734	2, 0 .4770	2, 0 .5758	2, 0 .6652	2, 0 .7425	2, 0 .8069	2, 0 .8586	2, 0 .8988	2, 0 .9293	2, 0 .9517
7	2, 1 .2161	2, 1 .3208	2, 1 .4310	2, 1 .5381	2, 1 .6359	2, 1 .7209	2, 1 .7917	2, 1 .8486	2, 1 .8927	2, 1 .9258	2, 1 .9500	2, 1 .9671
8	2, 2 .2487	2, 2 .3614	2, 2 .4757	2, 2 .5830	2, 2 .6781	2, 2 .7583	2, 2 .8235	2, 2 .8745	2, 2 .9131	2, 2 .9414	2, 2 .9615	2, 2 .9754
9	3, 0 .2836	3, 0 .4097	3, 0 .5346	3, 0 .6481	3, 0 .7441	3, 0 .8208	3, 0 .8789	3, 0 .9210	3, 0 .9502	3, 0 .9696	3, 0 .9821	3, 0 .9898
10	3, 1 .3186	3, 1 .4511	3, 1 .5776	3, 1 .6885	3, 1 .7793	3, 1 .8496	3, 1 .9012	3, 1 .9375	3, 1 .9619	3, 1 .9776	3, 1 .9873	3, 1 .9930
11	3, 2 .3491	3, 2 .4854	3, 2 .6117	3, 2 .7193	3, 2 .8052	3, 2 .8700	3, 2 .9166	3, 2 .9485	3, 2 .9694	3, 2 .9825	3, 2 .9904	3, 2 .9949
12	4, 0 .3808	4, 0 .5259	4, 0 .6567	4, 0 .7641	4, 0 .8457	4, 0 .9037	4, 0 .9426	4, 0 .9673	4, 0 .9822	4, 0 .9907	4, 0 .9954	4, 0 .9978
13	4, 1 .4117	4, 1 .5591	4, 1 .6879	4, 1 .7904	4, 1 .8662	4, 1 .9187	4, 1 .9529	4, 1 .9740	4, 1 .9863	4, 1 .9931	4, 1 .9967	4, 1 .9985
14	4, 2 .4388	4, 2 .5870	4, 2 .7130	4, 2 .8110	4, 2 .8818	4, 2 .9297	4, 2 .9603	4, 2 .9786	4, 2 .9891	4, 2 .9947	4, 2 .9975	4, 2 .9989
15	5, 0 .4674	5, 0 .6207	5, 0 .7473	5, 0 .8418	5, 0 .9067	5, 0 .9480	5, 0 .9726	5, 0 .9864	5, 0 .9936	5, 0 .9971	5, 0 .9988	5, 0 .9995

TABLE OPT2.3.3

P=3 K=3 OPTIMAL DESIGN AND ASSOCIATED CONFIDENCE COEFFICIENT FOR TWO-SIDED INTERVALS

b \ a/σ	.200	.300	.400	.500	.600	.700	.800	.900	1.000	1.100	1.200	1.300
16	4, 4 .0606	5, 1 .1712	5, 1 .3263	5, 1 .4941	5, 1 .6469	5, 1 .7697	5, 1 .8590	5, 1 .9187	5, 1 .9559	5, 1 .9774	5, 1 .9891	5, 1 .9950
17	5, 2 .0656	5, 2 .1848	5, 2 .3468	5, 2 .5176	5, 2 .6692	5, 2 .7880	5, 2 .8727	5, 2 .9282	5, 2 .9619	5, 2 .9810	5, 2 .9911	5, 2 .9961
18	5, 3 .0711	5, 3 .1976	6, 0 .3659	6, 0 .5431	6, 0 .6971	6, 0 .8140	6, 0 .8937	6, 0 .9433	6, 0 .9718	6, 0 .9869	6, 0 .9943	6, 0 .9977
19	5, 4 .0764	5, 4 .2097	6, 1 .3866	6, 1 .5659	6, 1 .7176	6, 1 .8300	6, 1 .9050	6, 1 .9505	6, 1 .9760	6, 1 .9891	6, 1 .9954	6, 1 .9982
20	5, 5 .0816	6, 2 .2231	6, 2 .4056	6, 2 .5860	6, 2 .7352	6, 2 .8434	6, 2 .9141	6, 2 .9562	6, 2 .9792	6, 2 .9908	6, 2 .9962	6, 2 .9986
21	6, 3 .0872	6, 3 .2358	7, 0 .4237	7, 0 .6087	7, 0 .7581	7, 0 .8629	7, 0 .9284	7, 0 .9655	7, 0 .9846	7, 0 .9937	7, 0 .9976	7, 0 .9991
22	6, 4 .0927	6, 4 .2478	7, 1 .4426	7, 1 .6280	7, 1 .7742	7, 1 .8745	7, 1 .9358	7, 1 .9698	7, 1 .9869	7, 1 .9947	7, 1 .9981	7, 1 .9993
23	6, 5 .0982	7, 2 .2607	7, 2 .4600	7, 2 .6452	7, 2 .7881	7, 2 .8842	7, 2 .9420	7, 2 .9733	7, 2 .9887	7, 2 .9956	7, 2 .9984	7, 2 .9995
24	7, 3 .1036	7, 3 .2731	8, 0 .4770	8, 0 .6652	8, 0 .8069	8, 0 .8988	8, 0 .9517	8, 0 .9789	8, 0 .9916	8, 0 .9969	8, 0 .9990	8, 0 .9997
25	7, 4 .1094	7, 4 .2849	8, 1 .4942	8, 1 .6815	8, 1 .8194	8, 1 .9072	8, 1 .9566	8, 1 .9815	8, 1 .9928	8, 1 .9974	8, 1 .9992	8, 1 .9998
26	7, 5 .1150	8, 2 .2973	8, 2 .5100	8, 2 .6960	8, 2 .8304	8, 2 .9143	8, 2 .9607	8, 2 .9836	8, 2 .9938	8, 2 .9979	8, 2 .9993	8, 2 .9998
27	7, 6 .1204	8, 3 .3094	9, 0 .5259	9, 0 .7136	9, 0 .8457	9, 0 .9253	9, 0 .9673	9, 0 .9871	9, 0 .9954	9, 0 .9985	9, 0 .9996	9, 0 .9999
28	8, 4 .1262	8, 4 .3208	9, 1 .5414	9, 1 .7274	9, 1 .8556	9, 1 .9313	9, 1 .9706	9, 1 .9887	9, 1 .9960	9, 1 .9988	9, 1 .9996	

TABLE OPT2.3.3

P=3 K=3 OPTIMAL DESIGN AND ASSOCIATED CONFIDENCE COEFFICIENT FOR TWO-SIDED INTERVALS

b \ a/σ	.100	.200	.300	.400	.500	.600	.700	.800	.900	1.000	1.100	1.200
29	7, 8 .0201	8, 5 .1320	9, 2 .3328	9, 2 .5557	9, 2 .7397	9, 2 .8642	9, 2 .9366	9, 2 .9734	9, 2 .9900	9, 2 .9966	9, 2 .9990	9, 2 .9997
30	8, 6 .0211	8, 6 .1375	9, 3 .3444	10, 0 .5705	10, 0 .7551	10, 0 .8766	10, 0 .9447	10, 0 .9779	10, 0 .9921	10, 0 .9975	10, 0 .9993	10, 0 .9998
31	8, 7 .0221	9, 4 .1433	9, 4 .3555	10, 1 .5845	10, 1 .7667	10, 1 .8844	10, 1 .9491	10, 1 .9801	10, 1 .9930	10, 1 .9978	10, 1 .9994	10, 1 .9998
32	8, 8 .0231	9, 5 .1491	10, 2 .3670	10, 2 .5974	10, 2 .7771	10, 2 .8913	10, 2 .9530	10, 2 .9819	10, 2 .9938	10, 2 .9981	10, 2 .9995	10, 2 .9999
33	9, 6 .0241	9, 6 .1547	10, 3 .3781	11, 0 .6112	11, 0 .7905	11, 0 .9013	11, 0 .9591	11, 0 .9850	11, 0 .9951	11, 0 .9986	11, 0 .9996	
34	9, 7 .0252	10, 4 .1604	11, 1 .3888	11, 1 .6237	11, 1 .8003	11, 1 .9074	11, 1 .9623	11, 1 .9865	11, 1 .9957	11, 1 .9988	11, 1 .9997	
35	9, 8 .0263	10, 5 .1662	11, 2 .3999	11, 2 .6353	11, 2 .8091	11, 2 .9129	11, 2 .9651	11, 2 .9877	11, 2 .9962	11, 2 .9990	11, 2 .9997	
36	9, 9 .0273	10, 6 .1719	11, 3 .4105	12, 0 .6481	12, 0 .8208	12, 0 .9210	12, 0 .9696	12, 0 .9898	12, 0 .9970	12, 0 .9992	12, 0 .9998	
37	10, 7 .0284	11, 4 .1775	12, 1 .4208	12, 1 .6593	12, 1 .8290	12, 1 .9258	12, 1 .9720	12, 1 .9908	12, 1 .9974	12, 1 .9993	12, 1 .9999	
38	10, 8 .0295	11, 5 .1834	12, 2 .4313	12, 2 .6697	12, 2 .8365	12, 2 .9301	12, 2 .9741	12, 2 .9916	12, 2 .9977	12, 2 .9994		
39	10, 9 .0306	11, 6 .1891	12, 3 .4414	13, 0 .6815	13, 0 .8467	13, 0 .9367	13, 0 .9775	13, 0 .9931	13, 0 .9981	13, 0 .9996		
40	10,10 .0317	11, 7 .1947	13, 1 .4514	13, 1 .6916	13, 1 .8536	13, 1 .9405	13, 1 .9792	13, 1 .9937	13, 1 .9984	13, 1 .9996		
41	11, 8 .0328	12, 5 .2005	13, 2 .4614	13, 2 .7010	13, 2 .8599	13, 2 .9440	13, 2 .9808	13, 2 .9963	13, 2 .9985	13, 2 .9997		

TABLE OPT2.4.3

P=4 K=3 OPTIMAL DESIGN AND ASSOCIATED CONFIDENCE COEFFICIENT FOR TWO-SIDED INTERVALS

b \ a/σ	.500	.600	.700	.800	.900	1.000	1.100	1.200	1.300	1.400	1.500	1.600
4	1, 0, 0 0, 0, 0 .0101	1, 0, 0 0, 0, 0 .0199	1, 0, 0 0, 0, 0 .0349	1, 0, 0 0, 0, 0 .0560	1, 0, 0 0, 0, 0 .0835	1, 0, 0 0, 0, 0 .1177	1, 0, 0 0, 0, 0 .1584	1, 0, 0 0, 0, 0 .2049	1, 0, 0 0, 0, 0 .2563	1, 0, 0 0, 0, 0 .3114	1, 0, 0 0, 0, 0 .3689	1, 0, 0 0, 0, 0 .4275
6	0, 1, 0 0, 0, 0 .0655	0, 1, 0 0, 0, 0 .1190	0, 1, 0 0, 0, 0 .1894	0, 1, 0 0, 0, 0 .2734	0, 1, 0 0, 0, 0 .3656	0, 1, 0 0, 0, 0 .4604	0, 1, 0 0, 0, 0 .5524	0, 1, 0 0, 0, 0 .6376	0, 1, 0 0, 0, 0 .7132	0, 1, 0 0, 0, 0 .7780	0, 1, 0 0, 0, 0 .8317	0, 1, 0 0, 0, 0 .8750
7	0, 0, 1 0, 0, 0 .0830	0, 0, 1 0, 0, 0 .1456	0, 0, 1 0, 0, 0 .2239	0, 0, 1 0, 0, 0 .3125	0, 0, 1 0, 0, 0 .4053	0, 0, 1 0, 0, 0 .4968	0, 0, 1 0, 0, 0 .5829	0, 0, 1 0, 0, 0 .6608	0, 0, 1 0, 0, 0 .7292	0, 0, 1 0, 0, 0 .7876	0, 0, 1 0, 0, 0 .8363	0, 0, 1 0, 0, 0 .8760
8												
10	0, 1, 0 0, 0, 1 .1545	0, 1, 0 0, 0, 1 .2539	0, 1, 0 0, 0, 1 .3655	0, 1, 0 0, 0, 1 .4786	0, 1, 0 0, 0, 1 .5847	0, 1, 0 0, 0, 1 .6787	0, 0, 0 0, 1, 0 .7606	0, 0, 0 0, 1, 0 .8284	0, 0, 0 0, 1, 0 .8806	0, 0, 0 0, 1, 0 .9194	0, 0, 0 0, 1, 0 .9470	0, 0, 0 0, 1, 0 .9662
11												
12	0, 2, 0 0, 0, 0 .1950	0, 2, 0 0, 0, 0 .3175	0, 2, 0 0, 0, 0 .4509	0, 2, 0 0, 0, 0 .5800	0, 2, 0 0, 0, 0 .6937	0, 2, 0 0, 0, 0 .7863	0, 2, 0 0, 0, 0 .8570	0, 2, 0 0, 0, 0 .9081	0, 2, 0 0, 0, 0 .9431	0, 2, 0 0, 0, 0 .9661	0, 2, 0 0, 0, 0 .9805	0, 2, 0 0, 0, 0 .9892
13	0, 1, 1 0, 0, 0 .2184	0, 1, 1 0, 0, 0 .3467	0, 1, 1 0, 0, 0 .4812	0, 1, 1 0, 0, 0 .6072	0, 1, 1 0, 0, 0 .7155	0, 1, 1 0, 0, 0 .8023	0, 1, 1 0, 0, 0 .8680	0, 1, 1 0, 0, 0 .9151	0, 1, 1 0, 0, 0 .9475	0, 1, 1 0, 0, 0 .9687	0, 1, 1 0, 0, 0 .9820	0, 1, 1 0, 0, 0 .9900
14	0, 0, 0 0, 1, 1 .2299	0, 0, 0 0, 1, 1 .3573	0, 0, 0 0, 1, 1 .4878	0, 0, 0 0, 1, 1 .6083								
15												

TABLE OPT2.4.3

P=4 K=3 OPTIMAL DESIGN AND ASSOCIATED CONFIDENCE COEFFICIENT FOR TWO-SIDED INTERVALS

b \ a/σ	.500	.600	.700	.800	.900	1.000	1.100	1.200	1.300	1.400	1.500	1.600
16	0, 2, 0 0, 0, 1 .2960	0, 2, 0 0, 0, 1 .4456	0, 2, 0 0, 0, 1 .5888	0, 2, 0 0, 0, 1 .7113	C, 2, 0 0, 0, 1 .8073	0, 2, 0 0, 0, 1 .8775	0, 2, 0 0, 0, 1 .9257	0, 2, 0 0, 0, 1 .9570	0, 1, 0 0, 1, C .9762	0, 1, 0 0, 1, 0 .9876	0, 1, 0 0, 1, 0 .9938	0, 1, 0 0, 1, 0 .9970
17	0, 1, 1 0, C, 1 .3038	0, 1, 1 0, 0, 1 .4503										
18	0, 3, 0 0, 0, 0 .3337	0, 3, 0 0, 0, 0 .4971	0, 3, 0 0, 0, 0 .6475	0, 3, 0 0, 0, 0 .7694	0, 3, 0 0, 0, 0 .8584	0, 3, 0 0, 0, 0 .9181	0, 3, 0 0, 0, 0 .9552	0, 3, 0 0, 0, 0 .9768	0, 3, 0 0, 0, 0 .9886	0, 3, 0 0, 0, 0 .9947	0, 3, 0 0, 0, 0 .9977	0, 3, 0 0, 0, 0 .9990
19	0, 2, 1 0, 0, 0 .3556	0, 2, 1 0, 0, 0 .5196	0, 2, 1 0, 0, 0 .6668	0, 2, 1 0, 0, 0 .7836	0, 2, 1 0, 0, 0 .8679	0, 2, 1 0, 0, 0 .9239	0, 2, 1 0, 0, 0 .9586	0, 2, 1 0, 0, 0 .9787	0, 2, 1 0, 0, 0 .9896	0, 2, 1 0, 0, 0 .9952	0, 2, 1 0, 0, 0 .9979	0, 2, 1 0, 0, 0 .9991
20	0, 2, 0 0, 0, 2 .3736	0, 2, 0 0, 0, 2 .5314	0, 2, 0 1, 0, 0 .6740	0, 2, 0 1, 0, 0 .7869	C, 2, 0 1, 0, 0 .8685							
21												
22	0, 3, 0 0, 0, 1 .4266	0, 3, 0 0, 0, 1 .5975	0, 3, 0 0, 0, 1 .7389	0, 3, 0 0, C, 1 .8425	0, 3, 0 0, 0, 1 .9113	0, 3, 0 0, 0, 1 .9533	0, 3, 0 0, 0, 1 .9769	0, 3, 0 0, 0, 1 .9893	0, 3, 0 0, 0, 1 .9953	0, 3, 0 0, 0, 1 .9981	0, 3, 0 0, 0, 1 .9993	0, 3, 0 0, 0, 1 .9997
23	0, 2, 1 0, 0, 1 .4359	C, 2, 1 0, 0, 1 .6035	0, 2, 1 0, 0, 1 .7411									
24	0, 4, 0 0, 0, 0 .4604	0, 4, 0 0, 0, 0 .6376	0, 4, 0 0, 0, 0 .7780	0, 4, 0 0, 0, 0 .8750	0, 4, 0 0, 0, 0 .9349	0, 4, 0 C, 0, 0 .9686	0, 4, 0 0, 0, 0 .9859	0, 4, 0 0, 0, 0 .9941	0, 4, 0 0, 0, 0 .9977	0, 4, 0 0, 0, 0 .9992	0, 4, 0 0, 0, 0 .9997	
25	0, 3, 1 0, 0, 0 .4785	0, 3, 1 0, C, 0 .6534	0, 3, 1 0, 0, 0 .7894	0, 3, 1 0, 0, 0 .8822	0, 3, 1 0, 0, 0 .9390	0, 3, 1 0, 0, 0 .9707	0, 3, 1 0, 0, 0 .9869	0, 3, 1 0, 0, 0 .9946	0, 3, 1 0, 0, 0 .9979			

TABLE CPT2.4.3

P=4 K=3 OPTIMAL DESIGN AND ASSOCIATED CONFIDENCE COEFFICIENT FOR TWO-SIDED INTERVALS

b \ a/σ	.100	.200	.300	.400	.500	.600	.700	.800	.900	1.000	1.100	1.200
26	0, 3, 0 0, 0, 2 .0028	0, 3, 0 0, 0, 2 .0376	0, 3, 0 0, 0, 2 .1417	0, 3, 0 0, 0, 2 .3089	0, 3, 0 0, 0, 2 .4966	0, 3, 0 0, 0, 2 .6648	0, 3, 0 1, 0, 0 .7946	0, 3, 0 1, 0, 0 .8846	0, 3, 0 1, 0, 0 .9399	0, 3, 0 1, 0, 0 .9710	0, 3, 0 1, 0, 0 .9870	
27	0, 2, 1 0, 0, 2 .0030	0, 2, 1 C, 0, 2 .0392	0, 2, 1 0, 0, 2 .1456	0, 2, 1 0, 0, 2 .3125	0, 2, 1 0, 0, 2 .4968							
28	C, 2, 0 0, 0, 4 .0032	0, 4, 0 0, C, 1 .0417	0, 4, 0 0, 0, 1 .1565	0, 4, 0 0, 0, 1 .3390	0, 4, 0 0, 0, 1 .5391	0, 4, 0 0, 0, 1 .7112	0, 4, 0 0, 0, 1 .8357	0, 4, 0 0, 0, 1 .9145	0, 4, 0 0, 0, 1 .9592	0, 4, 0 C, C, 1 .9821	0, 4, 0 0, 0, 1 .9927	0, 4, 0 0, 0, 1 .9973
29	0, 3, 1 0, 0, 1 .0034	0, 3, 1 0, 0, 1 .0443	0, 3, 1 0, 0, 1 .1637	0, 3, 1 0, 0, 1 .3488	C, 3, 1 0, C, 1 .5476	0, 3, 1 0, 0, 1 .7163	0, 3, 1 0, 0, 1 .8378	0, 3, 1 0, 0, 1 .9150				
30	0, 3, 0 C, 0, 3 .0038	0, 3, 0 0, 0, 3 .0481	0, 3, 0 0, 0, 3 .1725	0, 5, 0 0, 0, 0 .3604	0, 5, C 0, 0, 0 .5684	0, 5, 0 0, 0, 0 .7415	0, 5, 0 0, 0, 0 .8611	0, 5, 0 0, 0, 0 .9324	0, 5, 0 0, 0, 0 .9700	0, 5, 0 0, 0, 0 .9879	0, 5, 0 0, C, 0 .9955	0, 5, 0 0, 0, 0 .9985
31	0, 2, 1 0, 0, 3 .0039	0, 2, 1 0, 0, 3 .0492	0, 4, 1 0, 0, 0 .1770	0, 4, 1 0, 0, 0 .3748	0, 4, 1 C, 0, 0 .5827	0, 4, 1 0, 0, 0 .7523	0, 4, 1 0, 0, 0 .8678	0, 4, 1 0, 0, 0 .9361	0, 4, 1 0, 0, 0 .9719	0, 4, 1 0, 0, 0 .9887	0, 4, 1 0, 0, 0 .9958	0, 4, 1 0, 0, 0 .9986
32	0, 4, 0 0, 0, 2 .0042	0, 4, 0 0, 0, 2 .0536	0, 4, C 0, 0, 2 .1917	0, 4, 0 0, 0, 2 .3940	C, 4, 0 0, 0, 2 .5983	0, 4, 0 0, 0, 2 .7613	0, 4, C 0, 0, 2 .8715	0, 4, 0 1, 0, C .9376	0, 4, 0 1, 0, 0 .9724	0, 4, 0 1, 0, 0 .9889	0, 4, 0 1, 0, 0 .9959	
33	0, 3, 1 0, 0, 2 .0044	0, 3, 1 0, 0, 2 .0558	0, 3, 1 0, 0, 2 .1966	0, 3, 1 0, 0, 2 .3987	C, 3, 1 C, 0, 2 .6000							
34	0, 3, 0 0, 0, 4 .0048	0, 3, 0 C, 0, 4 .0586	0, 5, 0 0, 0, 1 .2063	0, 5, 0 0, 0, 1 .4211	0, 5, 0 0, 0, 1 .6326	0, 5, 0 0, 0, 1 .7942	0, 5, C 0, 0, 1 .8970	0, 5, 0 0, 0, 1 .9537	0, 5, C 0, 0, 1 .9812	0, 5, 0 0, 0, 1 .9931	0, 5, 0 0, 0, 1 .9977	0, 5, 0 0, 0, 1 .9993
35	0, 2, 1 0, 0, 4 .0048	0, 4, 1 0, 0, 1 .0609	0, 4, 1 0, 0, 1 .2138	0, 4, 1 0, C, 1 .4303	0, 4, 1 C, 0, 1 .6397	0, 4, 1 0, 0, 1 .7980	0, 4, 1 0, C, 1 .8986	0, 4, 1 0, 0, 1 .9541	0, 4, 1 0, 0, 1 .9813			

TABLE OPT2.5.3

P=5 K=3 OPTIMAL DESIGN AND ASSOCIATED CONFIDENCE COEFFICIENT FOR TWO-SIDED INTERVALS

b \ a/σ	.500	.600	.700	.800	.900	1.000	1.100	1.200	1.300	1.400	1.500	1.600
5	1, 0, 0 0, 0, 0 .0032	1, 0, 0 0, 0, 0 .0075	1, 0, 0 0, 0, 0 .0151	1, 0, 0 0, 0, 0 .0272	1, 0, 0 0, 0, 0 .0449	1, 0, 0 0, 0, 0 .0690	1, 0, 0 0, 0, 0 .0999	1, 0, 0 0, 0, 0 .1379	1, 0, 0 0, 0, 0 .1824	1, 0, 0 0, 0, 0 .2326	1, 0, 0 0, 0, 0 .2875	1, 0, 0 0, 0, 0 .3456
7	0, 1, 0 0, 0, 0 .0301	0, 1, 0 0, 0, 0 .0635	0, 1, 0 0, 0, 0 .1138	0, 1, 0 0, 0, 0 .1805	0, 1, 0 0, 0, 0 .2604	0, 1, 0 0, 0, 0 .3487	0, 1, 0 0, 0, 0 .4401	0, 1, 0 0, 0, 0 .5296	0, 1, 0 0, 0, 0 .6134	0, 1, 0 0, 0, 0 .6887	0, 1, 0 0, 0, 0 .7542	0, 1, 0 0, 0, 0 .8096
10	0, 0, 1 0, 0, 0 .0694	0, 0, 1 0, 0, 0 .1340	0, 0, 1 0, 0, 0 .2193	0, 0, 1 0, 0, 0 .3180	0, 0, 0 1, 0, 0 .4280	0, 0, 0 1, 0, 0 .5400	0, 0, 0 1, 0, 0 .6432	0, 0, 0 1, 0, 0 .7324	0, 0, 0 1, 0, 0 .8056	0, 0, 0 1, 0, 0 .8629	0, 0, 0 1, 0, 0 .9061	0, 0, 0 1, 0, 0 .9374
14	0, 2, 0 0, 0, 0 .1181	0, 2, 0 0, 0, 0 .2179	0, 2, 0 0, 0, 0 .3396	0, 2, 0 0, 0, 0 .4686	0, 2, 0 0, 0, 0 .5913	0, 2, 0 0, 0, 0 .6986	0, 2, 0 0, 0, 0 .7863	0, 2, 0 0, 0, 0 .8539	0, 2, 0 0, 0, 0 .9037	0, 2, 0 0, 0, 0 .9386	0, 2, 0 0, 0, 0 .9622	0, 2, 0 0, 0, 0 .9775
15												1, 0, 0 1, 0, 0 .9776
17	0, 0, 0 1, 1, 0 .1742	0, 0, 0 1, 1, 0 .3001	0, 1, 0 1, 0, 0 .4414	0, 1, 0 1, 0, 0 .5831	0, 1, 0 1, 0, 0 .7062	0, 1, 0 1, 0, 0 .8035	0, 1, 0 1, 0, 0 .8749	0, 1, 0 1, 0, 0 .9239	0, 1, 0 1, 0, 0 .9558	0, 1, 0 1, 0, 0 .9754	0, 1, 0 1, 0, 0 .9868	0, 1, 0 1, 0, 0 .9932
20	0, 0, 1 1, 0, 0 .2320	0, 0, 1 1, 0, 0 .3845	0, 0, 1 1, 0, 0 .5404	0, 0, 1 1, 0, 0 .6785	0, 0, 1 1, 0, 0 .7880	0, 0, 0 2, 0, 0 .8699	0, 0, 0 2, 0, 0 .9248	0, 0, 0 2, 0, 0 .9588	0, 0, 0 2, 0, 0 .9785	0, 0, 0 2, 0, 0 .9894	0, 0, 0 2, 0, 0 .9950	0, 0, 0 2, 0, 0 .9977
24	0, 2, 1 0, 0, 0 .2904	0, 2, 0 1, 0, 0 .4585	0, 2, 0 1, 0, 0 .6218	0, 2, 0 1, 0, 0 .7554	0, 2, 0 1, 0, 0 .8525	0, 2, 0 1, 0, 0 .9167	0, 2, 0 1, 0, 0 .9557	0, 2, 0 1, 0, 0 .9779	0, 2, 0 1, 0, 0 .9895	0, 2, 0 1, 0, 0 .9953	0, 2, 0 1, 0, 0 .9980	0, 2, 0 1, 0, 0 .9992
27	0, 0, 0 2, 1, 0 .3487	0, 0, 0 2, 1, 0 .5296	0, 0, 0 2, 1, 0 .6887	0, 1, 0 2, 0, 0 .8125	0, 1, 0 2, 0, 0 .8959	0, 1, 0 2, 0, 0 .9462	0, 1, 0 2, 0, 0 .9741	0, 1, 0 2, 0, 0 .9883	0, 1, 0 2, 0, 0 .9951	0, 1, 0 2, 0, 0 .9980	0, 1, 0 2, 0, 0 .9993	0, 1, 0 2, 0, 0 .9997
30	0, 0, 2 1, 0, 0 .4024	0, 0, 1 2, 0, 0 .5921	0, 0, 1 2, 0, 0 .7479	0, 0, 1 2, 0, 0 .8576	0, 0, 1 2, 0, 0 .9260	0, 0, 0 3, 0, 0 .9648	0, 0, 0 3, 0, 0 .9845	0, 0, 0 3, 0, 0 .9937	0, 0, 0 3, 0, 0 .9976	0, 0, 0 3, 0, 0 .9991	0, 0, 0 3, 0, 0 .9997	

TABLE OPT2.5.3

P=5 K=3 OPTIMAL DESIGN AND ASSOCIATED CONFIDENCE COEFFICIENT FOR TWO-SIDED INTERVALS

b \ a/σ	.100	.200	.300	.400	.500	.600	.700	.800	.900	1.000	1.100	1.200
34	0, 0, 1 1, 2, 0 .0008	0, 0, 0 2, 2, 0 .0193	0, 0, 0 2, 2, 0 .0987	0, 2, 1 1, 0, 0 .2557	0, 2, 1 1, 0, 0 .4565	0, 2, 1 1, 0, 0 .6459	0, 2, 0 2, 0, 0 .7949	0, 2, 0 2, 0, 0 .8925	0, 2, 0 2, 0, 0 .9485	0, 2, 0 2, 0, 0 .9774	0, 2, 0 2, 0, 0 .9909	0, 2, 0 2, 0, 0 .9966
37	0, 0, 2 1, 1, 0 .0010	0, 0, 2 1, 1, 0 .0237	0, 0, 1 2, 1, 0 .1170	0, 0, 1 2, 1, 0 .2935	0, 0, 0 3, 1, 0 .5054	0, 0, 0 3, 1, 0 .6959	0, 0, 0 3, 1, 0 .8329	0, 1, 0 3, 0, 0 .9182	0, 1, 0 3, 0, 0 .9638	0, 1, 0 3, 0, 0 .9854	0, 1, 0 3, 0, 0 .9946	0, 1, 0 3, 0, 0 .9982
40	0, 0, 4 0, 0, 0 .0012	0, 0, 3 1, 0, 0 .0284	0, 0, 3 1, 0, 0 .1366	0, 0, 2 2, 0, 0 .3305	0, 0, 2 2, 0, 0 .5510	0, 0, 1 3, 0, 0 .7382	0, 0, 1 3, 0, 0 .8653	0, 0, 1 3, 0, 0 .9380	0, 0, 1 3, 0, 0 .9743	0, 0, 0 4, 0, 0 .9904	0, 0, 0 4, 0, 0 .9968	0, 0, 0 4, 0, 0 .9990
44	0, 0, 1 2, 2, 0 .0015	0, 0, 1 2, 2, 0 .0337	0, 0, 0 3, 2, 0 .1563	0, 0, 0 3, 2, 0 .3664	0, 2, 1 2, 0, 0 .5939	0, 2, 1 2, 0, 0 .7744	0, 2, 0 3, 0, 0 .8909	0, 2, 0 3, 0, 0 .9533	0, 2, 0 3, 0, 0 .9821	0, 2, 0 3, 0, 0 .9938	0, 2, 0 3, 0, 0 .9981	0, 2, 0 3, 0, 0 .9995
47	0, 0, 3 1, 1, 0 .0017	0, 0, 3 1, 1, 0 .0392	0, 0, 2 2, 1, 0 .1769	0, 0, 1 3, 1, 0 .4020	0, 0, 0 4, 1, 0 .6324	0, 0, 0 4, 1, 0 .8070	0, 0, 0 4, 1, 0 .9115	0, 1, 0 4, 0, 0 .9645	0, 1, 0 4, 0, 0 .9874	0, 1, 0 4, 0, 0 .9960	0, 1, 0 4, 0, 0 .9989	0, 1, 0 4, 0, 0 .9997
50	0, 0, 5 0, 0, 0 .0021	0, 0, 4 1, 0, 0 .0451	0, 0, 3 2, 0, 0 .1976	0, 0, 2 3, 0, 0 .4355	0, 0, 2 3, 0, 0 .6682	0, 0, 1 4, 0, 0 .8345	0, 0, 1 4, 0, 0 .9287	0, 0, 1 4, 0, 0 .9731	0, 0, 1 4, 0, 0 .9911	0, 0, 1 4, 0, 0 .9974	0, 0, 1 4, 0, 0 .9993	0, 0, 1 4, 0, 0 .9998
54	0, 0, 2 2, 2, 0 .0024	0, 0, 2 2, 2, 0 .0514	0, 0, 1 3, 2, 0 .2186	0, 0, 0 4, 2, 0 .4686	0, 2, 1 3, 0, 0 .7010	0, 2, 1 3, 0, 0 .8578	0, 2, 0 4, 0, 0 .9424	0, 2, 0 4, 0, 0 .9797	0, 2, 0 4, 0, 0 .9937	0, 2, 0 4, 0, 0 .9983	0, 2, 0 4, 0, 0 .9996	
57	0, 0, 4 1, 1, 0 .0028	0, 0, 3 2, 1, 0 .0579	0, 0, 2 3, 1, 0 .2398	0, 0, 1 4, 1, 0 .5000	0, 0, 0 5, 1, 0 .7302	0, 0, 0 5, 1, 0 .8786	0, 0, 0 5, 1, 0 .9533	0, 1, 0 5, 0, 0 .9846	0, 1, 0 5, 0, 0 .9956	0, 1, 0 5, 0, 0 .9989	0, 1, 0 5, 0, 0 .9998	
60	0, 0, 6 0, 0, 0 .0032	0, 0, 5 1, 0, 0 .0648	0, 0, 4 2, 0, 0 .2608	0, 0, 3 3, 0, 0 .5296	0, 0, 2 4, 0, 0 .7573	0, 0, 1 5, 0, 0 .8961	0, 0, 1 5, 0, 0 .9624	0, 0, 1 5, 0, 0 .9883	0, 0, 1 5, 0, 0 .9969	0, 0, 1 5, 0, 0 .9993	0, 0, 1 5, 0, 0 .9999	
64	0, 0, 3 2, 2, 0 .0036	0, 0, 2 3, 2, 0 .0719	0, 0, 1 4, 2, 0 .2820	0, 0, 0 5, 2, 0 .5586	0, 2, 1 4, 0, 0 .7817	0, 2, 1 4, 0, 0 .9109	0, 2, 0 5, 0, 0 .9696	0, 2, 0 5, 0, 0 .9911	0, 2, 0 5, 0, 0 .9978	0, 2, 0 5, 0, 0 .9995		

TABLE OPT2.5.3

P=5 K=3 OPTIMAL DESIGN AND ASSOCIATED CONFIDENCE COEFFICIENT FOR TWO-SIDED INTERVALS

b \ a/σ	.100	.200	.300	.400	.500	.600	.700	.800	.900	1.000	1.100	1.200
67	0, 0, 5 1, 1, 0 .0040	0, 0, 4 2, 1, 0 .0793	0, 0, 3 3, 1, 0 .3027	0, 0, 1 5, 1, 0 .5856	0, 0, 0 6, 1, 0 .8035	0, 0, 0 6, 1, 0 .9239	0, 0, 0 6, 1, 0 .9754	0, 1, 0 6, 0, 0 .9933	0, 1, 0 6, 0, 0 .9984	0, 1, 0 6, 0, 0 .9997		
70	0, 0, 7 0, 0, 0 .0045	0, 0, 5 2, 0, 0 .0869	0, 0, 4 3, 0, 0 .3238	0, 0, 3 4, 0, 0 .6112	0, 0, 2 5, 0, 0 .8235	0, 0, 1 6, 0, 0 .9349	0, 0, 1 6, 0, 0 .9801	0, 0, 1 6, 0, 0 .9949	0, 0, 1 6, 0, 0 .9989	0, 0, 1 6, 0, 0 .9998		
74	0, 0, 4 2, 2, 0 .0050	0, 0, 3 3, 2, 0 .0948	0, 0, 1 5, 2, 0 .3442	0, 0, 0 6, 2, 0 .6358	0, 2, 1 5, 0, 0 .8415	0, 2, 1 5, 0, 0 .9443	0, 2, 0 6, 0, 0 .9839	0, 2, 0 6, 0, 0 .9961	0, 2, 0 6, 0, 0 .9992	0, 2, 0 6, 0, 0 .9999		
77	0, 0, 6 1, 1, 0 .0056	0, 0, 4 3, 1, 0 .1027	0, 0, 3 4, 1, 0 .3645	0, 0, 1 6, 1, 0 .6587	0, 0, 0 7, 1, 0 .8575	0, 0, 0 7, 1, 0 .9524	0, 0, 0 7, 1, 0 .9870	0, 0, 0 7, 1, 0 .9970	0, 1, 0 7, 0, 0 .9994			
80	0, 0, 7 1, 0, 0 .0062	0, 0, 6 2, 0, 0 .1110	0, 0, 4 4, 0, 0 .3845	0, 0, 3 5, 0, 0 .6804	0, 0, 2 6, 0, 0 .8721	0, 0, 1 7, 0, 0 .9593	0, 0, 1 7, 0, 0 .9895	0, 0, 1 7, 0, 0 .9978	0, 0, 1 7, 0, 0 .9996			
84	0, 0, 5 2, 2, 0 .0068	0, 0, 3 4, 2, 0 .1193	0, 0, 2 5, 2, 0 .4040	0, 0, 0 7, 2, 0 .7010	0, 2, 1 6, 0, 0 .8852	0, 2, 1 6, 0, 0 .9651	0, 2, 0 7, 0, 0 .9915	0, 2, 0 7, 0, 0 .9983	0, 2, 0 7, 0, 0 .9997			
87	0, 0, 7 1, 1, 0 .0074	0, 0, 5 3, 1, 0 .1279	0, 0, 3 5, 1, 0 .4234	0, 0, 1 7, 1, 0 .7201	0, 0, 0 8, 1, 0 .8969	0, 0, 0 8, 1, 0 .9702	0, 0, 0 8, 1, 0 .9931	0, 0, 0 8, 1, 0 .9987	0, 0, 0 8, 1, 0 .9998			
90	0, 0, 8 1, 0, 0 .0081	0, 0, 6 3, 0, 0 .1365	0, 0, 4 5, 0, 0 .4421	0, 0, 3 6, 0, 0 .7383	0, 0, 2 7, 0, 0 .9075	0, 0, 1 8, 0, 0 .9745	0, 0, 1 8, 0, 0 .9944	0, 0, 1 8, 0, 0 .9990	0, 0, 1 8, 0, 0 .9999			
94	0, 0, 6 2, 2, 0 .0088	0, 0, 4 4, 2, 0 .1454	0, 0, 2 6, 2, 0 .4606	0, 0, 0 8, 2, 0 .7554	0, 2, 1 7, 0, 0 .9171	0, 2, 1 7, 0, 0 .9782	0, 2, 0 8, 0, 0 .9955	0, 2, 0 8, 0, 0 .9992				

TABLE OPT2.6.3

P=6 K=3 OPTIMAL DESIGN AND ASSOCIATED CONFIDENCE COEFFICIENT FOR TWO-SIDED INTERVALS

b \ a/σ	.500	.600	.700	.800	.900	1.000	1.100	1.200	1.300	1.400	1.500	1.600
6	1, 0, 0 0, 0 .0010	1, 0, 0 0, 0 .0028	1, 0, 0 0, 0 .0065	1, 0, 0 0, 0 .0132	1, 0, 0 0, 0 .0241	1, 0, 0 0, 0 .0404	1, 0, 0 0, 0 .0631	1, 0, 0 0, 0 .0928	1, 0, 0 0, 0 .1297	1, 0, 0 0, 0 .1738	1, 0, 0 0, 0 .2240	1, 0, 0 0, 0 .2795
7	0, 1, 0 0, 0 .0120	0, 1, 0 0, 0 .0294	0, 1, 0 0, 0 .0593	0, 1, 0 0, 0 .1035	0, 1, 0 0, 0 .1614	0, 1, 0 0, 0 .2305	0, 1, 0 0, 0 .3071	0, 1, 0 0, 0 .3872	0, 1, 0 0, 0 .4669	0, 1, 0 0, 0 .5434	0, 1, 0 0, 0 .6145	0, 1, 0 0, 0 .6790
11	0, 0, 1 0, 0 .0334	0, 0, 1 0, 0 .0770	0, 0, 1 0, 0 .1454	0, 0, 1 0, 0 .2363	0, 0, 1 0, 0 .3426	0, 0, 1 0, 0 .4548	0, 0, 1 0, 0 .5638	0, 0, 1 0, 0 .6625	0, 0, 1 0, 0 .7470	0, 0, 1 0, 0 .8159	0, 0, 1 0, 0 .8697	0, 0, 1 0, 0 .9103
14	0, 2, 0 0, 0 .0620	0, 2, 0 0, 0 .1300	0, 2, 0 0, 0 .2232	0, 2, 0 0, 0 .3321	0, 2, 0 0, 0 .4454	0, 2, 0 0, 0 .5539	0, 2, 0 0, 0 .6513	0, 2, 0 0, 0 .7347	0, 2, 0 0, 0 .8033	0, 2, 0 0, 0 .8578	0, 2, 0 0, 0 .8998	0, 2, 0 0, 0 .9311
15		0, 0, 0 1, 0 .1334	0, 0, 0 1, 0 .2368	0, 0, 0 1, 0 .3616	0, 0, 0 1, 0 .4928	0, 0, 0 1, 0 .6164	0, 0, 0 1, 0 .7231	0, 0, 0 1, 0 .8085	0, 0, 0 1, 0 .8727	0, 0, 0 1, 0 .9185	0, 0, 0 1, 0 .9496	0, 0, 0 1, 0 .9698
17	0, 1, 0 0, 1 .0725	0, 1, 0 0, 1 .1410										
18	0, 1, 1 0, 0 .1059	0, 1, 1 0, 0 .2108	0, 1, 1 0, 0 .3428	0, 1, 1 0, 0 .4831	0, 1, 1 0, 0 .6145	0, 1, 1 0, 0 .7261	0, 1, 1 0, 0 .8140	0, 1, 1 0, 0 .8790	0, 1, 1 0, 0 .9244	0, 1, 1 0, 0 .9546	0, 1, 1 0, 0 .9738	0, 1, 1 0, 0 .9854
21	0, 0, 1 0, 1 .1403	0, 0, 1 0, 1 .2599	0, 0, 1 0, 1 .3972	0, 0, 1 0, 1 .5327	0, 0, 1 0, 1 .6534	0, 0, 1 0, 1 .7530	0, 0, 1 0, 1 .8307	1, 0, 0 1, 0 .8903	1, 0, 0 1, 0 .9345	1, 0, 0 1, 0 .9625	1, 0, 0 1, 0 .9794	1, 0, 0 1, 0 .9891
22	0, 0, 2 0, 0 .1512	0, 0, 2 0, 0 .2865	0, 0, 2 0, 0 .4436	0, 0, 2 0, 0 .5961	0, 0, 2 0, 0 .7256	0, 0, 2 0, 0 .8244	0, 0, 2 0, 0 .8938	0, 0, 2 0, 0 .9390	0, 0, 2 0, 0 .9667	0, 0, 2 0, 0 .9827	0, 0, 2 0, 0 .9914	0, 0, 2 0, 0 .9959
24	0, 2, 0 0, 1 .1546											

TABLE OPT2.6.3

P=6 K=3 OPTIMAL DESIGN AND ASSOCIATED CONFIDENCE COEFFICIENT FOR TWO-SIDED INTERVALS

b \ a/σ	.400	.500	.600	.700	.800	.900	1.000	1.100	1.200	1.300	1.400	1.500
25	0, 0, 0 1, 1 .0837	0, 0, 0 1, 1 .1984	0, 0, 0 1, 1 .3513	0, 0, 0 1, 1 .5131	0, 0, 0 1, 1 .6586	0, 0, 0 1, 1 .7749	0, 0, 0 1, 1 .8599	0, 0, 0 1, 1 .9174	0, 0, 0 1, 1 .9539	0, 0, 0 1, 1 .9755	0, 0, 0 1, 1 .9877	0, 0, 0 1, 1 .9941
26			0, 0, 1 1, 0 .3582	0, 0, 1 1, 0 .5302	0, 0, 1 1, 0 .6837	0, 0, 1 1, 0 .8026	0, 0, 1 1, 0 .8850	0, 0, 1 1, 0 .9372	0, 0, 1 1, 0 .9676	0, 0, 1 1, 0 .9843	0, 0, 1 1, 0 .9927	0, 0, 1 1, 0 .9968
27												
28	0, 4, 0 0, 0 .1035	0, 4, 0 0, 0 .2305	0, 4, 0 0, 0 .3872	0, 4, 0 0, 0 .5434								
29	0, 2, 0 1, 0 .1113	0, 2, 0 1, 0 .2531	0, 2, 0 1, 0 .4296	0, 2, 0 1, 0 .6024	0, 2, 0 1, 0 .7449	0, 2, 0 1, 0 .8481	0, 2, 0 1, 0 .9156	0, 2, 0 1, 0 .9561	0, 2, 0 1, 0 .9786	0, 2, 0 1, 0 .9902	0, 2, 0 1, 0 .9958	0, 2, 0 1, 0 .9983
30				0, 0, 0 2, 0 .6047	0, 0, 0 2, 0 .7522	0, 0, 0 2, 0 .8572	0, 0, 0 2, 0 .9237	0, 0, 0 2, 0 .9620	0, 0, 0 2, 0 .9822	0, 0, 0 2, 0 .9922	0, 0, 0 2, 0 .9968	0, 0, 0 2, 0 .9987
31	0, 0, 1 0, 2 .1152											
32	0, 0, 2 0, 1 .1377	0, 0, 2 0, 1 .2949	0, 0, 2 0, 1 .4760	0, 0, 2 0, 1 .6427	0, 0, 2 0, 1 .7743	0, 0, 2 0, 1 .8671	0, 0, 2 0, 1 .9269	0, 0, 2 0, 1 .9624				
33	0, 0, 3 0, 0 .1397	0, 0, 3 0, 0 .3053	0, 0, 3 0, 0 .4984	0, 0, 3 0, 0 .6738	0, 0, 3 0, 0 .8070	0, 0, 3 0, 0 .8951	0, 0, 3 0, 0 .9473	0, 0, 3 0, 0 .9754	0, 0, 3 0, 0 .9893	0, 0, 3 0, 0 .9957	0, 0, 3 0, 0 .9984	0, 0, 3 0, 0 .9994
34												

TABLE OPT2.6.3

P=6 K=3 OPTIMAL DESIGN AND ASSOCIATED CONFIDENCE COEFFICIENT FOR TWO-SIDED INTERVALS

b \ a/σ	.100	.200	.300	.400	.500	.600	.700	.800	.900	1.000	1.100	1.200
35	0, 0, 0 1, 2 .0001	0, 0, 0 1, 2 .0067	0, 0, 0 1, 2 .0492	0, 0, 0 1, 2 .1578	0, 0, 0 1, 2 .3214	0, 0, 0 1, 2 .4993						
36		0, 0, 1 1, 1 .0070	0, 0, 1 1, 1 .0521	0, 0, 1 1, 1 .1709	0, 0, 1 1, 1 .3528	0, 0, 1 1, 1 .5488	0, 0, 1 1, 1 .7160	0, 0, 1 1, 1 .8367	0, 0, 1 1, 1 .9137	0, 0, 1 1, 1 .9579	0, 0, 1 1, 1 .9810	0, 0, 1 1, 1 .9920
37					0, 0, 2 1, 0 .3554	0, 0, 2 1, 0 .5592	0, 0, 2 1, 0 .7317	0, 0, 2 1, 0 .8528	0, 0, 2 1, 0 .9264	0, 0, 2 1, 0 .9662	0, 0, 2 1, 0 .9857	0, 0, 2 1, 0 .9944
38	0, 4, 0 0, 1 .0002	0, 4, 0 0, 1 .0080	0, 4, 0 0, 1 .0557									
39		0, 4, 1 0, 0 .0089	0, 4, 1 0, 0 .0634	0, 4, 1 0, 0 .1968	0, 4, 1 0, 0 .3867	0, 4, 1 0, 0 .5798	0, 4, 1 0, 0 .7380					
40			0, 0, 0 2, 1 .0643	0, 0, 0 2, 1 .2038	0, 0, 0 2, 1 .4058	0, 0, 0 2, 1 .6102	0, 0, 0 2, 1 .7721	0, 0, 0 2, 1 .8798	0, 0, 0 2, 1 .9423	0, 0, 0 2, 1 .9747	0, 0, 0 2, 1 .9898	0, 0, 0 2, 1 .9962
41						0, 0, 1 2, 0 .6131	0, 0, 1 2, 0 .7789	0, 0, 1 2, 0 .8870	0, 0, 1 2, 0 .9478	0, 0, 1 2, 0 .9780	0, 0, 1 2, 0 .9915	0, 0, 1 2, 0 .9970
42		0, 0, 2 0, 2 .0108	0, 0, 2 0, 2 .0731	0, 0, 2 0, 2 .2158	0, 0, 2 0, 2 .4071							
43		0, 0, 3 0, 1 .0113	0, 0, 3 0, 1 .0778	0, 0, 3 0, 1 .2337	0, 0, 3 0, 1 .4439	0, 0, 3 0, 1 .6441	0, 0, 3 0, 1 .7960	0, 0, 3 0, 1 .8942	0, 0, 3 0, 1 .9501	0, 0, 3 0, 1 .9785		
44				0, 0, 4 0, 0 .2363	0, 0, 4 0, 0 .4548	0, 0, 4 0, 0 .6625	0, 0, 4 0, 0 .8159	0, 0, 4 0, 0 .9103	0, 0, 4 0, 0 .9606	0, 0, 4 0, 0 .9843	0, 0, 4 0, 0 .9943	0, 0, 4 0, 0 .9981

Selected Tables in Mathematical Statistics
Volume 8, 1985

EXPECTED VALUES AND VARIANCES AND COVARIANCES OF ORDER STATISTICS FOR A FAMILY OF SYMMETRIC DISTRIBUTIONS (STUDENT'S t)

M.L. Tiku and S. Kumra
Department of Mathematical Sciences
McMaster University
Hamilton, Ontario, Canada

ABSTRACT

The expected values and the variances and covariances of order statistics of random samples from a family of symmetric distributions (reducible to Student's t through a linear transformation) with kurtosis $\beta_2 = \mu_4/\mu_2^2 > 3$ are evaluated.

INTRODUCTION

The expected values and the variances and covariances of order statistics of random samples are needed in the construction of the best linear unbiased estimators (BLUE) of location and scale parameters or for constructing goodness-of-fit statistics based on spacings; see for example Lloyd (1952) and Tiku (1980a). The variances and covariances of order statistics of random samples from symmetric distributions are also needed in determining the efficiencies of some "robust" estimators of location; see Huber (1970) and Tiku (1930b). The expected values and the variances and covariances of order statistics are known only for a few symmetric distributions, normal and logistic for example; see Sarhan and Greenberg (1962), Tietjen, Kahaner and Beckman (1977), Gupta, Qureishi and Shah (1967) and Tiku and Jones (1971). In this paper, we have considered the family of symmetric distributions $C\{1+(x-\mu)^2/k\sigma^2\}^{-p}$, $-\infty < x < \infty$, $p \geq 2$, $k = 2p-3$, and have tabulated the expected values (Table I) and the variances and covariances (Table II) of order statistics of random samples of size $n = 2(1)20$ from this family; $p = 2(.5)10$.

Received by the editors August 1979 and in revised form December 1980.
AMS(MOS) Subject Classifications (1980): Primary 62Q05; Secondary 62E30 and 62G30.

Note that $t = \sqrt{v}(x-\mu)/\sigma\sqrt{k}$ has Student's t distribution with $v = 2p-1$ degrees of freedom (df) and, therefore, the expected values and the variances and covariances of order statistics of random samples from Student's t distribution with $v(\geq 3)$ df can be obtained by multiplying the values for $p = (v+1)/2$ given in Table I and Table II by $\sqrt{\{v/(v-2)\}}$ and $v/(v-2)$, respectively.

A SYMMETRIC FAMILY OF DISTRIBUTIONS

Consider the symmetric family of distributions

$$f(x) = (1/\sigma)C/\{1+(x-\mu)^2/k\sigma^2\}^p, \quad -\infty < x < \infty, \tag{1}$$

where $k = 2p-3$, $p \geq 2$ and $C = (1/\sqrt{2})\Gamma(p)/[\sqrt{\{\pi(p-3/2)\}}\Gamma(p-1/2)]$. The family (1) has mean μ and standard deviation σ, and kurtosis $\beta_2 = \mu_4/\mu_2^2 = 3(p-3/2)/(p-5/2)$ which assumes values between 3 and ∞. Note that for $p = \infty$, (1) reduces to the normal $N(\mu,\sigma)$ distribution.

Since the distribution of $t_v = \sqrt{(v/k)}y$, $y = (x-\mu)/\sigma$, is Student's t with $v = 2p-1$ df, the probability integral

$$F(z) = \Pr[y \leq z] = C\int_{-\infty}^{z} (1+y^2/k)^{-p} dy \tag{2}$$

is equal to $\Pr[t_v \leq t]$, $t = \sqrt{(v/k)}z$, where (Johnson and Kotz, 1970, p.96)

$$\Pr[t_v \leq t] = \frac{1}{2} + \frac{1}{\pi}[\theta+\{\cos(\theta)+ \frac{2}{3}\cos^3(\theta) + \cdots\cdots + \frac{(2)(4)\cdots(v-3)}{(3)(5)\cdots(v-2)}\cos^{v-2}(\theta)\}\sin(\theta)] \tag{3}$$

for v odd and greater than 1,

and

$$\Pr[t_v \leq t] = \frac{1}{2} + [\{1+ \frac{1}{2}\cos^2(\theta) + \frac{(1)(3)}{(2)(4)}\cos^4(\theta) + \cdots\cdots + \frac{(1)(3)\cdots(v-3)}{(2)(4)\cdots(v-2)}\cos^{v-2}(\theta)\}\sin(\theta)] \tag{4}$$

for v even;

$\theta = \tan^{-1}(t/\sqrt{v})$.

USEFULNESS OF THE TABLES

Estimation: Consider first the problem of estimating the location and scale parameters μ and σ, given a random sample $x_1, x_2, \ldots, x_n$ from the distribution (1); p known. The maximum likelihood (ML) equations for estimating μ and σ are given by

$$\partial \log L/\partial\mu = (2p/k\sigma^2) \sum_{i=1}^{n} [(x_i-\mu)/\{1+((x_i-\mu)^2/k\sigma^2\}] = 0$$

and

$$\partial \log L/\partial\sigma = -(n/\sigma) + (2p/k\sigma^3) \sum_{i=1}^{n} [(x_i-\mu)^2/\{1+(x_i-\mu)^2/k\sigma^2\}] = 0. \quad (5)$$

These equations have no explicit solutions and are difficult to solve by iterative methods. Lloyd's (1952) best linear unbiased estimators (BLUE) are given by

$$\mu^* = \underset{\sim}{I}'\underset{\sim}{V}^{-1}\underset{\sim}{X}/\underset{\sim}{I}'\underset{\sim}{V}^{-1}\underset{\sim}{I}$$

and

$$\sigma^* = \underset{\sim}{\alpha}'\underset{\sim}{V}^{-1}\underset{\sim}{X}/\underset{\sim}{\alpha}'\underset{\sim}{V}^{-1}\underset{\sim}{\alpha}. \quad (6)$$

The estimators μ^* and σ^* are uncorrelated and

$$\mathrm{Var}(\mu^*) = \sigma^2/\underset{\sim}{I}'\underset{\sim}{V}^{-1}\underset{\sim}{I}$$

and

$$\mathrm{Var}(\sigma^*) = \underset{\sim}{\sigma}^2/\underset{\sim}{\alpha}'\underset{\sim}{V}^{-1}\underset{\sim}{\alpha}; \quad (7)$$

see Lloyd (1952). In the expressions (6) and (7), $\underset{\sim}{X}' = (X_i)_{i\,=\,1,2,\ldots,n}$ is the vector of ordered observations, $\underset{\sim}{\alpha}' = (\mu_{r:n})_{r\,=\,1,2,\ldots,n}$ is the vector of the expected values $E(y_r)$ given in Table I, and $\underset{\sim}{V}$ is the variance-covariance matrix $\sigma_{rs:n} = \mathrm{Cov}(y_r, y_s)$, $r,s = 1,2,\ldots,n$, given in Table II; $y_i = (X_i-\mu)/\sigma$, $i = 1,2,\ldots,n$.

Note that the minimum variance bounds for estimating μ and σ in (1) are given by

$$\mathrm{MVB}(\mu) = 1/-E(\partial^2 \log L/\partial\mu^2) = (p-3/2)(p+1)\sigma^2/np(p-1/2)$$

and

$$\text{MVB}(\sigma) = 1/-E(\partial^2 \log L/\partial\sigma^2) = (p+1)\sigma^2/2n(p-1/2). \tag{8}$$

Consider the following 15 ordered observations and suppose that they are the order statistics of a random smaple of size 15 from the distribution (1) with p = 5 (effectively a logistic distribution which has the same first five moments as the distribution (1) with p = 5, Tiku and Jones (1971)):

8.921, 8.982, 9.048, 9.262, 9.689, 9.715, 9.774, 9.830, 10.128, 10.485, 10.591, 10.766, 10.840, 10.881, 11.263.

These observations represent the measurements of a certain characteristic in blood cells; see Elveback, Guillier and Kerting (1970). Substituting the values given in Table I and Table II for p = 5 and n = 15 in the equations (6) and (7), we obtain the BLUE

$$\mu^* = 10.010 \text{ and } \sigma^* = 0.843$$

with $V(\mu^*) = 0.063\sigma^2$ and $V(\sigma^*) = 0.048\sigma^2$. The MVB are $0.062\sigma^2$ (for μ) and $0.044\sigma^2$ (for σ). The estimators μ^* and σ^* have, therefore, efficiencies of 98% and 92%, respectively.

Incidentally, the ML estimates obtained by a laborious iterative method are $\hat{\mu} = 10.011$ and $\hat{\sigma} = 0.805$. There is close agreement between the BLUE and the ML estimates.

Goodness-of-fit: The following data represent the order statistics of 15 random observations on a variable x:

10.115, 10.185, 10.259, 10.502, 10.986, 11.016, 11.083, 11.146, 11.434 11.889, 12.009, 12.207, 12.291, 12.338, 12.771.

Suppose that one were to test from this data whether the random variable x follows, other than its location and scale parameters μ and σ, Student's t distribution with 9 df. Note that under the assumption of this t distribution, $E(x) = \mu$ and $V(x) = (9/7)\sigma^2$. Tiku's (1980a) goodness-of-fit statistic Z* which is location and scale invariant is given by (see also Tiku and Singh, 1981)

$$Z^* = 2\sum_{i=1}^{n-1}(n-1-i)G_i/(n-2)\sum_{i=1}^{n-1}G_i, \tag{9}$$

where the sample spacings G_i are given by $G_i = (X_{i+1}-X_i)/d_i$, $d_i = \mu_{i+1:n}-\mu_{i:n}$, $i = 1,2,\ldots,n-1$; $\mu_{r:n}$ being the expected value of the rth order statistic in a sample of size n from Student's t distribution with 9 df. The values of

$\mu_{r:n}$, r = 1,2,...,n, are obtained by multiplying the values for p = (9+1)/2 = 5 given in Table I by $\sqrt{(9/7)}$ = 1.13389. The values of d_i for n = 15 thus work out to be

.6414, .3526, .2600, .2161, .1922, .1793, .1734,

for i = 1,2,...,7, respectively; $d_{15-i} = d_i$. The values of the spacings G_i are then given by

.1078, .2122, .9333, 2.240, .1475, .3732, .3663, 1.949, 2.258, .6254, .9181, .3227, .1318, .6753,

for i = 1,2,...,14, respectively. Substituting these values in (9), the value of Z* is given by

$$Z^* = 2(72.5084)/13(11.2606) = 0.99$$

The null distribution of Z* is known to be approximately normal with mean 1 and variance (Tiku, 1980a, 1981)

$$V(Z^*) \simeq \{V(w_1) + V(w_2) - 2\,\mathrm{Cov}\,(w_1,w_2)\}/\{(n-1)(n-2)\sigma\}^2, \tag{10}$$

where

$$V(w_1) = 4(\underset{\sim}{\ell}_1'\underset{\sim}{V}\underset{\sim}{\ell})\sigma^2,\quad V(w_2) = (\underset{\sim}{\ell}_2'\underset{\sim}{V}\ell_2)\sigma^2$$

and

$$\mathrm{Cov}(w_1,w_2) = 2(\underset{\sim}{\ell}_1'\underset{\sim}{V}\underset{\sim}{\ell}_2)\sigma^2; \tag{11}$$

$\underset{\sim}{\ell}_1' = (n-2,n-3,\ldots,1,0)$ and $\underset{\sim}{\ell}_2' = (n-2,\ldots,n-2)$ are vectors each having n-1 elements and $\underset{\sim}{V}$ is the variance-covariance matrix of G_i/σ, i = 1,...,n-1. Note that $d_i d_j\,\mathrm{Cov}\,(G_i,G_j) = \mathrm{Cov}(X_{i+1}X_{j+1}) - \mathrm{Cov}\,(X_{i+1}X_j) - \mathrm{Cov}\,(X_iX_{j+1}) + \mathrm{Cov}(X_iX_j)$. The matrix $\underset{\sim}{V}$ can thus be worked out from the values of the variances and covariances of order statistics of random samples from Student's t distribution with 9 df. These variance and covariances are obtained by multiplying the values for p = (9+1)/2 = 5 given in Table II by 9/7 = 1.28571. Substituting the values for n = 15 in the equation (10), we obtain $V(Z^*) \simeq 0.023$. The normal approximation thus yields 0.75 and 1.25 for the lower and the upper 5% points of Z*, respectively. Since the calculated value of Z* is 0.99, there is no evidence that x does not follow Student's t distribution with 9 df; significance level α = 0.10.

Efficiencies of robust estimators: Let $X_1,X_2,\ldots,X_n$ be the n order statistics of a random sample of size n. The most prominent robust unbiased estimator of the location μ of symmetric distributions of the type $(1/\sigma)f((x-\mu)/\sigma)$ is the trimmed mean

$$\mu_{Trim} = \sum_{i=r+1}^{n-r} X_i/(n-2r); \tag{12}$$

see Stigler (1977). One of the competitors to μ_{Trim} is the unbiased estimator μ_c given by

$$\mu_c = \{\sum_{i=r+1}^{n-r} X_i + r\beta(X_{n-r}+X_{r+1})\}/(n-2r+2r\beta), \tag{13}$$

where $\beta = -f(t)\{t-f(t)/q\}/q$; $q = r/n$, $f(z) = (1/\sqrt{(2\pi)})\exp(-\frac{1}{2}z^2)$, $-\infty < z < \infty$, and t is determined by the equation $\int_{-\infty}^{t} f(z)dz = 1-q$; see Tiku (1967, 1968, 1978, 1980 b,c) and Tiku and Singh (1981).

The exact variances of the estimators μ_{Trim} and μ_c are given by $(\underset{\sim}{\ell}'\underset{\sim}{V}\underset{\sim}{\ell})\sigma^2$, where $\underset{\sim}{V}$ is the variance-covariance matrix of $y_i = (X_i-\mu)/\sigma$, $i = r+1, r+2,\ldots,n-r$; $(n-2r)\underset{\sim}{\ell}' = (1,1,\ldots,1)$ and $(n-2r+2r\beta)\underset{\sim}{\ell}' = (1+r\beta, 1,\ldots,1,1+r\beta)$ for μ_{Trim} and μ_c, respectively. Substituting the values of the variances and covariances (given in Table II) in $\underset{\sim}{\ell}'\underset{\sim}{V}\underset{\sim}{\ell}$, we obtain the following variances of μ_{Trim} and μ_c for the symmetric family of distributions (1), n = 10:

Values of $(1/\sigma)$ Variance

p	Kurtosis β_2	r = 1 μ_{Trim}	r = 1 μ_c	r = 2 μ_{Trim}	r = 2 μ_c	r = 3 μ_{Trim}	r = 3 μ_c
2	∞	.0599	.0653	.0553	.0578	.0559	.0563
2.5	∞	.0771	.0814	.0747	.0767	.0774	.0774
3	9	.0850	.0883	.0845	.0858	.0886	.0883
3.5	6	.0894	.0921	.0904	.0912	.0955	.0950
4	5	.0923	.0945	.0943	.0947	.1001	.0994
6	3.9	.0978	.0988	.1020	.1016	.1095	.1084
10	3.4	.1013	.1013	.1071	.1059	.1158	.1143
Normal	3	.1053	.1043	.1133	.1113	.1238	.1218

Realizing that the minimum variance bound for estimating μ in (1) is $(p-3/2)(p+1)\sigma^2/np(p-1/2)$, the estimators μ_{Trim} and μ_c have remarkably high efficiencies.

EXPECTED VALUES AND VARIANCES AND COVARIANCES

Let $x_1, x_2, \ldots, x_n$ be a random sample of size n from the distribution (1), p known. Let $X_1, X_2, \ldots, X_n$ be the n order statistics obtained by arranging x_i, $i = 1,2,\ldots,n$, in ascending order of magnitude. The expressions for the first and second moments of y_r and the first product moment of y_r and y_s are as follows ($r,s = 1,2,\ldots,n$, $r < s$):

$$E(y_r) = C \frac{n!}{(r-1)!(n-r)!} \int_{-\infty}^{\infty} y(1+\frac{y^2}{k})^{-p}\{F(y)\}^{r-1}\{1-F(y)\}^{n-r}dy$$

which under the transformation $y = \sqrt{k}\tan(\frac{1}{2}\pi u)$ reduces to

$$E(y_r) = C \frac{n!k\pi}{4(r-1)!(n-r)!} \int_{-1}^{1} \sin(\pi u)\cos^q(\tfrac{1}{2}\pi u)\{F(u)\}^{r-1}\{1-F(u)\}^{n-r}du, \quad (14)$$

and similarly,

$$E(y_r^2) = C \frac{n!k\sqrt{k}\pi}{2(r-1)!(n-r)!} \int_{-1}^{1} \sin^2(\tfrac{1}{2}\pi u)\cos^q(\tfrac{1}{2}\pi u)\{F(u)\}^{r-1}\{1-F(u)\}^{n-r}du \quad (15)$$

and

$$E(y_r y_s) = C^2 \frac{n!(k\pi)^2}{16(r-1)!(s-r-1)!(n-s)!} \int_{-1}^{1}\int_{-1}^{z} \sin(\pi u)\sin(\pi z)\cos^q(\tfrac{1}{2}\pi u)\cos^q(\tfrac{1}{2}\pi z)\,.$$

$$\{F(u)\}^{r-1}\{F(z)-F(u)\}^{s-r-1}\{1-F(z)\}^{n-s}dudz, \quad (16)$$

$q = 2p-4$ and $y_i = (X_i-\mu)/\sigma$, $i = 1,2,\ldots,n$. Note that $E(X_r) = \mu+\sigma\,\mu_{r:n}$, $\mu_{r:n} = E(y_r)$, $V(X_r) = \sigma^2\sigma_{rr:n}$, $\sigma_{rr:n} = V(y_r)$, and $Cov(X_r,X_s) = \sigma^2\sigma_{rs:n}$, $\sigma_{rs:n} = Cov(y_r,y_s)$.

The integrals (14) and (15) were numerically evaluated by the following traditional Simpson's method:

Let f(u) be a function integrable in the range $a \leq x \leq a+h$. Simpson's compound rule with 2m+1 points approximates $\int_a^{a+h} f(u)du$ by

$$S^{(m)} = (h/6m)\{f(a) + 4\sum_{j=1}^{m} f(u_{2j-1}) + 2\sum_{j=1}^{m-1} f(u_{2j}) + f(a+h)\} \quad (17)$$

where $u_j = a+jh/2m$; see Rowland and Varol (1972). A traditional method of applying Simpson's rule is to evaluate $S^{(1)}, S^{(2)}, S^{(4)}, \ldots$, and accept $S^{(2m)}$ as a sufficiently accurate approximation to $\int_a^{a+h} f(u)du$, if the "stopping inequality"

$$|S^{(m)}-S^{(2m)}| < \varepsilon \tag{18}$$

is satisfied; ε is a preassigned tolerance. Clenshaw and Curtis (1960) have, however, given an example where (18) is satisfied while the error is much greater than ε, and this happens when the fourth derivative $f^{(4)}$ is not of constant sign in the range of integration as is the case with the integrand functions (14) and (15); see also Rowland and Varol (1972). However, Rowland and Varol proved that if $f^{(4)}$ is not of constant sign then under some regularity conditions which the integrand functions (14) and (15) satisfy, the stopping inequality (18) is asymptotically valid, that is, there exists an integer m_0 such that the stopping inequality (13) is valid for all $m \geq m_0$ (threshold). The value of m used in the evaluation of (14) and (15) was between 100 and 1024.

To evaluate the range-dependent double integral (16), we notice that this integral is of the form

$$\int_{-1}^{1} g(z)\{\int_{-1}^{z} f(u,z)\,du\}dz = \int_{-1}^{1} G(z)\,dz, \tag{19}$$

where

$$g(z) = \sin(\pi z)\cos^q(\tfrac{1}{2}\pi z)\{1-F(z)\}^{n-s}$$

and

$$f(u,z) = \sin(\pi u)\cos^q(\tfrac{1}{2}\pi u)\{F(u)\ ^{r-1}\{F(z)-F(u)\}^{s-r-1}.$$

The value (calculated independently for each n, r and s) of the integral (16) was obtained by evaluating the inner integral $\int_{-1}^{z} f(u,z)\,du$ and the outer integral $\int_{-1}^{1} G(z)\,dz$ by using the 32-point Gaussian quadrature DQG 32 (Stroud and Secrest, 1966), used twice (once in each of the two intervals $[-1,z_1]$ and $[z_1,z]$, $z_1 = (z-1)/2$) in the inner interval $[-1,z]$ and, similarly, used twice in the outer interval $[-1,1]$, thus needing the evaluation of the function $f(u,z)$ at 4096 points (u,z); see also Davis and Robinowitz (1967). To provide a check on these values, the integral (16) was evaluated for a number of values of n, r and s, from the more expensive (in terms of the computing time) Simpson's formulae (17) and (18) and a perfect agreement (to at least 9 decimal places) between the two was found. Finally the values of the variances $V(y_r) = \sigma_{rr:n} = E(y_r^2) - E^2(y_r)$ and the covariances $Cov(y_r, y_s) = \sigma_{rs:n} = E(y_r y_s) - E(y_r)E(y_s)$ were calculated. The values of $\sigma_{rr:n}$ and $\sigma_{rs:n}$ were thoroughly checked to satisfy (to at least 9 decimal places) the recurrence relations given by Govindarajulu (1963). The values of $\sigma_{rs:n}$ were also thoroughly checked to satisfy (to as least 8 decimal places) the well known equation $\Sigma_{r=1}^{n} \Sigma_{s=1}^{n} \sigma_{rs:n} = n$ (for a given p). The values of $\mu_{r:n}$, $\sigma_{rr:n}$, and $\sigma_{rs:n}$ are given in Table I and II to 8 decimal places and we are confident that these values are accurate to within a unit in the last decimal place. The missing values in Tables I and II may be obtained from the following equations:

(1) $\mu_{n-r+1:n} = -\mu_{r:n}$

(2) $\sigma_{n-r+1,\ n-r+1:n} = \sigma_{rr:n}$

and

(3) $\sigma_{n-r+1,\ n-s+1:n} = \sigma_{n-s+1,\ n-r+1:n} = \sigma_{rs:n}$.

The calculations were carried out in double precision on the CDC 6400 Computer of McMaster University.

ACKNOWLEDGEMENT

Thanks are due to my colleague Dr. P.C. Chakravarty for advice on the numerical evaluation and to the referees for useful comments. Thanks are also due to the National Research Council of Canada and McMaster University Research Board for research grants. Thanks are due to my student Mr. W.K. Wong for assistance in the computations.

REFERENCES

Clenshaw, C.W. and Curtis, A.R. (1960). A method for numerical integration on an automatic computer. Numerische Mathematik 2, 197-205.

Davis, P. and Rabinowitz, (1967). Numerical Integration. London: Blaisdell Publishing Company.

Elveback, L.R., Guillier, C.L. and Kerting, F.R. (1970). Health, Normality and the Ghost of Gauss. J. American Medical Assoc. 211, 69-75.

Govindarajulu, Z. (1963). On moments of order statistics and quasi-ranges from normal populations. Ann. Math. Statist. 34, 633-51.

Gupta, S.S., Qureishi, A.S. and Shah, B.K. (1967). Best linear unbiased estimators of the parameters of the logistic distribution using order statistics. Technometrics 9, 43-56.

Huber, Peter J. (1970). Studentizing robust estimates. Nonparametric Techniques in Statistical Inferences. (Ed., M.L. Puri). England: Cambridge University Press.

Johnson, Norman L. and Kotz, S. (1970). Continuous Univariate Distributions 2. Boston: Houghton Mifflin Company.

Lloyd, E.H. (1952). Least square estimation of location and scale parameters using order statistics. Biometrika 39, 88-95.

Rowland, J.H. and Varol, Y.L. (1972). Exit criteria for Simpson's Compound rule. Math. Computation 26, No. 119, 699-703.

Sarhan, A.E., and Greenberg, B.G. (1962). Contributions to Order Statistics. New York: John Wiley and Sons.

Stigler, M.S. (1977). Do robust estimators work with real data? Ann. Statist. 6, 1055-98.

Stround, A.H. and Secrest, D. (1966). Gaussian Quandrature Formulas. Toronto: Prentice-Hall of Canada.

Tietjen, G.L., Kahaner, D.K. and Beckman, R.J. (1977). Variances and co-variances of the normal order statistics for sample sizes 2 to 50. Selected Tables in Mathematical Statistics V. Rhode Island: American Mathematical Society.

Tiku, M.L. (1967). Estimating the mean and standard deviation from a censored normal sample. Biometrika 54, 155-65.

Tiku, M.L. (1968). Estimating the parameters of log-normal distribution from censored samples. J. Amer. Statist. Assoc. 63, 134-40.

Tiku, M.L. (1978). Linear regression model with censored observations. Commun. Statist. A7 (13), 1219-32.

Tiku, M.L. (1980a). Goodness of fit statistics based on the spacings of complete or censored samples. Aust. J. Stat. 22 (3), to appear. Abstract No. 79-24. Canadian J. Stat. 7 (1979), p. 108.

Tiku, M.L. (1980b). Robustness of MML estimators based on censored samples and robust test statistics. J. Statistical Planning and Inference 4, 123-43.

Tiku, M.L. (1980c). Robust two-sample test and robust regression and analysis-of-variance via MML estimators. Proceedings of the International Symposium on Statistics and Related Topics: Ottawa, May 5-7 (Eds., A.K.E. Saleh. M. Csorgo, E.A. Dawson and J.N.K. Rao). North Holland, 297-314.

Tiku, M.L. (1981). A goodness of fit statistic based on the sample spacings for testing a symmetric distribution against symmetric alternatives. Aust. J. Stat. 23 (3), 148-58.

Tiku, M.L. and Jones, P.W. (1971). Best linear unbiased estimators for a distribution similar to the logistic. Proceedings of "Statistics '71 Canada". Eds: Carter, C.S., Dwividi, T.T., Fellegi, I.P., Fraser, D.A.S., Mcgregor, J.R. and Sprott, D.A..

Tiku, M.L. and Singh, M. (1981). Testing the two parameter Weibull distribution. Commun. Statist. A10 (9), 907-18.

Tiku, M.L. and Singh, M. (1981). Robust test for means when population variances are unequal. Commun. Statist. A10 (20), 2057-71.

APPENDIX

The random variable x is assumed to have the symmetric distribution $f(x) = (1/\sigma)C\{1+(x-\mu)^2/k\sigma^2\}^{-p}$, $-\infty < x < \infty$, $p \geq 2$, and $X_1, X_2, \ldots, X_n$ denote the n order statistics of a random sample of size n from the distribution f(x).

Table I. Gives the means (expected values) $\mu_{r:n}$ of $y_r = (X_r-\mu)/\sigma$, $r = 1,2,\ldots,n$; $n = 2(1)20$ and $p = 2(.5)10$.

Table II. Gives the variances $\sigma_{rr:n}$ of $y_r = (X_r-\mu)/\sigma$, and covariances $\sigma_{rs:n}$ of y_r and $y_s = (X_s-\mu)/\sigma$, r, s = 1,2,...,n, r < s; n = 2(1)20 and p = 2(.5)10. Note that $\sigma_{rr:n} = \mathrm{Cov}(y_r, y_s)$, r = s.

NOTATION IN TABLES

P = p(df of t is v = 2p-1)

N = sample size

R = rank of order statistic

S = rank of second order statistic

TABLE I

P = 2.0

N	R	MEAN	N	R	MEAN	N	R	MEAN	N	R	MEAN
1	1	.00000000	10	4	-.25068692	15	1	-1.74937845	18	4	-.59145424
2	1	-.47746483	10	5	-.08058994	15	2	-.98025337	18	5	-.44880268
3	1	-.71619724	11	1	-1.52539105	15	3	-.68102658	18	6	-.33138540
3	2	.00000000	11	2	-.80305669	15	4	-.48976976	18	7	-.22850481
4	1	-.88437943	11	3	-.50811150	15	5	-.34327477	18	8	-.13411921
4	2	-.21165069	11	4	-.30950902	15	6	-.21930586	18	9	-.04423515
5	1	-1.01728649	11	5	-.14774823	15	7	-.10708621	19	1	-1.93196988
5	2	-.35275116	11	6	.00000000	15	8	.00000000	19	2	-1.11981053
5	3	.00000000	12	1	-1.58658375	16	1	-1.79814445	19	3	-.81243602
6	1	-1.12862362	12	2	-.85227137	16	2	-1.01788841	19	4	-.62134789
6	2	-.46060088	12	3	-.55698330	16	3	-.71680808	19	5	-.47935305
6	3	-.13705172	12	4	-.36149609	16	4	-.52597339	19	6	-.36326163
7	1	-1.22524338	12	5	-.20553488	16	5	-.38115886	19	7	-.26232023
7	2	-.54890502	12	6	-.06684692	16	6	-.25992979	19	8	-.17053552
7	3	-.23984051	13	1	-1.64397866	16	7	-.15159931	19	9	-.08404679
7	4	.00000000	13	2	-.89784486	16	8	-.04985510	19	10	.00000000
8	1	-1.31109920	13	3	-.60161719	17	1	-1.84468191	20	1	-1.97308558
8	2	-.62425266	13	4	-.40820368	17	2	-1.05354519	20	2	-1.15077149
8	3	-.32286211	13	5	-.25640401	17	3	-.75046255	20	3	-.84116182
8	4	-.10147118	13	6	-.12414428	17	4	-.55975389	20	4	-.64965650
9	1	-1.38869489	13	7	.00000000	17	5	-.41618677	20	5	-.50811345
9	2	-.69033367	14	1	-1.69810344	17	6	-.29709187	20	6	-.39307184
9	3	-.39296911	14	2	-.94035646	17	7	-.19179930	20	7	-.29370447
9	4	-.18264813	14	3	-.64277521	17	8	-.09417074	20	8	-.20403522
9	5	.00000000	14	4	-.45070443	17	9	.00000000	20	9	-.12028596
10	1	-1.45972429	14	5	-.30195180	18	1	-1.88922465	20	10	-.03975447
10	2	-.74943029	14	6	-.17441800	18	2	-1.08745532			
10	3	-.45394719	14	7	-.05711265	18	3	-.78226421			

TABLE I

P = 2.5

N	R	MEAN	N	R	MEAN	N	R	MEAN	N	R	MEAN
2	1	-.52065034	10	5	-.09546680	15	2	-1.10075735	18	5	-.52754219
3	1	-.78097552	11	1	-1.59819915	15	3	-.78587880	18	6	-.39194512
3	2	.00000000	11	2	-.91202433	15	4	-.57320844	18	7	-.27136328
4	1	-.95974569	11	3	-.59154114	15	5	-.40512544	18	8	-.15967017
4	2	-.24466499	11	4	-.36457275	15	6	-.26012392	18	9	-.05272453
5	1	-1.09773837	11	5	-.17502248	15	7	-.12736369	19	1	-1.97110274
5	2	-.40777498	11	6	.00000000	15	8	.00000000	19	2	-1.24531504
5	3	.00000000	12	1	-1.65575992	16	1	-1.85075711	19	3	-.92991622
6	1	-1.21102536	12	2	-.96503073	16	2	-1.14009106	19	4	-.72229833
6	2	-.53130341	12	3	-.64699229	16	3	-.82542141	19	5	-.56267292
6	3	-.16071810	12	4	-.42518769	16	4	-.61452749	19	6	-.42917615
7	1	-1.30764257	12	5	-.24334286	16	5	-.44925129	19	7	-.31127790
7	2	-.63132211	12	6	-.07937394	16	6	-.30804855	19	8	-.20293823
7	3	-.28125668	13	1	-1.70926389	16	7	-.18024954	19	9	-.10017660
7	4	.00000000	13	2	-1.01371228	16	8	-.05936759	19	10	.00000000
8	1	-1.39220410	13	3	-.69728221	17	1	-1.89285960	20	1	-2.00763923
8	2	-.71571191	13	4	-.47935925	17	2	-1.17711730	20	2	-1.27690945
8	3	-.37815270	13	5	-.30330168	17	3	-.86239420	20	3	-.96096533
8	4	-.11976330	13	6	-.14740874	17	4	-.65288170	20	4	-.75397120
9	1	-1.46761313	13	7	.00000000	17	5	-.48987634	20	5	-.59560684
9	2	-.78893186	14	1	-1.75930161	17	6	-.35175118	20	6	-.46387117
9	3	-.45944208	14	2	-1.05877354	17	7	-.22792707	20	7	-.34822111
9	4	-.21557395	14	3	-.74334473	17	8	-.11213878	20	8	-.24266909
9	5	.00000000	14	4	-.52838631	17	9	.00000000	20	9	-.14334193
10	1	-1.53581962	14	5	-.35679160	18	1	-1.93290339	20	10	-.04741898
10	2	-.85375466	14	6	-.20701983	18	2	-1.21211516			
10	3	-.52964067	14	7	-.06792730	18	3	-.89713444			
10	4	-.29564537	15	1	-1.80634048	18	4	-.68869298			

TABLE I

P = 3.0

N	R	MEAN	N	R	MEAN	N	R	MEAN	N	R	MEAN
1	1	.00000000	10	4	-.31676798	15	1	-1.80640960	18	4	-.73268310
2	1	-.53601420	10	5	-.10253965	15	2	-1.14844467	18	5	-.56417381
3	1	-.80402131	11	1	-1.61080183	15	3	-.83199946	18	6	-.42060975
3	2	.00000000	11	2	-.95742977	15	4	-.61155575	18	7	-.29187203
4	1	-.98558241	11	3	-.62932050	15	5	-.43423162	18	8	-.17197687
4	2	-.25933801	11	4	-.39035576	15	6	-.27959475	18	9	-.05682601
5	1	-1.12392050	11	5	-.18798936	15	7	-.13710584	19	1	-1.95852735
5	2	-.43223001	11	6	.00000000	15	8	.00000000	19	2	-1.29238835
5	3	.00000000	12	1	-1.66529083	16	1	-1.84764914	19	3	-.98007242
6	1	-1.23620253	12	2	-1.01142279	16	2	-1.18781648	19	4	-.76772433
6	2	-.56251039	12	3	-.68746467	16	3	-.87284202	19	5	-.60127852
6	3	-.17166925	12	4	-.45488799	16	4	-.65501502	19	6	-.46028062
7	1	-1.33101193	12	5	-.26129132	16	5	-.48117792	19	7	-.33465620
7	2	-.66734608	12	6	-.08536662	16	6	-.33094975	19	8	-.21852774
7	3	-.30042119	13	1	-1.71566718	16	7	-.19400308	19	9	-.10796943
7	4	.00000000	13	2	-1.06077465	16	8	-.06395226	19	10	.00000000
8	1	-1.41326431	13	3	-.73998756	17	1	-1.88658105	20	1	-1.99194604
8	2	-.75524532	13	4	-.51238837	17	2	-1.22473863	20	2	-1.32357224
8	3	-.40364835	13	5	-.32551214	17	3	-.91090033	20	3	-1.01173335
8	4	-.12837592	13	6	-.15853801	17	4	-.69523659	20	4	-.80066049
9	1	-1.48604128	13	7	.00000000	17	5	-.52429490	20	5	-.63597968
9	2	-.83104849	14	1	-1.76254527	17	6	-.37769718	20	6	-.49717502
9	3	-.48993420	14	2	-1.10625198	17	7	-.24524614	20	7	-.37419369
9	4	-.23107665	14	3	-.78791072	17	8	-.12079871	20	8	-.26122944
9	5	.00000000	14	4	-.56426931	17	9	.00000000	20	9	-.15447519
10	1	-1.55140437	14	5	-.38268600	18	1	-1.92346740	20	10	-.05112905
10	2	-.89777354	14	6	-.22259919	18	2	-1.25951299			
10	3	-.56414830	14	7	-.07312312	18	3	-.94654377			

TABLE I

P = 3.5

N	R	MEAN	N	R	MEAN	N	R	MEAN	N	R	MEAN
2	1	-.54360444	10	5	-.10669103	15	2	-1.17340523	18	5	-.58539496
3	1	-.81540666	11	1	-1.61241908	15	3	-.85782598	18	6	-.43738889
3	2	.00000000	11	2	-.98203480	15	4	-.63361011	18	7	-.30395581
4	1	-.99800304	11	3	-.65086063	15	5	-.45121075	18	8	-.17925632
4	2	-.26761752	11	4	-.40535271	15	6	-.29104555	18	9	-.05925650
5	1	-1.13599650	11	5	-.19560021	15	7	-.14285973	19	1	-1.94331969
5	2	-.44602920	11	6	.00000000	15	8	.00000000	19	2	-1.31609430
5	3	.00000000	12	1	-1.66478867	16	1	-1.83865147	19	3	-1.00762923
6	1	-1.24718657	12	2	-1.03635363	16	2	-1.21256711	19	4	-.79351208
6	2	-.58004615	12	3	-.71044063	16	3	-.89927204	19	5	-.62358797
6	3	-.17799532	12	4	-.47212065	16	4	-.67822639	19	6	-.47845451
7	1	-1.34047302	12	5	-.27181683	16	5	-.49976129	19	7	-.34841338
7	2	-.68746788	12	6	-.08889696	16	6	-.34439956	19	8	-.22774284
7	3	-.31149180	13	1	-1.71303355	16	7	-.20212218	19	9	-.11258736
7	4	.00000000	13	2	-1.08585011	16	8	-.06666515	19	10	.00000000
8	1	-1.42094412	13	3	-.76412298	17	1	-1.87549208	20	1	-1.97471399
8	2	-.77717531	13	4	-.53149945	17	2	-1.24920173	20	2	-1.34682796
8	3	-.41834560	13	5	-.33851837	17	3	-.93780748	20	3	-1.03949136
8	4	-.13340214	13	6	-.16509435	17	4	-.71944000	20	4	-.82707716
9	1	-1.49178241	13	7	.00000000	17	5	-.54428216	20	5	-.65925175
9	2	-.85423784	14	1	-1.75778013	17	6	-.39291120	20	6	-.51659666
9	3	-.50745647	14	2	-1.13132800	17	7	-.25546156	20	7	-.38945618
9	4	-.24012386	14	3	-.81298281	17	8	-.12592307	20	8	-.27219102
9	5	.00000000	14	4	-.58497028	17	9	.00000000	20	9	-.16167057
10	1	-1.55511142	14	5	-.39782235	18	1	-1.91030783	20	10	-.05333010
10	2	-.92182131	14	6	-.23177122	18	2	-1.28362429			
10	3	-.58390393	14	7	-.07619186	18	3	-.97382126			
10	4	-.32907908	15	1	-1.79952120	18	4	-.75773858			

TABLE I

P = 4.0

N	R	MEAN	N	R	MEAN	N	R	MEAN	N	R	MEAN
2	1	-.54805716	10	5	-.10942298	15	2	-1.18858494	18	5	-.59923790
3	1	-.82208574	11	1	-1.61136655	15	3	-.87429197	18	6	-.44841078
3	2	.00000000	11	2	-.99737815	15	4	-.64792711	18	7	-.31192823
4	1	-1.00513883	11	3	-.66476408	15	5	-.46233855	18	8	-.18407157
4	2	-.27292647	11	4	-.41516273	15	6	-.29859096	18	9	-.06086621
5	1	-1.14270417	11	5	-.20060880	15	7	-.14666204	19	1	-1.93054256
5	2	-.45487745	11	6	.00000000	15	8	.00000000	19	2	-1.33009368
5	3	.00000000	12	1	-1.66223548	16	1	-1.83016001	19	3	-1.02496678
6	1	-1.25299324	12	2	-1.05180831	16	2	-1.22751591	19	4	-.81010490
6	2	-.59125883	12	3	-.72522732	16	3	-.91606812	19	5	-.63811649
6	3	-.18211469	12	4	-.48337437	16	4	-.69326202	19	6	-.49037783
7	1	-1.34511177	12	5	-.27873944	16	5	-.51192239	19	7	-.35748217
7	2	-.70028208	12	6	-.09122591	16	6	-.35325410	19	8	-.23383577
7	3	-.31870070	13	1	-1.70898018	16	7	-.20748573	19	9	-.11564579
7	4	.00000000	13	2	-1.10129918	16	8	-.06846016	19	10	.00000000
8	1	-1.42425985	13	3	-.77960854	17	1	-1.86555305	20	1	-1.96054749
8	2	-.79107518	13	4	-.54395660	17	2	-1.26387130	20	2	-1.36044900
8	3	-.42790277	13	5	-.34706437	17	3	-.95485049	20	3	-1.05689578
8	4	-.13669726	13	6	-.16941955	17	4	-.73508372	20	4	-.84403581
9	1	-1.49368477	13	7	.00000000	17	5	-.55734147	20	5	-.67438129
9	2	-.86886049	14	1	-1.75223410	17	6	-.40291660	20	6	-.52932210
9	3	-.51882663	14	2	-1.14667921	17	7	-.26220620	20	7	-.39950786
9	4	-.24605506	14	3	-.82901900	17	8	-.12931363	20	8	-.27943446
9	5	.00000000	14	4	-.59843683	17	9	.00000000	20	9	-.16543774
10	1	-1.55554942	14	5	-.40775602	18	1	-1.89893999	20	10	-.05478896
10	2	-.93690286	14	6	-.23781939	18	2	-1.29797506			
10	3	-.59669099	14	7	-.07821976	18	3	-.99104122			
10	4	-.33714312	15	1	-1.79249475	18	4	-.77389682			

TABLE I

P = 4.5

N	R	MEAN	N	R	MEAN	N	R	MEAN	N	R	MEAN
2	1	-.55096109	10	5	-.11135741	15	2	-1.19872936	18	5	-.60897785
3	1	-.82644163	11	1	-1.60967639	15	3	-.88568799	18	6	-.45620475
3	2	.00000000	11	2	-1.00782860	15	4	-.65796579	18	7	-.31758356
4	1	-1.00971652	11	3	-.67447332	15	5	-.47019461	18	8	-.18749371
4	2	-.27661697	11	4	-.42207913	15	6	-.30393863	18	9	-.06201121
5	1	-1.14688858	11	5	-.20415525	15	7	-.14936236	19	1	-1.92029306
5	2	-.46102828	11	6	.00000000	15	8	.00000000	19	2	-1.33923506
5	3	.00000000	12	1	-1.65943897	16	1	-1.82301998	19	3	-1.03684903
6	1	-1.25645878	12	2	-1.06228805	16	2	-1.23745286	19	4	-.82166467
6	2	-.59903758	12	3	-.73553138	16	3	-.92766488	19	5	-.64832646
6	3	-.18500967	12	4	-.49129914	16	4	-.70378814	19	6	-.49880173
7	1	-1.34767760	12	5	-.28363910	16	5	-.52049876	19	7	-.36391131
7	2	-.70914584	12	6	-.09287787	16	6	-.35952547	19	8	-.23816456
7	3	-.32376692	13	1	-1.70508166	16	7	-.21129392	19	9	-.11782129
7	4	.00000000	13	2	-1.11172666	16	8	-.06973608	19	10	.00000000
8	1	-1.42582341	13	3	-.79037568	17	1	-1.85736072	20	1	-1.94929353
8	2	-.80065692	13	4	-.55271704	17	2	-1.27356802	20	2	-1.36928413
8	3	-.43461261	13	5	-.35310887	17	3	-.96658920	20	3	-1.06879344
8	4	-.13902412	13	6	-.17248747	17	4	-.74601806	20	4	-.85583071
9	1	-1.49418852	13	7	.00000000	17	5	-.56654088	20	5	-.68500050
9	2	-.87890255	14	1	-1.74724252	17	6	-.40999769	20	6	-.53830433
9	3	-.52679720	14	2	-1.15699051	17	7	-.26699306	20	7	-.40662899
9	4	-.25024341	14	3	-.84014355	17	8	-.13172371	20	8	-.28457847
9	5	.00000000	14	4	-.60789348	17	9	.00000000	20	9	-.16854370
10	1	-1.55496296	14	5	-.41477595	18	1	-1.88971106	20	10	-.05582724
10	2	-.94721855	14	6	-.24210813	18	2	-1.30740495			
10	3	-.60563854	14	7	-.07965993	18	3	-1.00287255			
10	4	-.34283408	15	1	-1.78642203	18	4	-.78517241			

TABLE I

P = 5.0

N	R	MEAN	N	R	MEAN	N	R	MEAN	N	R	MEAN
2	1	-.55299553	10	5	-.11279904	15	2	-1.20596224	18	5	-.61620193
3	1	-.82949330	11	1	-1.60793523	15	3	-.89403571	18	6	-.46200738
3	2	.00000000	11	2	-1.01539157	15	4	-.66539211	18	7	-.32180392
4	1	-1.01288116	11	3	-.68163419	15	5	-.47603630	18	8	-.19005112
4	2	-.27932972	11	4	-.42721695	15	6	-.30792675	18	9	-.06286745
5	1	-1.14971407	11	5	-.20679824	15	7	-.15137927	19	1	-1.91205564
5	2	-.46554953	11	6	.00000000	15	8	.00000000	19	2	-1.34563161
5	3	.00000000	12	1	-1.65685242	16	1	-1.81713238	19	3	-1.04548760
6	1	-1.25870752	12	2	-1.06984615	16	2	-1.24450768	19	4	-.83017452
6	2	-.60474677	12	3	-.74311868	16	3	-.93614414	19	5	-.65589223
6	3	-.18715505	12	4	-.49718072	16	4	-.71156584	19	6	-.50506912
7	1	-1.34921929	12	5	-.28728941	16	5	-.52687093	19	7	-.36870694
7	2	-.71563695	12	6	-.09411060	16	6	-.36420013	19	8	-.24139874
7	3	-.32752134	13	1	-1.70165513	16	7	-.21413779	19	9	-.11944815
7	4	.00000000	13	2	-1.11921988	16	8	-.07068973	19	10	.00000000
8	1	-1.42658554	13	3	-.79829061	17	1	-1.85067683	20	1	-1.94029898
8	2	-.80765549	13	4	-.55921225	17	2	-1.28042120	20	2	-1.37543222
8	3	-.43958133	13	5	-.35760978	17	3	-.97515625	20	3	-1.07742613
8	4	-.14075468	13	6	-.17477684	17	4	-.75408761	20	4	-.86450261
9	1	-1.49413170	13	7	.00000000	17	5	-.57337011	20	5	-.69286217
9	2	-.88621626	14	1	-1.74298460	17	6	-.41527289	20	6	-.54498241
9	3	-.53269278	14	2	-1.16437203	17	7	-.27056672	20	7	-.41193810
9	4	-.25335843	14	3	-.84830699	17	8	-.13352504	20	8	-.28842050
9	5	.00000000	14	4	-.61489723	17	9	.00000000	20	9	-.17086609
10	1	-1.55406762	14	5	-.41999979	18	1	-1.88224385	20	10	-.05660399
10	2	-.95470841	14	6	-.24530776	18	2	-1.31403750			
10	3	-.61224767	14	7	-.08073561	18	3	-1.01149080			
10	4	-.34706469	15	1	-1.78134334	18	4	-.79348351			

TABLE I

P = 5.5

N	R	MEAN	N	R	MEAN	N	R	MEAN	N	R	MEAN
2	1	-.55449618	10	5	-.11391488	15	2	-1.21136813	18	5	-.62177260
3	1	-.83174428	11	1	-1.60631756	15	3	-.90041039	18	6	-.46649512
3	2	.00000000	11	2	-1.02111248	15	4	-.67110714	18	7	-.32507397
4	1	-1.01518999	11	3	-.68713201	15	5	-.48054997	18	8	-.19203485
4	2	-.28140714	11	4	-.43118368	15	6	-.31101525	18	9	-.06353195
5	1	-1.15173451	11	5	-.20884394	15	7	-.15294308	19	1	-1.90535034
5	2	-.46901189	11	6	.00000000	15	8	.00000000	19	2	-1.35033881
5	3	.00000000	12	1	-1.65456937	16	1	-1.81226148	19	3	-1.05204510
6	1	-1.26025868	12	2	-1.07554767	16	2	-1.24976216	19	4	-.83669829
6	2	-.60911366	12	3	-.74893652	16	3	-.94260993	19	5	-.66172218
6	3	-.18880835	12	4	-.50171848	16	4	-.71754573	19	6	-.50991376
7	1	-1.35020292	12	5	-.29011409	16	5	-.53179137	19	7	-.37242140
7	2	-.72059328	12	6	-.09506573	16	6	-.36781890	19	8	-.24390695
7	3	-.33041461	13	1	-1.69871215	16	7	-.21634250	19	9	-.12071071
7	4	.00000000	13	2	-1.12485604	16	8	-.07142953	19	10	.00000000
8	1	-1.42694786	13	3	-.80435163	17	1	-1.84518366	20	1	-1.93300375
8	2	-.81298833	13	4	-.56421947	17	2	-1.28550653	20	2	-1.37993548
8	3	-.44340816	13	5	-.36109125	17	3	-.98167935	20	3	-1.08396884
8	4	-.14209204	13	6	-.17655064	17	4	-.76028596	20	4	-.87114394
9	1	-1.49384427	13	7	.00000000	17	5	-.57863997	20	5	-.69891569
9	2	-.89177659	14	1	-1.73938946	17	6	-.41935474	20	6	-.55014167
9	3	-.53722940	14	2	-1.16990709	17	7	-.27333654	20	7	-.41604864
9	4	-.25576567	14	3	-.85454974	17	8	-.13492245	20	8	-.29139939
9	5	.00000000	14	4	-.62029190	17	9	.00000000	20	9	-.17266829
10	1	-1.55311710	14	5	-.42403840	18	1	-1.87613920	20	10	-.05720700
10	2	-.96038876	14	6	-.24778638	18	2	-1.31893947			
10	3	-.61732792	14	7	-.08156964	18	3	-1.01804297			
10	4	-.35033287	15	1	-1.77710527	18	4	-.79986121			

TABLE I

P = 6.0

N	R	MEAN	N	R	MEAN	N	R	MEAN	N	R	MEAN
2	1	-.55564684	10	5	-.11480412	15	2	-1.21555582	18	5	-.62619873
3	1	-.83347026	11	1	-1.60486576	15	3	-.90543582	18	6	-.47006928
3	2	.00000000	11	2	-1.02558823	15	4	-.67564061	18	7	-.32768218
4	1	-1.01694412	11	3	-.69148501	15	5	-.48414204	18	8	-.19361847
4	2	-.28304866	11	4	-.43433853	15	6	-.31347763	18	9	-.06406265
5	1	-1.15324322	11	5	-.21047421	15	7	-.15419106	19	1	-1.89981141
5	2	-.47174776	11	6	.00000000	15	8	.00000000	19	2	-1.35393798
5	3	.00000000	12	1	-1.65258099	16	1	-1.80819254	19	3	-1.05718962
6	1	-1.26137966	12	2	-1.07999824	16	2	-1.25382076	19	4	-.84185729
6	2	-.61256098	12	3	-.75353820	16	3	-.94770122	19	5	-.66635166
6	3	-.19012132	12	4	-.50532544	16	4	-.72228576	19	6	-.51377053
7	1	-1.35085953	12	5	-.29236471	16	5	-.53570516	19	7	-.37538322
7	2	-.72450045	12	6	-.09582751	16	6	-.37070316	19	8	-.24590896
7	3	-.33271231	13	1	-1.69619231	16	7	-.21810174	19	9	-.12171904
7	4	.00000000	13	2	-1.12924510	16	8	-.07202017	19	10	.00000000
8	1	-1.42709871	13	3	-.80914054	17	1	-1.84061568	20	1	-1.92699281
8	2	-.81718532	13	4	-.56819706	17	2	-1.28942225	20	2	-1.38336496
8	3	-.44644582	13	5	-.36386429	17	3	-.98680960	20	3	-1.08909514
8	4	-.14315647	13	6	-.17796538	17	4	-.76519540	20	4	-.87639165
9	1	-1.49346797	13	7	.00000000	17	5	-.58282944	20	5	-.70371985
9	2	-.89614459	14	1	-1.73634507	17	6	-.42260691	20	6	-.55424708
9	3	-.54082790	14	2	-1.17420648	17	7	-.27554629	20	7	-.41932524
9	4	-.25768164	14	3	-.85947678	17	8	-.13603810	20	8	-.29377661
9	5	.00000000	14	4	-.62457433	17	9	.00000000	20	9	-.17410748
10	1	-1.55220417	14	5	-.42725390	18	1	-1.87108123	20	10	-.05768871
10	2	-.96484219	14	6	-.24976300	18	2	-1.32270131			
10	3	-.62135415	14	7	-.08223523	18	3	-1.02318978			
10	4	-.35293333	15	1	-1.77354430	18	4	-.80490874			

TABLE I

P = 6.5

N	R	MEAN	N	R	MEAN	N	R	MEAN	N	R	MEAN
2	1	-.55655615	10	5	-.11552941	15	2	-1.21889226	18	5	-.62979993
3	1	-.83483422	11	1	-1.60357800	15	3	-.90949839	18	6	-.47298293
3	2	.00000000	11	2	-1.02918383	15	4	-.67932425	18	7	-.32981097
4	1	-1.01831954	11	3	-.69501660	15	5	-.48706848	18	8	-.19491194
4	2	-.28437825	11	4	-.43690748	15	6	-.31548672	18	9	-.06449626
5	1	-1.15440849	11	5	-.21180391	15	7	-.15521010	19	1	-1.89517102
5	2	-.47396375	11	6	.00000000	15	8	.00000000	19	2	-1.35677377
5	3	.00000000	12	1	-1.65085173	16	1	-1.80475555	19	3	-1.06133177
6	1	-1.26221998	12	2	-1.08356692	16	2	-1.25704643	19	4	-.84603852
6	2	-.61535105	12	3	-.75726841	16	3	-.95181305	19	5	-.67011652
6	3	-.19118916	12	4	-.50826117	16	4	-.72613488	19	6	-.51691349
7	1	-1.35131346	12	5	-.29420009	16	5	-.53889239	19	7	-.37780006
7	2	-.72765905	12	6	-.09644927	16	6	-.37305587	19	8	-.24754397
7	3	-.33458103	13	1	-1.69402626	16	7	-.21953813	19	9	-.12254290
7	4	.00000000	13	2	-1.13275743	16	8	-.07250263	19	10	.00000000
8	1	-1.42713344	13	3	-.81301910	17	1	-1.83676989	20	1	-1.92196645
8	2	-.82057360	13	4	-.57143276	17	2	-1.29252611	20	2	-1.38605772
8	3	-.44891540	13	5	-.36612510	17	3	-.99094880	20	3	-1.09321821
8	4	-.14402375	13	6	-.17912006	17	4	-.76917956	20	4	-.88064193
9	1	-1.49306693	13	7	.00000000	17	5	-.58623965	20	5	-.70762487
9	2	-.89966556	14	1	-1.73374830	17	6	-.42525895	20	6	-.55759145
9	3	-.54375173	14	2	-1.17763974	17	7	-.27735024	20	7	-.42199824
9	4	-.25924274	14	3	-.86346357	17	8	-.13694941	20	8	-.29571773
9	5	.00000000	14	4	-.62805605	17	9	.00000000	20	9	-.17528332
10	1	-1.55136035	14	5	-.42987456	18	1	-1.86683432	20	10	-.05808238
10	2	-.96842615	14	6	-.25137607	18	2	-1.32567461			
10	3	-.62462320	14	7	-.08277872	18	3	-1.02733810			
10	4	-.35505164	15	1	-1.77052373	18	4	-.80900231			

TABLE I

P = 7.0

N	R	MEAN	N	R	MEAN	N	R	MEAN	N	R	MEAN
2	1	-.55729227	10	5	-.11613225	15	2	-1.22161126	18	5	-.63278701
3	1	-.83593840	11	1	-1.60243858	15	3	-.91285009	18	6	-.47540364
3	2	.00000000	11	2	-1.03213473	15	4	-.68237633	18	7	-.33158140
4	1	-1.01942554	11	3	-.69793897	15	5	-.48949854	18	8	-.19598830
4	2	-.28547700	11	4	-.43903979	15	6	-.31715712	18	9	-.06485719
5	1	-1.15533317	11	5	-.21290912	15	7	-.15605791	19	1	-1.89123325
5	2	-.47579500	11	6	.00000000	15	8	.00000000	19	2	-1.35906268
5	3	.00000000	12	1	-1.64934289	16	1	-1.80182059	19	3	-1.06473755
6	1	-1.26286877	12	2	-1.08649109	16	2	-1.25966964	19	4	-.84949548
6	2	-.61765520	12	3	-.76035296	16	3	-.95520263	19	5	-.67323810
6	3	-.19207461	12	4	-.51069699	16	4	-.72932241	19	6	-.51952394
7	1	-1.35163605	12	5	-.29572539	16	5	-.54153808	19	7	-.37980964
7	2	-.73026505	12	6	-.09696634	16	6	-.37501155	19	8	-.24890440
7	3	-.33613057	13	1	-1.69215225	16	7	-.22073307	19	9	-.12322867
7	4	.00000000	13	2	-1.13563057	16	8	-.07290413	19	10	.00000000
8	1	-1.42710322	13	3	-.81622396	17	1	-1.83349410	20	1	-1.91770739
8	2	-.82336589	13	4	-.57411629	17	2	-1.29504452	20	2	-1.38822461
8	3	-.45096253	13	5	-.36800355	17	3	-.99435808	20	3	-1.09660524
8	4	-.14474398	13	6	-.18008034	17	4	-.77247718	20	4	-.88415401
9	1	-1.49267068	13	7	.00000000	17	5	-.58906940	20	5	-.71086136
9	2	-.90256351	14	1	-1.73151450	17	6	-.42746289	20	6	-.56036834
9	3	-.54617421	14	2	-1.18044311	17	7	-.27885075	20	7	-.42422033
9	4	-.26053917	14	3	-.86675533	17	8	-.13770781	20	8	-.29733265
9	5	.00000000	14	4	-.63094225	17	9	.00000000	20	9	-.17626202
10	1	-1.55059277	14	5	-.43205140	18	1	-1.86322427	20	10	-.05841012
10	2	-.97137187	14	6	-.25271743	18	2	-1.32808108			
10	3	-.62733010	14	7	-.08323088	18	3	-1.03075196			
10	4	-.35681046	15	1	-1.76793616	18	4	-.81238866			

TABLE I

P = 7.5

N	R	MEAN	N	R	MEAN	N	R	MEAN	N	R	MEAN
2	1	-.55790006	10	5	-.11664123	15	2	-1.22386865	18	5	-.63530461
3	1	-.83685009	11	1	-1.60142879	15	3	-.91566214	18	6	-.47744669
3	2	.00000000	11	2	-1.03459952	15	4	-.68494630	18	7	-.33307690
4	1	-1.02033339	11	3	-.70039708	15	5	-.49154858	18	8	-.19689799
4	2	-.28640016	11	4	-.44083801	15	6	-.31856779	18	9	-.06516231
5	1	-1.15608334	11	5	-.21384225	15	7	-.15677429	19	1	-1.88785330
5	2	-.47733360	11	6	.00000000	15	8	.00000000	19	2	-1.36094712
5	3	.00000000	12	1	-1.64801957	16	1	-1.79928884	19	3	-1.06758673
6	1	-1.26338200	12	2	-1.08893024	16	2	-1.26184353	19	4	-.85240110
6	2	-.61959005	12	3	-.76294594	16	3	-.95804449	19	5	-.67586819
6	3	-.19282070	12	4	-.51275049	16	4	-.73200528	19	6	-.52172660
7	1	-1.35187040	12	5	-.29701307	16	5	-.54376938	19	7	-.38150689
7	2	-.73245158	12	6	-.09740311	16	6	-.37666284	19	8	-.25005406
7	3	-.33743623	13	1	-1.69051923	16	7	-.22174270	19	9	-.12380838
7	4	.00000000	13	2	-1.13802366	16	8	-.07324348	19	10	.00000000
8	1	-1.42703668	13	3	-.81891639	17	1	-1.83067393	20	1	-1.91405590
8	2	-.82570644	13	4	-.57637779	17	2	-1.29712741	20	2	-1.39000385
8	3	-.45268698	13	5	-.36958906	17	3	-.99721442	20	3	-1.09943653
8	4	-.14535164	13	6	-.18089149	17	4	-.77525146	20	4	-.88710454
9	1	-1.49229252	13	7	.00000000	17	5	-.59145519	20	5	-.71358731
9	2	-.90499000	14	1	-1.72957652	17	6	-.42932342	20	6	-.56271084
9	3	-.54821400	14	2	-1.18277445	17	7	-.28011843	20	7	-.42609671
9	4	-.26163295	14	3	-.86951897	17	8	-.13834880	20	8	-.29869722
9	5	.00000000	14	4	-.63337358	17	9	.00000000	20	9	-.17708933
10	1	-1.54989886	14	5	-.43388832	18	1	-1.86012140	20	10	-.05868722
10	2	-.97383544	14	6	-.25385039	18	2	-1.33006708			
10	3	-.62960824	14	7	-.08361296	18	3	-1.03361005			
10	4	-.35829410	15	1	-1.76569851	18	4	-.81523627			

TABLE I

P = 8.0

N	R	MEAN	N	R	MEAN	N	R	MEAN	N	R	MEAN
2	1	-.55841018	10	5	-.11707668	15	2	-1.22577210	18	5	-.63745530
3	1	-.83761526	11	1	-1.60053082	15	3	-.91805495	18	6	-.47919404
3	2	.00000000	11	2	-1.03668883	15	4	-.68713996	18	7	-.33435690
4	1	-1.02109146	11	3	-.70249332	15	5	-.49330126	18	8	-.19767693
4	2	-.28718666	11	4	-.44237493	15	6	-.31977493	18	9	-.06542362
5	1	-1.15670322	11	5	-.21464058	15	7	-.15738761	19	1	-1.88492257
5	2	-.47864444	11	6	.00000000	15	8	.00000000	19	2	-1.36252443
5	3	.00000000	12	1	-1.64685222	16	1	-1.79708472	19	3	-1.07000509
6	1	-1.26379632	12	2	-1.09099538	16	2	-1.26367366	19	4	-.85487738
6	2	-.62123770	12	3	-.76515606	16	3	-.96046123	19	5	-.67811431
6	3	-.19345791	12	4	-.51450509	16	4	-.73429444	19	6	-.52361006
7	1	-1.35204367	12	5	-.29811462	16	5	-.54567653	19	7	-.38295934
7	2	-.73431224	12	6	-.09777694	16	6	-.37807567	19	8	-.25103842
7	3	-.33855135	13	1	-1.68908597	16	7	-.22260703	19	9	-.12430488
7	4	.00000000	13	2	-1.14004726	16	8	-.07353407	19	10	.00000000
8	1	-1.42695041	13	3	-.82121003	17	1	-1.82822265	20	1	-1.91089273
8	2	-.82769651	13	4	-.57830950	17	2	-1.29887788	20	2	-1.39148958
8	3	-.45415944	13	5	-.37094517	17	3	-.99964198	20	3	-1.10183810
8	4	-.14587119	13	6	-.18158574	17	4	-.77761770	20	4	-.88961807
9	1	-1.49193781	13	7	.00000000	17	5	-.59349384	20	5	-.71591461
9	2	-.90705120	14	1	-1.72788157	17	6	-.43091499	20	6	-.56471342
9	3	-.54995509	14	2	-1.18474315	17	7	-.28120357	20	7	-.42770220
9	4	-.26256814	14	3	-.87187196	17	8	-.13889768	20	8	-.29986545
9	5	.00000000	14	4	-.63544964	17	9	.00000000	20	9	-.17779786
10	1	-1.54927246	14	5	-.43545915	18	1	-1.85742793	20	10	-.05892457
10	2	-.97592601	14	6	-.25482000	18	2	-1.33173292			
10	3	-.63155194	14	7	-.08394006	18	3	-1.03603756			
10	4	-.35956244	15	1	-1.76374653	18	4	-.81766410			

TABLE I

P = 8.5

N	R	MEAN	N	R	MEAN	N	R	MEAN	N	R	MEAN
2	1	-.55884429	10	5	-.11745346	15	2	-1.22739838	18	5	-.63931379
3	1	-.83826643	11	1	-1.59972890	15	3	-.92011565	18	6	-.48070555
3	2	.00000000	11	2	-1.03848217	15	4	-.68903426	18	7	-.33546483
4	1	-1.02173367	11	3	-.70430206	15	5	-.49481688	18	8	-.19835141
4	2	-.28786473	11	4	-.44370362	15	6	-.32081962	18	9	-.06564993
5	1	-1.15722344	11	5	-.21533135	15	7	-.15791860	19	1	-1.88235836
5	2	-.47977455	11	6	.00000000	15	8	.00000000	19	2	-1.36386327
5	3	.00000000	12	1	-1.64581642	16	1	-1.79514982	19	3	-1.07208325
6	1	-1.26413661	12	2	-1.09276616	16	2	-1.26523509	19	4	-.85701283
6	2	-.62265761	12	3	-.76706221	16	3	-.96254144	19	5	-.68005479
6	3	-.19400844	12	4	-.51602160	16	4	-.73627056	19	6	-.52523900
7	1	-1.35217359	12	5	-.29906767	16	5	-.54732536	19	7	-.38421640
7	2	-.73591474	12	6	-.09810051	16	6	-.37929820	19	8	-.25189071
7	3	-.33951478	13	1	-1.68781941	16	7	-.22335531	19	9	-.12473487
7	4	.00000000	13	2	-1.14178050	16	8	-.07378569	19	10	.00000000
8	1	-1.42685422	13	3	-.82318728	17	1	-1.82607362	20	1	-1.90812733
8	2	-.82940920	13	4	-.57997863	17	2	-1.30036905	20	2	-1.39274801
8	3	-.45543135	13	5	-.37211830	17	3	-1.00173039	20	3	-1.10390062
8	4	-.14632049	13	6	-.18218666	17	4	-.77965968	20	4	-.89178486
9	1	-1.49160803	13	7	.00000000	17	5	-.59525594	20	5	-.71792470
9	2	-.90882367	14	1	-1.72638803	17	6	-.43229197	20	6	-.56644506
9	3	-.55145858	14	2	-1.18642735	17	7	-.28214295	20	7	-.42909152
9	4	-.26337688	14	3	-.87389937	17	8	-.13937298	20	8	-.30087688
9	5	.00000000	14	4	-.63724296	17	9	.00000000	20	9	-.17841146
10	1	-1.54870647	14	5	-.43681779	18	1	-1.85506915	20	10	-.05913015
10	2	-.97772215	14	6	-.25565921	18	2	-1.33314959			
10	3	-.63322976	14	7	-.08422326	18	3	-1.03812476			
10	4	-.36065916	15	1	-1.76203015	18	4	-.81975850			

TABLE I

P = 9.0

N	R	MEAN	N	R	MEAN	N	R	MEAN	N	R	MEAN
2	1	-.55921812	10	5	-.11778268	15	2	-1.22880364	18	5	-.64093582
3	1	-.83882718	11	1	-1.59900954	15	3	-.92190880	18	6	-.48202592
3	2	.00000000	11	2	-1.04003812	15	4	-.69068654	18	7	-.33643321
4	1	-1.02228446	11	3	-.70587860	15	5	-.49614046	18	8	-.19894114
4	2	-.28845533	11	4	-.44486370	15	6	-.32173257	18	9	-.06584784
5	1	-1.15766586	11	5	-.21593492	15	7	-.15838281	19	1	-1.88009680
5	2	-.48075888	11	6	.00000000	15	8	.00000000	19	2	-1.36501342
5	3	.00000000	12	1	-1.64489212	16	1	-1.79343849	19	3	-1.07388813
6	1	-1.26442025	12	2	-1.09430114	16	2	-1.26658263	19	4	-.85887322
6	2	-.62389390	12	3	-.76872301	16	3	-.96435074	19	5	-.68174799
6	3	-.19448885	12	4	-.51734538	16	4	-.73799371	19	6	-.52666173
7	1	-1.35227208	12	5	-.29990034	16	5	-.54876501	19	7	-.38531499
7	2	-.73730927	12	6	-.09838332	16	6	-.38036644	19	8	-.25263586
7	3	-.34035548	13	1	-1.68669301	16	7	-.22400945	19	9	-.12511089
7	4	.00000000	13	2	-1.14328147	16	8	-.07400571	19	10	.00000000
8	1	-1.42675400	13	3	-.82490932	17	1	-1.82417501	20	1	-1.90568995
8	2	-.83089867	13	4	-.58143527	17	2	-1.30165418	20	2	-1.39382703
8	3	-.45654104	13	5	-.37314313	17	3	-1.00354596	20	3	-1.10569098
8	4	-.14671289	13	6	-.18271188	17	4	-.78143973	20	4	-.89367196
9	1	-1.49130275	13	7	.00000000	17	5	-.59679418	20	5	-.71967825
9	2	-.91036402	14	1	-1.72506291	17	6	-.43349501	20	6	-.56795723
9	3	-.55276996	14	2	-1.18788433	17	7	-.28296407	20	7	-.43030556
9	4	-.26408320	14	3	-.87566434	17	8	-.13978856	20	8	-.30176108
9	5	.00000000	14	4	-.63880758	17	9	.00000000	20	9	-.17894803
10	1	-1.54819396	14	5	-.43800450	18	1	-1.85298715	20	10	-.05930994
10	2	-.97928184	14	6	-.25639267	18	2	-1.33436866			
10	3	-.63469272	14	7	-.08447083	18	3	-1.03993841			
10	4	-.36161687	15	1	-1.76051000	18	4	-.82158370			

TABLE I

P = 9.5

N	R	MEAN	N	R	MEAN	N	R	MEAN	N	R	MEAN
2	1	-.55954336	10	5	-.11807280	15	2	-1.23002987	18	5	-.64236379
3	1	-.83931503	11	1	-1.59836136	15	3	-.92348326	18	6	-.48318925
3	2	.00000000	11	2	-1.04140080	15	4	-.69214036	18	7	-.33728683
4	1	-1.02276193	11	3	-.70726493	15	5	-.49730633	18	8	-.19946113
4	2	-.28897434	11	4	-.44588534	15	6	-.32253723	18	9	-.06602236
5	1	-1.15804644	11	5	-.21646681	15	7	-.15879208	19	1	-1.87808783
5	2	-.48162390	11	6	.00000000	15	8	.00000000	19	2	-1.36601180
5	3	.00000000	12	1	-1.64406291	16	1	-1.79191466	19	3	-1.07547020
6	1	-1.26465973	12	2	-1.09564437	16	2	-1.26775717	19	4	-.86050843
6	2	-.62498000	12	3	-.77018292	16	3	-.96593874	19	5	-.68323834
6	3	-.19491172	12	4	-.51851097	16	4	-.73950951	19	6	-.52791507
7	1	-1.35234738	12	5	-.30063409	16	5	-.55003290	19	7	-.38628331
7	2	-.73853380	12	6	-.09863262	16	6	-.38130787	19	8	-.25329287
7	3	-.34109550	13	1	-1.68568533	16	7	-.22458615	19	9	-.12544249
7	4	.00000000	13	2	-1.14459380	16	8	-.07419971	19	10	.00000000
8	1	-1.42665332	13	3	-.82642252	17	1	-1.82248602	20	1	-1.90352603
8	2	-.83220583	13	4	-.58271757	17	2	-1.30277296	20	2	-1.39476204
8	3	-.45751768	13	5	-.37404611	17	3	-1.00513879	20	3	-1.10725961
8	4	-.14705854	13	6	-.18317486	17	4	-.78300517	20	4	-.89533019
9	1	-1.49102061	13	7	.00000000	17	5	-.59814864	20	5	-.72122140
9	2	-.91171498	14	1	-1.72387982	17	6	-.43455511	20	6	-.56928916
9	3	-.55392384	14	2	-1.18915699	17	7	-.28368795	20	7	-.43137551
9	4	-.26470537	14	3	-.87721468	17	8	-.14015501	20	8	-.30254065
9	5	.00000000	14	4	-.64018462	17	9	.00000000	20	9	-.17942120
10	1	-1.54772858	14	5	-.43904996	18	1	-1.85113646	20	10	-.05946851
10	2	-.98064882	14	6	-.25703917	18	2	-1.33542847			
10	3	-.63597959	14	7	-.08468911	18	3	-1.04152887			
10	4	-.36246042	15	1	-1.75915482	18	4	-.82318841			

TABLE I

P = 10.0

N	R	MEAN	N	R	MEAN	N	R	MEAN	N	R	MEAN
2	1	-.55982885	10	5	-.11833041	15	2	-1.23110910	18	5	-.64363056
3	1	-.83974328	11	1	-1.59777478	15	3	-.92487670	18	6	-.48422197
3	2	.00000000	11	2	-1.04260404	15	4	-.69342944	18	7	-.33804495
4	1	-1.02317970	11	3	-.70849350	15	5	-.49834107	18	8	-.19992307
4	2	-.28943404	11	4	-.44679192	15	6	-.32325177	18	9	-.06617742
5	1	-1.15837710	11	5	-.21693908	15	7	-.15915562	19	1	-1.87629174
5	2	-.48239006	11	6	.00000000	15	8	.00000000	19	2	-1.36688633
5	3	.00000000	12	1	-1.64331525	16	1	-1.79054948	19	3	-1.07686825
6	1	-1.26486419	12	2	-1.09682959	16	2	-1.26878988	19	4	-.86195699
6	2	-.62594169	12	3	-.77147629	16	3	-.96734365	19	5	-.68456021
6	3	-.19528680	12	4	-.51954510	16	4	-.74085324	19	6	-.52902755
7	1	-1.35240528	12	5	-.30128555	16	5	-.55115802	19	7	-.38714322
7	2	-.73961760	12	6	-.09885402	16	6	-.38214380	19	8	-.25387649
7	3	-.34175190	13	1	-1.68477896	16	7	-.22509840	19	9	-.12573710
7	4	.00000000	13	2	-1.14575085	16	8	-.07437206	19	10	.00000000
8	1	-1.42655430	13	3	-.82776267	17	1	-1.82097411	20	1	-1.90159237
8	2	-.83336219	13	4	-.58385504	17	2	-1.30375553	20	2	-1.39557979
8	3	-.45838384	13	5	-.37484774	17	3	-1.00654746	20	3	-1.10864521
8	4	-.14736533	13	6	-.18358604	17	4	-.78439258	20	4	-.89679877
9	1	-1.49075991	13	7	.00000000	17	5	-.59935040	20	5	-.72258986
9	2	-.91290940	14	1	-1.72281748	17	6	-.43549630	20	6	-.57047125
9	3	-.55494696	14	2	-1.19027811	17	7	-.28433088	20	7	-.43232559
9	4	-.26525760	14	3	-.87858725	17	8	-.14048056	20	8	-.30323310
9	5	.00000000	14	4	-.64140587	17	9	.00000000	20	9	-.17984158
10	1	-1.54730471	14	5	-.43997797	18	1	-1.84948093	20	10	-.05960941
10	2	-.98185667	14	6	-.25761331	18	2	-1.33635811			
10	3	-.63712034	14	7	-.08488300	18	3	-1.04293489			
10	4	-.36320907	15	1	-1.75793951	18	4	-.82461030			

TABLE II

P = 2.0

N	R	S	COVARIANCE	N	R	S	COVARIANCE	N	R	S	COVARIANCE	N	R	S	COVARIANCE
1	1	1	1.00000000	7	1	2	.22603474	9	1	1	1.48980591	10	2	2	.20497537
2	1	1	.77202734	7	1	3	.11705431	9	1	2	.25047687	10	2	3	.10523512
2	1	2	.22797266	7	1	4	.07978931	9	1	3	.12667835	10	2	4	.07018908
3	1	1	.86701595	7	1	5	.06264667	9	1	4	.08418666	10	2	5	.05296563
3	1	2	.18997722	7	1	6	.05533950	9	1	5	.06368993	10	2	6	.04311685
3	1	3	.13298405	7	1	7	.06177982	9	1	6	.05230981	10	2	7	.03715460
3	2	2	.24009112	7	2	2	.18812609	9	1	7	.04597915	10	2	8	.03379952
4	1	1	.97778191	7	2	3	.10072143	9	1	8	.04390418	10	2	9	.03314190
4	1	2	.19277492	7	2	4	.06981786	9	1	9	.05210886	10	3	3	.10401825
4	1	3	.11118863	7	2	5	.05538796	9	2	2	.19875359	10	3	4	.07016282
4	1	4	.09820898	7	2	6	.04928383	9	2	3	.10312445	10	3	5	.05331256
4	2	2	.19529511	7	3	3	.10880482	9	2	4	.06940038	10	3	6	.04360446
4	2	3	.12078690	7	3	4	.07684537	9	2	5	.05290135	10	3	7	.03770530
5	1	1	1.08754495	7	3	5	.06170359	9	2	6	.04367050	10	3	8	.03439330
5	1	2	.20235939	7	4	4	.09418982	9	2	7	.03852840	10	4	4	.07535550
5	1	3	.11008100	8	1	1	1.39452438	9	2	8	.03689736	10	4	5	.05770622
5	1	4	.08094564	8	1	2	.23829559	9	3	3	.10409552	10	4	6	.04745572
5	1	5	.08027739	8	1	3	.12175153	9	3	4	.07096412	10	4	7	.04120314
5	2	2	.18544395	8	1	4	.08175200	9	3	5	.05452811	10	5	5	.06569028
5	2	3	.10606605	8	1	5	.06270670	9	3	6	.04526319	10	5	6	.05438108
5	2	4	.08012363	8	1	6	.05270151	9	3	7	.04009883	11	1	1	1.67174604
5	3	3	.13541181	8	1	7	.04871435	9	4	4	.07843333	11	1	2	.27430646
6	1	1	1.19368683	8	1	8	.05630669	9	4	5	.06083435	11	1	3	.13671385
6	1	2	.21389402	8	2	2	.19300192	9	4	6	.05083448	11	1	4	.08962923
6	1	3	.11290101	8	2	3	.10149608	9	5	5	.07218502	11	1	5	.06673372
6	1	4	.07888799	8	2	4	.06913430	10	1	1	1.58211310	11	1	6	.05360672
6	1	5	.06509409	8	2	5	.05349129	10	1	2	.26249245	11	1	7	.04543796
6	1	6	.06927511	8	2	6	.04522416	10	1	3	.13169000	11	1	8	.04028451
6	2	2	.18495693	8	2	7	.04198897	10	1	4	.08685110	11	1	9	.03743941
6	2	3	.10158714	8	3	3	.10537144	10	1	5	.06510101	11	1	10	.03734633
6	2	4	.07245334	8	3	4	.07287515	10	1	6	.05275870	11	1	11	.04603772
6	2	5	.06056952	8	3	5	.05692910	10	1	7	.04531623	11	2	2	.21145015
6	3	3	.11662863	8	3	6	.04845673	10	1	8	.04112182	11	2	3	.10763213
6	3	4	.08524779	8	4	4	.08389342	10	1	9	.04024005	11	2	4	.07128806
7	1	1	1.29593263	8	4	5	.06631295	10	1	10	.04876959	11	2	5	.05340023

TABLE II

P = 2.0

N	R	S	COVARIANCE	N	R	S	COVARIANCE	N	R	S	COVARIANCE	N	R	S	COVARIANCE
11	2	6	.04306907	12	2	4	.07258043	13	1	8	.04045933	13	5	7	.03918601
11	2	7	.03661161	12	2	5	.05406090	13	1	9	.03624101	13	5	8	.03425885
11	2	8	.03253041	12	2	6	.04331162	13	1	10	.03350910	13	5	9	.03088043
11	2	9	.03028579	12	2	7	.03649031	13	1	11	.03220454	13	6	6	.05108666
11	2	10	.03025487	12	2	8	.03198764	13	1	12	.03304386	13	6	7	.04326001
11	3	3	.10466304	12	2	9	.02908352	13	1	13	.04180653	13	6	8	.03792332
11	3	4	.07001927	12	2	10	.02757224	13	2	2	.22471325	13	7	7	.04918481
11	3	5	.05277026	12	2	11	.02796305	13	2	3	.11288240	14	1	1	1.92702964
11	3	6	.04273704	12	3	3	.10576794	13	2	4	.07399707	14	1	2	.30846992
11	3	7	.03643876	12	3	4	.07029202	13	2	5	.05486656	14	1	3	.15155644
11	3	8	.03245175	12	3	5	.05264341	13	2	6	.04373423	14	1	4	.09815488
11	3	9	.03026887	12	3	6	.04233161	13	2	7	.03661269	14	1	5	.07216270
11	4	4	.07361302	12	3	7	.03575983	13	2	8	.03181394	14	1	6	.05711043
11	4	5	.05585142	12	3	8	.03141106	13	2	9	.02853723	14	1	7	.04747634
11	4	6	.04544216	12	3	9	.02860548	13	2	10	.02641627	14	1	8	.04093004
11	4	7	.03887805	12	3	10	.02715535	13	2	11	.02541231	14	1	9	.03634730
11	4	8	.03471656	12	4	4	.07268691	13	2	12	.02609669	14	1	10	.03315728
11	5	5	.06175338	12	4	5	.05475825	13	3	3	.10717800	14	1	11	.03111140
11	5	6	.05051813	12	4	6	.04421016	13	3	4	.07084103	14	1	12	.03026912
11	5	7	.04339794	12	4	7	.03745741	13	3	5	.05278823	14	1	13	.03139253
11	6	6	.05850754	12	4	8	.03297736	13	3	6	.04221811	14	1	14	.04012445
12	1	1	1.75896530	12	4	9	.03008698	13	3	7	.03542843	14	2	2	.23138208
12	1	2	.28590674	12	5	5	.05924932	13	3	8	.03084099	14	2	3	.11562559
12	1	3	.14171163	12	5	6	.04805731	13	3	9	.02770428	14	2	4	.07549485
12	1	4	.09245978	12	5	7	.04085700	13	3	10	.02567537	14	2	5	.05576831
12	1	5	.06848735	12	5	8	.03606688	13	3	11	.02472413	14	2	6	.04427437
12	1	6	.05466869	12	6	6	.05403903	13	4	4	.07229177	14	2	7	.03688762
12	1	7	.04593832	12	6	7	.04613248	13	4	5	.05415337	14	2	8	.03185433
12	1	8	.04019047	13	1	1	1.84399535	13	4	6	.04346537	14	2	9	.02832449
12	1	9	.03648486	13	1	2	.29729281	13	4	7	.03657053	14	2	10	.02586563
12	1	10	.03454487	13	1	3	.14666279	13	4	8	.03189898	14	2	11	.02429077
12	1	11	.03499602	13	1	4	.09530838	13	4	9	.02870035	14	2	12	.02365078
12	1	12	.04375260	13	1	5	.07030750	13	4	10	.02663342	14	2	13	.02454492
12	2	2	.21805409	13	1	6	.05585327	13	5	5	.05762304	14	3	3	.10879630
12	2	3	.11020373	13	1	7	.04664613	13	5	6	.04643617	14	3	4	.07157987

TABLE II

P = 2.0

N	R	S	COVARIANCE	N	R	S	COVARIANCE	N	R	S	COVARIANCE	N	R	S	COVARIANCE
14	3	5	.05311837	15	1	8	.04152217	15	4	7	.03570986	16	1	14	.02724415
14	3	6	.04229951	15	1	9	.03665053	15	4	8	.03079075	16	1	15	.02874080
14	3	7	.03531952	15	1	10	.03314487	15	4	9	.02727806	16	1	16	.03735048
14	3	8	.03055070	15	1	11	.03067958	15	4	10	.02474208	16	2	2	.24464463
14	3	9	.02720063	15	1	12	.02913502	15	4	11	.02295815	16	2	3	.12120366
14	3	10	.02486553	15	1	13	.02863877	15	4	12	.02184802	16	2	4	.07863089
14	3	11	.02337215	15	1	14	.02997436	15	5	5	.05591073	16	2	5	.05774615
14	3	12	.02277365	15	1	15	.03865240	15	5	6	.04463989	16	2	6	.04556850
14	4	4	.07225827	15	2	2	.23803194	15	5	7	.03729333	16	2	7	.03770785
14	4	5	.05387793	15	2	3	.11840553	15	5	8	.03221363	16	2	8	.03229440
14	4	6	.04304318	15	2	4	.07704573	15	5	9	.02857915	16	2	9	.02840956
14	4	7	.03602484	15	2	5	.05673503	15	5	10	.02595228	16	2	10	.02555986
14	4	8	.03121650	15	2	6	.04489396	15	5	11	.02410453	16	2	11	.02347023
14	4	9	.02783275	15	2	7	.03726325	15	6	6	.04764380	16	2	12	.02199856
14	4	10	.02547279	15	2	8	.03202953	15	6	7	.03990748	16	2	13	.02111339
14	4	11	.02396619	15	2	9	.02830571	15	6	8	.03454169	16	2	14	.02094573
14	5	5	.05657026	15	2	10	.02562300	15	6	9	.03069430	16	2	15	.02210607
14	5	6	.04535500	15	2	11	.02373613	15	6	10	.02791024	16	3	3	.11242493
14	5	7	.03805889	15	2	12	.02255640	15	7	7	.04363349	16	3	4	.07342148
14	5	8	.03304534	15	2	13	.02218529	15	7	8	.03785567	16	3	5	.05413372
14	5	9	.02951058	15	2	14	.02323238	15	7	9	.03370303	16	3	6	.04283013
14	5	10	.02704396	15	3	3	.11055956	15	8	8	.04242350	16	3	7	.03550833
14	6	6	.04906206	15	3	4	.07245256	16	1	1	2.08775819	16	3	8	.03045348
14	6	7	.04129270	15	3	5	.05357899	16	1	2	.33023043	16	3	9	.02681955
14	6	8	.03593645	15	3	6	.04251632	16	1	3	.16115278	16	3	10	.02415068
14	6	9	.03215241	15	3	7	.03536103	16	1	4	.10379823	16	3	11	.02219246
14	7	7	.04592295	15	3	8	.03044070	16	1	5	.07590895	16	3	12	.02081376
14	7	8	.04007618	15	3	9	.02693360	16	1	6	.05973602	16	3	13	.01998692
15	1	1	2.00823544	15	3	10	.02440437	16	1	7	.04933511	16	3	14	.01983762
15	1	2	.31944608	15	3	11	.02262525	16	1	8	.04219104	16	4	4	.07288675
15	1	3	.15638716	15	3	12	.02151532	16	1	9	.03707387	16	4	5	.05395645
15	1	4	.10098726	15	3	13	.02117393	16	1	10	.03332499	16	4	6	.04280577
15	1	5	.07403384	15	4	4	.07247996	16	1	11	.03057785	16	4	7	.03555756
15	1	6	.05841075	15	4	5	.05383362	16	1	12	.02864262	16	4	8	.03054086
15	1	7	.04838197	15	4	6	.04284439	16	1	13	.02747531	16	4	9	.02692778

TABLE II

P = 2.0

N	R	S	COVARIANCE	N	R	S	COVARIANCE	N	R	S	COVARIANCE	N	R	S	COVARIANCE
16	4	10	.02427096	17	1	12	.02847209	17	4	5	.05420313	18	1	1	2.24224783
16	4	11	.02232040	17	1	13	.02693627	17	4	6	.04288514	18	1	2	.35126141
16	4	12	.02094761	17	1	14	.02605976	17	4	7	.03552370	18	1	3	.17048856
16	4	13	.02012703	17	1	15	.02603566	17	4	8	.03041646	18	1	4	.10934047
16	5	5	.05553115	17	1	16	.02765609	17	4	9	.02671860	18	1	5	.07964301
16	5	6	.04418323	17	1	17	.03618857	17	4	10	.02396995	18	1	6	.06241848
16	5	7	.03677948	17	2	2	.25120861	17	4	11	.02190606	18	1	7	.05131995
16	5	8	.03164140	17	2	3	.12400737	17	4	12	.02037545	18	1	8	.04366177
16	5	9	.02793373	17	2	4	.08023738	17	4	13	.01930542	18	1	9	.03812641
16	5	10	.02520381	17	2	5	.05878760	17	4	14	.01870190	18	1	10	.03399977
16	5	11	.02319835	17	2	6	.04628163	17	5	5	.05535691	18	1	11	.03086859
16	5	12	.02178763	17	2	7	.03820093	17	5	6	.04391498	18	1	12	.02848644
16	6	6	.04664196	17	2	8	.03262128	17	5	7	.03644723	18	1	13	.02671174
16	6	7	.03891730	17	2	9	.02859545	17	5	8	.03125319	18	1	14	.02548434
16	6	8	.03354097	17	2	10	.02561054	17	5	9	.02748548	18	1	15	.02483653
16	6	9	.02965325	17	2	11	.02337288	17	5	10	.02468115	18	1	16	.02497680
16	6	10	.02678672	17	2	12	.02171433	17	5	11	.02257366	18	1	17	.02669329
16	6	11	.02467959	17	2	13	.02055326	17	5	12	.02101031	18	1	18	.03514342
16	7	7	.04197884	17	2	14	.01989312	17	5	13	.01991835	18	2	2	.25771669
16	7	8	.03625397	17	2	15	.01988249	17	6	6	.04593739	18	2	3	.12680791
16	7	9	.03210463	17	2	16	.02112764	17	6	7	.03820658	18	2	4	.08185609
16	7	10	.02904052	17	3	3	.11436262	17	6	8	.03281505	18	2	5	.05984952
16	8	8	.03993797	17	3	4	.07446049	17	6	9	.02889624	18	2	6	.04702203
16	8	9	.03543529	17	3	5	.05475758	17	6	10	.02597523	18	2	7	.03872868
17	1	1	2.16572505	17	3	6	.04321500	17	6	11	.02377806	18	2	8	.03299222
17	1	2	.34083244	17	3	7	.03573226	17	6	12	.02214781	18	2	9	.02883846
17	1	3	.16585297	17	3	8	.03055329	17	7	7	.04076005	18	2	10	.02573770
17	1	4	.10658347	17	3	9	.02681009	17	7	8	.03507208	18	2	11	.02338271
17	1	5	.07778034	17	3	10	.02403127	17	7	9	.03092873	18	2	12	.02159006
17	1	6	.06107451	17	3	11	.02194643	17	7	10	.02783548	18	2	13	.02025444
17	1	7	.05031824	17	3	12	.02040069	17	7	11	.02550647	18	2	14	.01933158
17	1	8	.04290988	17	3	13	.01931930	17	8	8	.03811330	18	2	15	.01884692
17	1	9	.03757444	17	3	14	.01870681	17	8	9	.03366701	18	2	16	.01895951
17	1	10	.03362376	17	3	15	.01870401	17	8	10	.03034180	18	2	17	.02026866
17	1	11	.03066466	17	4	4	.07343077	17	9	9	.03729583	18	3	3	.11635141

TABLE II

P = 2.0

N	R	S	COVARIANCE	N	R	S	COVARIANCE	N	R	S	COVARIANCE	N	R	S	COVARIANCE
18	3	4	.07555099	18	6	6	.04545167	19	1	16	.02376761	19	3	17	.01688256
18	3	5	.05543290	18	6	7	.03770018	19	1	17	.02404014	19	4	4	.07480508
18	3	6	.04365290	18	6	8	.03228787	19	1	18	.02583169	19	4	5	.05495546
18	3	7	.03601301	18	6	9	.02834117	19	1	19	.03419680	19	4	6	.04328974
18	3	8	.03071664	18	6	10	.02537950	19	2	2	.26416457	19	4	7	.03570034
18	3	9	.02687512	18	6	11	.02312178	19	2	3	.12959923	19	4	8	.03042202
18	3	10	.02400389	18	6	12	.02139948	19	2	4	.08348055	19	4	9	.02657829
18	3	11	.02182131	18	6	13	.02011632	19	2	5	.06092484	19	4	10	.02368892
18	3	12	.02015902	18	7	7	.03985256	19	2	6	.04778166	19	4	11	.02147225
18	3	13	.01892049	18	7	8	.03418726	19	2	7	.03928149	19	4	12	.01975624
18	3	14	.01806551	18	7	9	.03004756	19	2	8	.03339511	19	4	13	.01843557
18	3	15	.01761875	18	7	10	.02693627	19	2	9	.02912239	19	4	14	.01745154
18	3	16	.01772963	18	7	11	.02456192	19	2	10	.02591838	19	4	15	.01678700
18	4	4	.07407833	18	7	12	.02274957	19	2	11	.02346479	19	4	16	.01647963
18	4	5	.05454339	18	8	8	.03674195	19	2	12	.02156774	19	5	5	.05543523
18	4	6	.04305338	18	8	9	.03234070	19	2	13	.02010864	19	5	6	.04376815
18	4	7	.03557830	18	8	10	.02902721	19	2	14	.01902108	19	5	7	.03615482
18	4	8	.03038441	18	8	11	.02649555	19	2	15	.01828473	19	5	8	.03084804
18	4	9	.02661085	18	9	9	.03533908	19	2	16	.01793932	19	5	9	.02697707
18	4	10	.02378686	18	9	10	.03176305	19	2	17	.01814997	19	5	10	.02406349
18	4	11	.02163827	19	1	1	2.31742541	19	2	18	.01950762	19	5	11	.02182611
18	4	12	.02000101	19	1	2	.36152630	19	3	3	.11837587	19	5	12	.02009294
18	4	13	.01878111	19	1	3	.17506104	19	3	4	.07667946	19	5	13	.01875865
18	4	14	.01793988	19	1	4	.11206793	19	3	5	.05614689	19	5	14	.01776471
18	4	15	.01750268	19	1	5	.08149369	19	3	6	.04413090	19	5	15	.01709449
18	5	5	.05533695	19	1	6	.06376269	19	3	7	.03633663	19	6	6	.04513118
18	5	6	.04378770	19	1	7	.05233252	19	3	8	.03092748	19	6	7	.03734723
18	5	7	.03624980	19	1	8	.04443555	19	3	9	.02699485	19	6	8	.03190872
18	5	8	.03099986	19	1	9	.03871335	19	3	10	.02404227	19	6	9	.02793456
18	5	9	.02717881	19	1	10	.03442799	19	3	11	.02177917	19	6	10	.02493921
18	5	10	.02431549	19	1	11	.03114950	19	3	12	.02002832	19	6	11	.02263670
18	5	11	.02213494	19	1	12	.02861637	19	3	13	.01868125	19	6	12	.02085184
18	5	12	.02047243	19	1	13	.02666869	19	3	14	.01767736	19	6	13	.01947734
18	5	13	.01923374	19	1	14	.02521677	19	3	15	.01699850	19	6	14	.01845374
18	5	14	.01838056	19	1	15	.02423247	19	3	16	.01668220	19	7	7	.03917452

TABLE II

P = 2.0

N	R	S	COVARIANCE	N	R	S	COVARIANCE	N	R	S	COVARIANCE	N	R	S	COVARIANCE
19	7	8	.03351983	20	2	2	.27054985	20	4	5	.05542337	20	7	7	.03867003
19	7	9	.02937972	20	2	3	.13237710	20	4	6	.04357909	20	7	8	.03301588
19	7	10	.02625469	20	2	4	.08510605	20	4	7	.03587467	20	7	9	.02887095
19	7	11	.02384984	20	2	5	.06200845	20	4	8	.03051322	20	7	10	.02573330
19	7	12	.02198427	20	2	6	.04855471	20	4	9	.02660285	20	7	11	.02330552
19	7	13	.02054717	20	2	7	.03985249	20	4	10	.02365440	20	7	12	.02140311
19	8	8	.03569422	20	2	8	.03382146	20	4	11	.02137990	20	7	13	.01990931
19	8	9	.03132699	20	2	9	.02943630	20	4	12	.01960156	20	7	14	.01875228
19	8	10	.02802519	20	2	10	.02613768	20	4	13	.01820714	20	8	8	.03488504
19	8	11	.02548125	20	2	11	.02359765	20	4	14	.01712748	20	8	9	.03054202
19	8	12	.02350620	20	2	12	.02161435	20	4	15	.01632642	20	8	10	.02724940
19	9	9	.03385188	20	2	13	.02006056	20	4	16	.01580090	20	8	11	.02469872
19	9	10	.03032144	20	2	14	.01885784	20	4	17	.01559648	20	8	12	.02269829
19	9	11	.02759772	20	2	15	.01796486	20	5	5	.05562563	20	8	13	.02112669
19	10	10	.03327370	20	2	16	.01737714	20	5	6	.04383239	20	9	9	.03269962
20	1	1	2.39134568	20	2	17	.01714386	20	5	7	.03613920	20	9	10	.02920657
20	1	2	.37163555	20	2	18	.01743350	20	5	8	.03077427	20	9	11	.02649714
20	1	3	.17957224	20	2	19	.01882793	20	5	9	.02685510	20	9	12	.02437020
20	1	4	.11476531	20	3	3	.12042463	20	5	10	.02389637	20	10	10	.03169328
20	1	5	.08333030	20	3	4	.07783588	20	5	11	.02161179	20	10	11	.02878362
20	1	6	.06510359	20	3	5	.05689011	20	5	12	.01982432				
20	1	7	.05335067	20	3	6	.04463951	20	5	13	.01842212				
20	1	8	.04522362	20	3	7	.03669304	20	5	14	.01733629				
20	1	9	.03932434	20	3	8	.03117445	20	5	15	.01653099				
20	1	10	.03489242	20	3	9	.02715569	20	5	16	.01600367				
20	1	11	.03148304	20	3	10	.02412909	20	6	6	.04493808				
20	1	12	.02882284	20	3	11	.02179644	20	6	7	.03711209				
20	1	13	.02673971	20	3	12	.01997385	20	6	8	.03164246				
20	1	14	.02512752	20	3	13	.01854535	20	6	9	.02764005				
20	1	15	.02393015	20	3	14	.01743944	20	6	10	.02461452				
20	1	16	.02314082	20	3	15	.01661862	20	6	11	.02227603				
20	1	17	.02282443	20	3	16	.01607923	20	6	12	.02044503				
20	1	18	.02320464	20	3	17	.01586722	20	6	13	.01900801				
20	1	19	.02505511	20	3	18	.01613889	20	6	14	.01789509				
20	1	20	.03333421	20	4	4	.07559306	20	6	15	.01707009				

TABLE II

P = 2.5

N	R	S	COVARIANCE	N	R	S	COVARIANCE	N	R	S	COVARIANCE	N	R	S	COVARIANCE
2	1	1	.72892322	7	1	3	.13151861	9	1	2	.24412439	10	2	3	.12684076
2	1	2	.27107678	7	1	4	.09231832	9	1	3	.13573256	10	2	4	.08816446
3	1	1	.73293439	7	1	5	.07237337	9	1	4	.09392895	10	2	5	.06783627
3	1	2	.22857143	7	1	6	.06172214	9	1	5	.07223621	10	2	6	.05551474
3	1	3	.15277990	7	1	7	.06084587	9	1	6	.05930282	10	2	7	.04749044
3	2	2	.31428571	7	2	2	.22316051	9	1	7	.05120350	10	2	8	.04224292
4	1	1	.76460249	7	2	3	.12930060	9	1	8	.04672730	10	2	9	.03945927
4	1	2	.22232669	7	2	4	.09194224	9	1	9	.04866141	10	3	3	.13251193
4	1	3	.13376080	7	2	5	.07266011	9	2	2	.22356408	10	3	4	.09291247
4	1	4	.10788145	7	2	6	.06232021	9	2	3	.12680523	10	3	5	.07187510
4	2	2	.25442476	7	3	3	.14629467	9	2	4	.08862663	10	3	6	.05903741
4	2	3	.16091633	7	3	4	.10550540	9	2	5	.06856847	10	3	7	.05064166
5	1	1	.80196347	7	3	5	.08414724	9	2	6	.05652086	10	3	8	.04514224
5	1	2	.22369843	7	4	4	.13070212	9	2	7	.04894751	10	4	4	.10258503
5	1	3	.13021681	8	1	1	.91394901	9	2	8	.04477473	10	4	5	.07983667
5	1	4	.09515949	8	1	2	.23837045	9	3	3	.13503216	10	4	6	.06585408
5	1	5	.08462613	8	1	3	.13346878	9	3	4	.09531734	10	4	7	.05666809
5	2	2	.23431897	8	1	4	.09291144	9	3	5	.07420289	10	5	5	.09194365
5	2	3	.14151146	8	1	5	.07196688	9	3	6	.06142879	10	5	6	.07623000
5	2	4	.10551145	8	1	6	.05978256	9	3	7	.05336974	11	1	1	1.01569761
5	3	3	.18481518	8	1	7	.05302650	9	4	4	.10794563	11	1	2	.25569559
6	1	1	.84011087	8	1	8	.05391813	9	4	5	.08463991	11	1	3	.14065952
6	1	2	.22770696	8	2	2	.22265438	9	4	6	.07042832	11	1	4	.09653866
6	1	3	.13019289	8	2	3	.12746986	9	5	5	.10105756	11	1	5	.07361152
6	1	4	.09260827	8	2	4	.08973305	10	1	1	.98296398	11	1	6	.05975480
6	1	5	.07451432	8	2	5	.06998125	10	1	2	.24992411	11	1	7	.05064362
6	1	6	.07043112	8	2	6	.05840668	10	1	3	.13815688	11	1	8	.04441499
6	2	2	.22620819	8	2	7	.05199046	10	1	4	.09517149	11	1	9	.04026163
6	2	3	.13323988	8	3	3	.13921924	10	1	5	.07283447	11	1	10	.03819220
6	2	4	.09626203	8	3	4	.09914763	10	1	6	.05939012	11	1	11	.04117590
6	2	5	.07823246	8	3	5	.07789440	10	1	7	.05067268	11	2	2	.22759500
6	3	3	.15898488	8	3	6	.06535040	10	1	8	.04498235	11	2	3	.12732103
6	3	4	.11698378	8	4	4	.11635887	10	1	9	.04194795	11	2	4	.08810676
7	1	1	.87759174	8	4	5	.09224052	10	1	10	.04452462	11	2	5	.06751282
7	1	2	.23280356	9	1	1	.94906843	10	2	2	.22531714	11	2	6	.05498390

TABLE II

P = 2.5

N	R	S	COVARIANCE	N	R	S	COVARIANCE	N	R	S	COVARIANCE	N	R	S	COVARIANCE
11	2	7	.04671007	12	2	5	.06745259	13	1	9	.03950135	13	5	8	.04772961
11	2	8	.04103885	12	2	6	.05474149	13	1	10	.03594165	13	5	9	.04265644
11	2	9	.03725485	12	2	7	.04629329	13	1	11	.03356135	13	6	6	.07165767
11	2	10	.03538338	12	2	8	.04039261	13	1	12	.03265354	13	6	7	.06094868
11	3	3	.13103174	12	2	9	.03621109	13	1	13	.03606576	13	6	8	.05331415
11	3	4	.09138519	12	2	10	.03340370	13	2	2	.23303887	13	7	7	.06949672
11	3	5	.07036042	12	2	11	.03215870	13	2	3	.12906919	14	1	1	1.10770601
11	3	6	.05748967	12	3	3	.13023797	13	2	4	.08870785	14	1	2	.27253011
11	3	7	.04895492	12	3	4	.09043286	13	2	5	.06757018	14	1	3	.14825681
11	3	8	.04309008	12	3	5	.06936821	13	2	6	.05468714	14	1	4	.10093720
11	3	9	.03917517	12	3	6	.05646070	13	2	7	.04609732	14	1	5	.07640785
11	4	4	.09899390	12	3	7	.04784841	13	2	8	.04004348	14	1	6	.06154193
11	4	5	.07661587	12	3	8	.04181718	13	2	9	.03564921	14	1	7	.05165118
11	4	6	.06282680	12	3	9	.03753673	13	2	10	.03246776	14	1	8	.04466583
11	4	7	.05364286	12	3	10	.03466395	13	2	11	.03034217	14	1	9	.03954586
11	4	8	.04731510	12	4	4	.09651510	13	2	12	.02954287	14	1	10	.03573210
11	5	5	.08600540	12	4	5	.07437439	13	3	3	.12991666	14	1	11	.03293637
11	5	6	.07082458	12	4	6	.06072678	13	3	4	.08987496	14	1	12	.03109313
11	5	7	.06066400	12	4	7	.05158336	13	3	5	.06873012	14	1	13	.03054722
11	6	6	.08236072	12	4	8	.04516232	13	3	6	.05577399	14	1	14	.03406445
12	1	1	1.04734683	12	4	9	.04059817	13	3	7	.04710377	14	2	2	.23600986
12	1	2	.26139928	12	5	5	.08190648	13	3	8	.04097748	14	2	3	.13018151
12	1	3	.14319367	12	5	6	.06711753	13	3	9	.03652287	14	2	4	.08922792
12	1	4	.09797498	12	5	7	.05716480	13	3	10	.03329502	14	2	5	.06781207
12	1	5	.07449148	12	5	8	.05015394	13	3	11	.03114047	14	2	6	.05476150
12	1	6	.06027230	12	6	6	.07606050	13	4	4	.09477791	14	2	7	.04604594
12	1	7	.05086093	12	6	7	.06499014	13	4	5	.07277905	14	2	8	.03987413
12	1	8	.04430587	13	1	1	1.07799105	13	4	6	.05922634	14	2	9	.03534184
12	1	9	.03966811	13	1	2	.26701426	13	4	7	.05012254	14	2	10	.03196162
12	1	10	.03655373	13	1	3	.14573168	13	4	8	.04367259	14	2	11	.02948255
12	1	11	.03515858	13	1	4	.09944784	13	4	9	.03897414	14	2	12	.02785032
12	1	12	.03840348	13	1	5	.07543200	13	4	10	.03556675	14	2	13	.02737716
12	2	2	.23020833	13	1	6	.06087973	13	5	5	.07896482	14	3	3	.12993121
12	2	3	.12809522	13	1	7	.05121517	13	5	6	.06446215	14	3	4	.08959739
12	2	4	.08831600	13	1	8	.04442265	13	5	7	.05468029	14	3	5	.06834186

TABLE II

P = 2.5

N	R	S	COVARIANCE	N	R	S	COVARIANCE	N	R	S	COVARIANCE	N	R	S	COVARIANCE
14	3	6	.05532457	15	1	9	.03970930	15	4	8	.04181222	16	1	15	.02719476
14	3	7	.04660135	15	1	10	.03572271	15	4	9	.03701911	16	1	16	.03080792
14	3	8	.04040889	15	1	11	.03269439	15	4	10	.03338347	16	2	2	.24218394
14	3	9	.03585339	15	1	12	.03045950	15	4	11	.03061522	16	2	3	.13266813
14	3	10	.03245194	15	1	13	.02902043	15	4	12	.02857182	16	2	4	.09051811
14	3	11	.02995631	15	1	14	.02874924	15	5	5	.07516909	16	2	5	.06853925
14	3	12	.02831551	15	1	15	.03232911	15	5	6	.06102375	16	2	6	.05515888
14	4	4	.09355770	15	2	2	.23906995	15	5	7	.05147561	16	2	7	.04621344
14	4	5	.07163093	15	2	3	.13139075	15	5	8	.04464680	16	2	8	.03985206
14	4	6	.05813532	15	2	4	.08983976	15	5	9	.03957313	16	2	9	.03513367
14	4	7	.04905988	15	2	5	.06814311	15	5	10	.03571947	16	2	10	.03153539
14	4	8	.04260103	15	2	6	.05492706	15	5	11	.03278285	16	2	11	.02875188
14	4	9	.03784090	15	2	7	.04609414	15	6	6	.06603432	16	2	12	.02660719
14	4	10	.03428243	15	2	8	.03982248	15	6	7	.05581726	16	2	13	.02502264
14	4	11	.03167066	15	2	9	.03518771	15	6	8	.04848979	16	2	14	.02403827
14	5	5	.07679649	15	2	10	.03168095	15	6	9	.04303433	16	2	15	.02399569
14	5	6	.06250180	15	2	11	.02901491	15	6	10	.03888452	16	3	3	.13063213
14	5	7	.05285301	15	2	12	.02704715	15	7	7	.06149216	16	3	4	.08960502
14	5	8	.04596736	15	2	13	.02578241	15	7	8	.05351851	16	3	5	.06806461
14	5	9	.04088268	15	2	14	.02555347	15	7	9	.04756814	16	3	6	.05489351
14	5	10	.03707676	15	3	3	.13019063	15	8	8	.06010592	16	3	7	.04606114
14	6	6	.06844752	15	3	4	.08952468	16	1	1	1.16462139	16	3	8	.03976622
14	6	7	.05801648	15	3	5	.06813543	16	1	2	.28325055	16	3	9	.03508938
14	6	8	.05055005	15	3	6	.05504586	16	1	3	.15323165	16	3	10	.03151837
14	6	9	.04502441	15	3	7	.04626921	16	1	4	.10391926	16	3	11	.02875353
14	7	7	.06488260	15	3	8	.04002290	16	1	5	.07840797	16	3	12	.02662221
14	7	8	.05665450	15	3	9	.03539891	16	1	6	.06295404	16	3	13	.02504781
15	1	1	1.13656160	15	3	10	.03189599	16	1	7	.05265740	16	3	14	.02407198
15	1	2	.27794247	15	3	11	.02923084	16	1	8	.04535310	16	4	4	.09214048
15	1	3	.15075885	15	3	12	.02726360	16	1	9	.03994518	16	4	5	.07021485
15	1	4	.10243040	15	3	13	.02600153	16	1	10	.03582663	16	4	6	.05675012
15	1	5	.07740325	15	4	4	.09271063	16	1	11	.03264371	16	4	7	.04769336
15	1	6	.06223780	15	4	5	.07080406	16	1	12	.03019258	16	4	8	.04122411
15	1	7	.05213811	15	4	6	.05733565	16	1	13	.02838136	16	4	9	.03640966
15	1	8	.04498533	15	4	7	.04827559	16	1	14	.02725358	16	4	10	.03272895

TABLE II

P = 2.5

N	R	S	COVARIANCE	N	R	S	COVARIANCE	N	R	S	COVARIANCE	N	R	S	COVARIANCE
16	4	11	.02987665	17	1	13	.02809505	17	4	6	.05632610	18	1	2	.29356038
16	4	12	.02767686	17	1	14	.02661152	17	4	7	.04726223	18	1	3	.15807668
16	4	13	.02605225	17	1	15	.02572819	17	4	8	.04078371	18	1	4	.10686453
16	5	5	.07393390	17	1	16	.02583604	17	4	9	.03595229	18	1	5	.08042054
16	5	6	.05989335	17	1	17	.02946186	17	4	10	.03224154	18	1	6	.06441558
16	5	7	.05041930	17	2	2	.24532729	17	4	11	.02933826	18	1	7	.05374996
16	5	8	.04363605	17	2	3	.13399323	17	4	12	.02705197	18	1	8	.04617201
16	5	9	.03857894	17	2	4	.09124504	17	4	13	.02527369	18	1	9	.04054028
16	5	10	.03470755	17	2	5	.06898359	17	4	14	.02396566	18	1	10	.03621835
16	5	11	.03170466	17	2	6	.05543977	17	5	5	.07299140	18	1	11	.03282716
16	5	12	.02938760	17	2	7	.04638481	17	5	6	.05902127	18	1	12	.03013174
16	6	6	.06417899	17	2	8	.03993985	17	5	7	.04959987	18	1	13	.02798638
16	6	7	.05412694	17	2	9	.03514871	17	5	8	.04285103	18	1	14	.02631003
16	6	8	.04691154	17	2	10	.03147783	17	5	9	.03780960	18	1	15	.02508502
16	6	9	.04152185	17	2	11	.02861096	17	5	10	.03393258	18	1	16	.02439683
16	6	10	.03738983	17	2	12	.02635616	17	5	11	.03089628	18	1	17	.02463716
16	6	11	.03418150	17	2	13	.02460331	17	5	12	.02850368	18	1	18	.02826101
16	7	7	.05891868	17	2	14	.02331297	17	5	13	.02664223	18	2	2	.24848261
16	7	8	.05114719	17	2	15	.02254685	17	6	6	.06272869	18	2	3	.13535125
16	7	9	.04532964	17	2	16	.02264871	17	6	7	.05280384	18	2	4	.09200754
16	7	10	.04086238	17	3	3	.13121120	17	6	8	.04567768	18	2	5	.06946393
16	8	8	.05658141	17	3	4	.08980179	17	6	9	.04034480	18	2	6	.05575751
16	8	9	.05022197	17	3	5	.06809698	17	6	10	.03623795	18	2	7	.04659511
17	1	1	1.19194274	17	3	6	.05483688	17	6	11	.03301831	18	2	8	.04007048
17	1	2	.28845568	17	3	7	.04594601	17	6	12	.03047944	18	2	9	.03521329
17	1	3	.15567158	17	3	8	.03960449	17	7	7	.05691690	18	2	10	.03148091
17	1	4	.10539846	17	3	9	.03488276	17	7	8	.04930658	18	2	11	.02854937
17	1	5	.07941525	17	3	10	.03126068	17	7	9	.04360011	18	2	12	.02621756
17	1	6	.06368192	17	3	11	.02842937	17	7	10	.03919884	18	2	13	.02436072
17	1	7	.05319727	17	3	12	.02620114	17	7	11	.03574440	18	2	14	.02290958
17	1	8	.04575225	17	3	13	.02446848	17	8	8	.05389193	18	2	15	.02184974
17	1	9	.04022741	17	3	14	.02319348	17	8	9	.04771776	18	2	16	.02125641
17	1	10	.03600004	17	3	15	.02243854	17	8	10	.04294761	18	2	17	.02147171
17	1	11	.03270185	17	4	4	.09178057	17	9	9	.05294967	18	3	3	.13189548
17	1	12	.03010959	17	4	5	.06980576	18	1	1	1.21857730	18	3	4	.09008839

TABLE II

P = 2.5

N	R	S	COVARIANCE	N	R	S	COVARIANCE	N	R	S	COVARIANCE	N	R	S	COVARIANCE
18	3	5	.06820916	18	6	7	.05175400	19	1	17	.02322379	19	4	4	.09151375
18	3	6	.05485407	18	6	8	.04469820	19	1	18	.02357054	19	4	5	.06937545
18	3	7	.04590205	18	6	9	.03941252	19	1	19	.02718197	19	4	6	.05582462
18	3	8	.03951442	18	6	10	.03533102	19	2	2	.25163749	19	4	7	.04672159
18	3	9	.03475200	18	6	11	.03211316	19	2	3	.13673133	19	4	8	.04021304
18	3	10	.03108815	18	6	12	.02954642	19	2	4	.09279598	19	4	9	.03534949
18	3	11	.02820784	18	6	13	.02749882	19	2	5	.06997128	19	4	10	.03159669
18	3	12	.02591524	18	7	7	.05533027	19	2	6	.05610317	19	4	11	.02863324
18	3	13	.02408883	18	7	8	.04784894	19	2	7	.04683491	19	4	12	.02625685
18	3	14	.02266131	18	7	9	.04223425	19	2	8	.04023331	19	4	13	.02433756
18	3	15	.02161927	18	7	10	.03789249	19	2	9	.03531446	19	4	14	.02279431
18	3	16	.02103785	18	7	11	.03446563	19	2	10	.03152751	19	4	15	.02158554
18	4	4	.09158327	18	7	12	.03172993	19	2	11	.02854237	19	4	16	.02071448
18	4	5	.06953590	18	8	8	.05178636	19	2	12	.02615184	19	5	5	.07172798
18	4	6	.05602655	18	8	9	.04576309	19	2	13	.02422307	19	5	6	.05782357
18	4	7	.04694676	18	8	10	.04109800	19	2	14	.02267310	19	5	7	.04845879
18	4	8	.04045510	18	8	11	.03741139	19	2	15	.02145904	19	5	8	.04175019
18	4	9	.03560780	18	9	9	.05016979	19	2	16	.02058293	19	5	9	.03672971
18	4	10	.03187429	18	9	10	.04510558	19	2	17	.02012842	19	5	10	.03285132
18	4	11	.02893655	19	1	1	1.24457156	19	2	18	.02043376	19	5	11	.02978590
18	4	12	.02659667	19	1	2	.29856784	19	3	3	.13266098	19	5	12	.02732598
18	4	13	.02473178	19	1	3	.16044603	19	3	4	.09044514	19	5	13	.02533821
18	4	14	.02327404	19	1	4	.10831525	19	3	5	.06838388	19	5	14	.02373942
18	4	15	.02221056	19	1	5	.08142074	19	3	6	.05492909	19	5	15	.02248719
18	5	5	.07227254	19	1	6	.06515100	19	3	7	.04591353	19	6	6	.06066577
18	5	6	.05834594	19	1	7	.05431019	19	3	8	.03947963	19	6	7	.05091271
18	5	7	.04895986	19	1	8	.04660527	19	3	9	.03467875	19	6	8	.04391177
18	5	8	.04223546	19	1	9	.04087374	19	3	10	.03097846	19	6	9	.03866430
18	5	9	.03720646	19	1	10	.03646670	19	3	11	.02805905	19	6	10	.03460555
18	5	10	.03332826	19	1	11	.03299620	19	3	12	.02571955	19	6	11	.03139444
18	5	11	.03027379	19	1	12	.03021911	19	3	13	.02383101	19	6	12	.02881566
18	5	12	.02783922	19	1	13	.02797969	19	3	14	.02231292	19	6	13	.02673070
18	5	13	.02589799	19	1	14	.02618069	19	3	15	.02112385	19	6	14	.02505320
18	5	14	.02438043	19	1	15	.02477158	19	3	16	.02026634	19	7	7	.05405421
18	6	6	.06158124	19	1	16	.02375398	19	3	17	.01982330	19	7	8	.04667641

TABLE II

P = 2.5

N	R	S	COVARIANCE
19	7	9	.04113711
19	7	10	.03684692
19	7	11	.03344911
19	7	12	.03071815
19	7	13	.02850883
19	8	8	.05010464
19	8	9	.04420480
19	8	10	.03962866
19	8	11	.03600017
19	8	12	.03308115
19	9	9	.04798507
19	9	10	.04305970
19	9	11	.03914922
19	10	10	.04731550
20	1	1	1.26996740
20	1	2	.30348162
20	1	3	.16277945
20	1	4	.10974923
20	1	5	.08241372
20	1	6	.06588535
20	1	7	.05487429
20	1	8	.04704714
20	1	9	.04122110
20	1	10	.03673559
20	1	11	.03319480
20	1	12	.03034925
20	1	13	.02803658
20	1	14	.02615056
20	1	15	.02462543
20	1	16	.02343115
20	1	17	.02258237
20	1	18	.02218167
20	1	19	.02261467
20	1	20	.02620621
20	2	2	.25478296

N	R	S	COVARIANCE
20	2	3	.13812538
20	2	4	.09360317
20	2	5	.07049891
20	2	6	.05647014
20	2	7	.04709731
20	2	8	.04042072
20	2	9	.03544321
20	2	10	.03160627
20	2	11	.02857452
20	2	12	.02613620
20	2	13	.02415334
20	2	14	.02253561
20	2	15	.02122714
20	2	16	.02020263
20	2	17	.01947514
20	2	18	.01913357
20	2	19	.01951105
20	3	3	.13348958
20	3	4	.09085718
20	3	5	.06860816
20	3	6	.05504995
20	3	7	.04596886
20	3	8	.03948823
20	3	9	.03465008
20	3	10	.03091651
20	3	11	.02796389
20	3	12	.02558761
20	3	13	.02365418
20	3	14	.02207619
20	3	15	.02079963
20	3	16	.01980020
20	3	17	.01909112
20	3	18	.01875988
20	4	4	.09154596
20	4	5	.06930220

N	R	S	COVARIANCE
20	4	6	.05570039
20	4	7	.04656806
20	4	8	.04003921
20	4	9	.03515840
20	4	10	.03138787
20	4	11	.02840349
20	4	12	.02600003
20	4	13	.02404347
20	4	14	.02244602
20	4	15	.02115347
20	4	16	.02014166
20	4	17	.01942440
20	5	5	.07132151
20	5	6	.05742220
20	5	7	.04806729
20	5	8	.04136714
20	5	9	.03635125
20	5	10	.03247209
20	5	11	.02939905
20	5	12	.02692245
20	5	13	.02490527
20	5	14	.02325770
20	5	15	.02192436
20	5	16	.02088074
20	6	6	.05993164
20	6	7	.05023397
20	6	8	.04327513
20	6	9	.03805790
20	6	10	.03401835
20	6	11	.03081526
20	6	12	.02823195
20	6	13	.02612665
20	6	14	.02440642
20	6	15	.02301402
20	7	7	.05301629

N	R	S	COVARIANCE
20	7	8	.04572163
20	7	9	.04024403
20	7	10	.03599757
20	7	11	.03262706
20	7	12	.02990655
20	7	13	.02768808
20	7	14	.02587459
20	8	8	.04873999
20	8	9	.04294157
20	8	10	.03844028
20	8	11	.03486360
20	8	12	.03197417
20	8	13	.02961636
20	9	9	.04623208
20	9	10	.04142204
20	9	11	.03759541
20	9	12	.03450106
20	10	10	.04506694
20	10	11	.04093790

****

TABLE II

P = 3.0

N	R	S	COVARIANCE
1	1	1	1.00000000
2	1	1	.71268877
2	1	2	.28731123
3	1	1	.67862187
3	1	2	.24380410
3	1	3	.15884206
3	2	2	.34985574
4	1	1	.67877158
4	1	2	.23200922
4	1	3	.14231626
4	1	4	.10943900
4	2	2	.28259954
4	2	3	.18053891
5	1	1	.68872640
5	1	2	.22815060
5	1	3	.13729891
5	1	4	.09990720
5	1	5	.08389781
5	2	2	.25620372
5	2	3	.15891375
5	2	4	.11758135
5	3	3	.21009959
6	1	1	.70221376
6	1	2	.22743107
6	1	3	.13536340
6	1	4	.09712946
6	1	5	.07705816
6	1	6	.06840676
6	2	2	.24307202
6	2	3	.14827821
6	2	4	.10779209
6	2	5	.08624094
6	3	3	.18062925
6	3	4	.13333248
7	1	1	.71702217

N	R	S	COVARIANCE
7	1	2	.22821561
7	1	3	.13476826
7	1	4	.09603707
7	1	5	.07519128
7	1	6	.06289039
7	1	7	.05802896
7	2	2	.23583274
7	2	3	.14223150
7	2	4	.10246563
7	2	5	.08077290
7	2	6	.06788712
7	3	3	.16500319
7	3	4	.12028820
7	3	5	.09555869
7	4	4	.14989093
8	1	1	.73223159
8	1	2	.22979945
8	1	3	.13487329
8	1	4	.09567875
8	1	5	.07447647
8	1	6	.06145506
8	1	7	.05328142
8	1	8	.05059086
8	2	2	.23169082
8	2	3	.13853680
8	2	4	.09921304
8	2	5	.07767836
8	2	6	.06435519
8	2	7	.05596650
8	3	3	.15554318
8	3	4	.11248688
8	3	5	.08862149
8	3	6	.07374691
8	4	4	.13341055
8	4	5	.10590719

N	R	S	COVARIANCE
9	1	1	.74742243
9	1	2	.23183348
9	1	3	.13538143
9	1	4	.09571335
9	1	5	.07424612
9	1	6	.06092205
9	1	7	.05207616
9	1	8	.04634648
9	1	9	.04499378
9	2	2	.22935771
9	2	3	.13620237
9	2	4	.09711114
9	2	5	.07571828
9	2	6	.06234730
9	2	7	.05343198
9	2	8	.04765109
9	3	3	.14935529
9	3	4	.10738624
9	3	5	.08417169
9	3	6	.06956244
9	3	7	.05978010
9	4	4	.12324746
9	4	5	.09719525
9	4	6	.08067573
9	5	5	.11644973
10	1	1	.76239500
10	1	2	.23412810
10	1	3	.13613551
10	1	4	.09598145
10	1	5	.07427269
10	1	6	.06075429
10	1	7	.05163621
10	1	8	.04527639
10	1	9	.04110824
10	1	10	.04062474

N	R	S	COVARIANCE
10	2	2	.22815936
10	2	3	.13472034
10	2	4	.09571598
10	2	5	.07441102
10	2	6	.06105755
10	2	7	.05201181
10	2	8	.04568602
10	2	9	.04154075
10	3	3	.14510644
10	3	4	.10386002
10	3	5	.08111357
10	3	6	.06676770
10	3	7	.05700898
10	3	8	.05016745
10	4	4	.11643136
10	4	5	.09139872
10	4	6	.07550494
10	4	7	.06464387
10	5	5	.10593535
10	5	6	.08789660
11	1	1	.77705482
11	1	2	.23657338
11	1	3	.13704503
11	1	4	.09639704
11	1	5	.07445257
11	1	6	.06077589
11	1	7	.05149738
11	1	8	.04488642
11	1	9	.04012452
11	1	10	.03701174
11	1	11	.03711556
11	2	2	.22771047
11	2	3	.13380387
11	2	4	.09478516
11	2	5	.07351867

TABLE II

P = 3.0

N	R	S	COVARIANCE	N	R	S	COVARIANCE	N	R	S	COVARIANCE	N	R	S	COVARIANCE
11	2	6	.06018408	12	2	4	.09417457	13	1	8	.04476336	13	5	7	.06288662
11	2	7	.05110039	12	2	5	.07290726	13	1	9	.03963866	13	5	8	.05484008
11	2	8	.04461009	12	2	6	.05957903	13	1	10	.03574146	13	5	9	.04877436
11	2	9	.03992775	12	2	7	.05048075	13	1	11	.03283856	13	6	6	.08265840
11	2	10	.03686997	12	2	8	.04392653	13	1	12	.03101428	13	6	7	.07047422
11	3	3	.14209742	12	2	9	.03906657	13	1	13	.03181778	13	6	8	.06156887
11	3	4	.10133088	12	2	10	.03548905	13	2	2	.22820431	13	7	7	.08048896
11	3	5	.07891769	12	2	11	.03318755	13	2	3	.13303409	14	1	1	.81887227
11	3	6	.06478429	12	3	3	.13992638	13	2	4	.09379303	14	1	2	.24425939
11	3	7	.05511873	12	3	4	.09947188	13	2	5	.07249477	14	1	3	.14024399
11	3	8	.04819414	12	3	5	.07729393	13	2	6	.05915875	14	1	4	.09810773
11	3	9	.04319111	12	3	6	.06332258	13	2	7	.05004932	14	1	5	.07546129
11	4	4	.11160023	12	3	7	.05375067	13	2	8	.04346373	14	1	6	.06136632
11	4	5	.08730192	12	3	8	.04683736	13	2	9	.03852756	14	1	7	.05177198
11	4	6	.07188817	12	3	9	.04170204	13	2	10	.03476892	14	1	8	.04484225
11	4	7	.06130289	12	3	10	.03791869	13	2	11	.03196795	14	1	9	.03963209
11	4	8	.05369753	12	4	4	.10804159	13	2	12	.03021150	14	1	10	.03561654
11	5	5	.09882537	12	4	5	.08428314	13	3	3	.13834618	14	1	11	.03250228
11	5	6	.08167097	12	4	6	.06923553	13	3	4	.09808395	14	1	12	.03016723
11	5	7	.06983477	12	4	7	.05888704	13	3	5	.07606918	14	1	13	.02875110
11	6	6	.09519097	12	4	8	.05139234	13	3	6	.06221808	14	1	14	.02976508
12	1	1	.79136065	12	4	9	.04581467	13	3	7	.05272494	14	2	2	.22889496
12	1	2	.23910255	12	5	5	.09373107	13	3	8	.04584513	14	2	3	.13299479
12	1	3	.13805458	12	5	6	.07723398	13	3	9	.04067922	14	2	4	.09358005
12	1	4	.09690936	12	5	7	.06584099	13	3	10	.03674095	14	2	5	.07222787
12	1	5	.07473040	12	5	8	.05756461	13	3	11	.03380497	14	2	6	.05887141
12	1	6	.06090930	12	6	6	.08790351	13	4	4	.10534650	14	2	7	.04974816
12	1	7	.05151112	12	6	7	.07514325	13	4	5	.08199075	14	2	8	.04314208
12	1	8	.04475944	13	1	1	.80529926	13	4	6	.06722392	14	2	9	.03816599
12	1	9	.03976244	13	1	2	.24167359	13	4	7	.05706800	14	2	10	.03432561
12	1	10	.03608740	13	1	3	.13912872	13	4	8	.04968908	14	2	11	.03134458
12	1	11	.03371936	13	1	4	.09748656	13	4	9	.04413805	14	2	12	.02910913
12	1	12	.03423194	13	1	5	.07507352	13	4	10	.03990091	14	2	13	.02775699
12	2	2	.22777582	13	1	6	.06111399	13	5	5	.08992827	14	3	3	.13719658
12	2	3	.13327894	13	1	7	.05161354	13	5	6	.07393056	14	3	4	.09703891

TABLE II

P = 3.0

N	R	S	COVARIANCE	N	R	S	COVARIANCE	N	R	S	COVARIANCE	N	R	S	COVARIANCE
14	3	5	.07513347	15	1	8	.04496796	15	4	7	.05454416	16	1	14	.02603204
14	3	6	.06136940	15	1	9	.03969189	15	4	8	.04735495	16	1	15	.02517460
14	3	7	.05193817	15	1	10	.03559965	15	4	9	.04190831	16	1	16	.02645625
14	3	8	.04509336	15	1	11	.03237419	15	4	10	.03766733	16	2	2	.23080395
14	3	9	.03992857	15	1	12	.02983700	15	4	11	.03431517	16	2	3	.13334134
14	3	10	.03593752	15	1	13	.02793141	15	4	12	.03167390	16	2	4	.09350514
14	3	11	.03283705	15	1	14	.02682868	15	5	5	.08469834	16	2	5	.07199667
14	3	12	.03051181	15	1	15	.02799674	15	5	6	.06939056	16	2	6	.05856969
14	4	4	.10326348	15	2	2	.22977800	15	5	7	.05884323	16	2	7	.04940566
14	4	5	.08021076	15	2	3	.13310945	15	5	8	.05114963	16	2	8	.04276450
14	4	6	.06566058	15	2	4	.09349396	15	5	9	.04531039	16	2	9	.03774343
14	4	7	.05565884	15	2	5	.07207034	15	5	10	.04075748	16	2	10	.03383081
14	4	8	.04838281	15	2	6	.05868290	15	5	11	.03715515	16	2	11	.03071977
14	4	9	.04288288	15	2	7	.04954148	15	6	6	.07566836	16	2	12	.02822348
14	4	10	.03862729	15	2	8	.04291790	15	6	7	.06428088	16	2	13	.02623933
14	4	11	.03531858	15	2	9	.03791636	15	6	8	.05595316	16	2	14	.02475318
14	5	5	.08700225	15	2	10	.03403176	15	6	9	.04962014	16	2	15	.02394622
14	5	6	.07139099	15	2	11	.03096693	15	6	10	.04467470	16	3	3	.13578853
14	5	7	.06062324	15	2	12	.02855466	15	7	7	.07111589	16	3	4	.09565876
14	5	8	.05276998	15	2	13	.02674309	15	7	8	.06200125	16	3	5	.07385850
14	5	9	.04682226	15	2	14	.02569810	15	7	9	.05505432	16	3	6	.06019536
14	5	10	.04221365	15	3	3	.13636929	15	8	8	.06971760	16	3	7	.05084423
14	6	6	.07872011	15	3	4	.09625054	16	1	1	.84496373	16	3	8	.04405361
14	6	7	.06698154	15	3	5	.07441349	16	1	2	.24941021	16	3	9	.03891155
14	6	8	.05839595	15	3	6	.06071026	16	1	3	.14253898	16	3	10	.03489982
14	6	9	.05187955	15	3	7	.05132561	16	1	4	.09942946	16	3	11	.03170704
14	7	7	.07514197	15	3	8	.04451105	16	1	5	.07632030	16	3	12	.02914347
14	7	8	.06563191	15	3	9	.03935687	16	1	6	.06195961	16	3	13	.02710514
15	1	1	.83208924	15	3	10	.03534873	16	1	7	.05218909	16	3	14	.02557888
15	1	2	.24684218	15	3	11	.03218356	16	1	8	.04512462	16	4	4	.10033389
15	1	3	.14138443	15	3	12	.02969098	16	1	9	.03979274	16	4	5	.07768188
15	1	4	.09875866	15	3	13	.02781934	16	1	10	.03564345	16	4	6	.06342964
15	1	5	.07588002	15	4	4	.10162942	16	1	11	.03234757	16	4	7	.05364843
15	1	6	.06165146	15	4	5	.07880539	16	1	12	.02970490	16	4	8	.04653093
15	1	7	.05196769	15	4	6	.06442304	16	1	13	.02760520	16	4	9	.04113287

TABLE II

P = 3.0

N	R	S	COVARIANCE	N	R	S	COVARIANCE	N	R	S	COVARIANCE	N	R	S	COVARIANCE
16	4	10	.03691628	17	1	12	.02966936	17	4	5	.07677552	18	1	1	.86974737
16	4	11	.03355731	17	1	13	.02746890	17	4	6	.06262379	18	1	2	.25447349
16	4	12	.03085849	17	1	14	.02570872	17	4	7	.05292000	18	1	3	.14486145
16	4	13	.02871190	17	1	15	.02439795	17	4	8	.04586103	18	1	4	.10080436
16	5	5	.08285151	17	1	16	.02373560	17	4	9	.04050501	18	1	5	.07724085
16	5	6	.06778475	17	1	17	.02510122	17	4	10	.03631419	18	1	6	.06261987
16	5	7	.05741483	17	2	2	.23193692	17	4	11	.03296208	18	1	7	.05268082
16	5	8	.04985265	17	2	3	.13366365	17	4	12	.03024353	18	1	8	.04549529
16	5	9	.04410781	17	2	4	.09359201	17	4	13	.02803147	18	1	9	.04006667
16	5	10	.03961450	17	2	5	.07198839	17	4	14	.02626093	18	1	10	.03582990
16	5	11	.03603152	17	2	6	.05851500	17	5	5	.08135011	18	1	11	.03244259
16	5	12	.03315066	17	2	7	.04932433	17	5	6	.06647633	18	1	12	.02968775
16	6	6	.07324551	17	2	8	.04266438	17	5	7	.05625021	18	1	13	.02742530
16	6	7	.06213905	17	2	9	.03762603	17	5	8	.04879639	18	1	14	.02556836
16	6	8	.05402087	17	2	10	.03369288	17	5	9	.04313201	18	1	15	.02407690
16	6	9	.04784244	17	2	11	.03055258	17	5	10	.03869451	18	1	16	.02297670
16	6	10	.04300314	17	2	12	.02800935	17	5	11	.03514166	18	1	17	.02247168
16	6	11	.03914003	17	2	13	.02594195	17	5	12	.03225818	18	1	18	.02389923
16	7	7	.06798494	17	2	14	.02428783	17	5	13	.02991073	18	2	2	.23315036
16	7	8	.05918549	17	2	15	.02305659	17	6	6	.07128483	18	2	3	.13405644
16	7	9	.05247507	17	2	16	.02243725	17	6	7	.06040624	18	2	4	.09373851
16	7	10	.04721073	17	3	3	.13539981	17	6	8	.05245985	18	2	5	.07203179
16	8	8	.06562771	17	3	4	.09522041	17	6	9	.04641107	18	2	6	.05850651
16	8	9	.05826310	17	3	5	.07343223	17	6	10	.04166616	18	2	7	.04928572
17	1	1	.85751112	17	3	6	.05979265	17	6	11	.03786322	18	2	8	.04260541
17	1	2	.25195569	17	3	7	.05046414	17	6	12	.03477429	18	2	9	.03755039
17	1	3	.14369984	17	3	8	.04369121	17	7	7	.06548856	18	2	10	.03360027
17	1	4	.10011306	17	3	9	.03855980	17	7	8	.05694428	18	2	11	.03043903
17	1	5	.07677551	17	3	10	.03454937	17	7	9	.05042845	18	2	12	.02786605
17	1	6	.06228401	17	3	11	.03134448	17	7	10	.04530970	18	2	13	.02575170
17	1	7	.05242853	17	3	12	.02874713	17	7	11	.04120237	18	2	14	.02401564
17	1	8	.04530250	17	3	13	.02663473	17	8	8	.06244109	18	2	15	.02262111
17	1	9	.03992052	17	3	14	.02494427	17	8	9	.05535937	18	2	16	.02159303
17	1	10	.03572453	17	3	15	.02368654	17	8	10	.04978693	18	2	17	.02112370
17	1	11	.03237773	17	4	4	.09929951	17	9	9	.06148730	18	3	3	.13516295

TABLE II

P = 3.0

N	R	S	COVARIANCE	N	R	S	COVARIANCE	N	R	S	COVARIANCE	N	R	S	COVARIANCE
18	3	4	.09490375	18	6	6	.06967361	19	1	16	.02265759	19	3	17	.02066742
18	3	5	.07310807	18	6	7	.05898177	19	1	17	.02172882	19	4	4	.09780548
18	3	6	.05947879	18	6	8	.05117742	19	1	18	.02135219	19	4	5	.07544028
18	3	7	.05016388	18	6	9	.04523751	19	1	19	.02282507	19	4	6	.06142360
18	3	8	.04340294	18	6	10	.04057461	19	2	2	.23442434	19	4	7	.05182848
18	3	9	.03827969	18	6	11	.03682929	19	2	3	.13450461	19	4	8	.04485489
18	3	10	.03427183	18	6	12	.03377207	19	2	4	.09393247	19	4	9	.03956362
18	3	11	.03106157	18	6	13	.03125424	19	2	5	.07211653	19	4	10	.03541793
18	3	12	.02844688	18	7	7	.06345834	19	2	6	.05853497	19	4	11	.03209002
18	3	13	.02629712	18	7	8	.05512339	19	2	7	.04928115	19	4	12	.02936990
18	3	14	.02453137	18	7	9	.04876898	19	2	8	.04257889	19	4	13	.02711907
18	3	15	.02311287	18	7	10	.04377399	19	2	9	.03750704	19	4	14	.02524623
18	3	16	.02206770	18	7	11	.03975758	19	2	10	.03354160	19	4	15	.02369592
18	4	4	.09847046	18	7	12	.03647625	19	2	11	.03036374	19	4	16	.02244845
18	4	5	.07603975	18	8	8	.05989428	19	2	12	.02776978	19	5	5	.07909240
18	4	6	.06196502	18	8	9	.05304342	19	2	13	.02562565	19	5	6	.06449936
18	4	7	.05232225	18	8	10	.04765016	19	2	14	.02384303	19	5	7	.05448631
18	4	8	.04531066	18	8	11	.04330826	19	2	15	.02236813	19	5	8	.04719622
18	4	9	.03999006	18	9	9	.05825810	19	2	16	.02118133	19	5	9	.04165728
18	4	10	.03582323	18	9	10	.05238490	19	2	17	.02031755	19	5	10	.03731286
18	4	11	.03248273	19	1	1	.88168829	19	2	18	.01996967	19	5	11	.03382239
18	4	12	.02976007	19	1	2	.25696032	19	3	3	.13504773	19	5	12	.03096739
18	4	13	.02752035	19	1	3	.14601983	19	3	4	.09468492	19	5	13	.02860363
18	4	14	.02568007	19	1	4	.10149961	19	3	5	.07286609	19	5	14	.02663601
18	4	15	.02420159	19	1	5	.07771279	19	3	6	.05923635	19	5	15	.02500683
18	5	5	.08011589	19	1	6	.06296371	19	3	7	.04992757	19	6	6	.06833293
18	5	6	.06539745	19	1	7	.05294224	19	3	8	.04317366	19	6	7	.05779546
18	5	7	.05528854	19	1	8	.04569871	19	3	9	.03805583	19	6	8	.05010941
18	5	8	.04792426	19	1	9	.04022570	19	3	10	.03405021	19	6	9	.04426124
18	5	9	.04232795	19	1	10	.03595177	19	3	11	.03083744	19	6	10	.03966901
18	5	10	.03794018	19	1	11	.03252997	19	3	12	.02821320	19	6	11	.03597601
18	5	11	.03441931	19	1	12	.02973903	19	3	13	.02604286	19	6	12	.03295310
18	5	12	.03154756	19	1	13	.02743348	19	3	14	.02423774	19	6	13	.03044881
18	5	13	.02918389	19	1	14	.02551754	19	3	15	.02274384	19	6	14	.02836329
18	5	14	.02724105	19	1	15	.02393278	19	3	16	.02154175	19	7	7	.06178056

TABLE II

P = 3.0

N	R	S	COVARIANCE
19	7	8	.05361930
19	7	9	.04739993
19	7	10	.04251014
19	7	11	.03857390
19	7	12	.03534923
19	7	13	.03267607
19	8	8	.05781718
19	8	9	.05115722
19	8	10	.04591389
19	8	11	.04168835
19	8	12	.03822354
19	9	9	.05567374
19	9	10	.05000979
19	9	11	.04543958
19	10	10	.05499414
20	1	1	.89334920
20	1	2	.25941419
20	1	3	.14717211
20	1	4	.10219604
20	1	5	.07818871
20	1	6	.06331295
20	1	7	.05321012
20	1	8	.04590972
20	1	9	.04039389
20	1	10	.03608517
20	1	11	.03263260
20	1	12	.02981160
20	1	13	.02747321
20	1	14	.02551680
20	1	15	.02387553
20	1	16	.02251009
20	1	17	.02141150
20	1	18	.02062403
20	1	19	.02035328
20	1	20	.02185881

N	R	S	COVARIANCE
20	2	2	.23574370
20	2	3	.13499654
20	2	4	.09416448
20	2	5	.07223463
20	2	6	.05859331
20	2	7	.04930402
20	2	8	.04257837
20	2	9	.03748921
20	2	10	.03350915
20	2	11	.03031697
20	2	12	.02770676
20	2	13	.02554179
20	2	14	.02372960
20	2	15	.02220877
20	2	16	.02094328
20	2	17	.01992519
20	2	18	.01919601
20	2	19	.01894754
20	3	3	.13503103
20	3	4	.09454566
20	3	5	.07269106
20	3	6	.05905209
20	3	7	.04974323
20	3	8	.04299207
20	3	9	.03787703
20	3	10	.03387267
20	3	11	.03065838
20	3	12	.02802836
20	3	13	.02584579
20	3	14	.02401809
20	3	15	.02248378
20	3	16	.02120685
20	3	17	.02017962
20	3	18	.01944446
20	4	4	.09727342

N	R	S	COVARIANCE
20	4	5	.07495125
20	4	6	.06097720
20	4	7	.05141875
20	4	8	.04447529
20	4	9	.03920796
20	4	10	.03508028
20	4	11	.03176434
20	4	12	.02904940
20	4	13	.02679515
20	4	14	.02490665
20	4	15	.02332080
20	4	16	.02200076
20	4	17	.02093893
20	5	5	.07823790
20	5	6	.06374607
20	5	7	.05381158
20	5	8	.04658302
20	5	9	.04109239
20	5	10	.03678534
20	5	11	.03332245
20	5	12	.03048527
20	5	13	.02812826
20	5	14	.02615280
20	5	15	.02449340
20	5	16	.02311191
20	6	6	.06720587
20	6	7	.05679685
20	6	8	.04920997
20	6	9	.04343940
20	6	10	.03890790
20	6	11	.03526137
20	6	12	.03227161
20	6	13	.02978638
20	6	14	.02770251
20	6	15	.02595146

N	R	S	COVARIANCE
20	7	7	.06037566
20	7	8	.05235990
20	7	9	.04625438
20	7	10	.04145430
20	7	11	.03758802
20	7	12	.03441564
20	7	13	.03177695
20	7	14	.02956330
20	8	8	.05609506
20	8	9	.04959476
20	8	10	.04447789
20	8	11	.04035221
20	8	12	.03696412
20	8	13	.03414406
20	9	9	.05356246
20	9	10	.04807256
20	9	11	.04364106
20	9	12	.03999840
20	10	10	.05237996
20	10	11	.04758590

TABLE II

P = 3.5

N	R	S	COVARIANCE
2	1	1	.70449421
2	1	2	.29550579
3	1	1	.64979729
3	1	2	.25174825
3	1	3	.16139152
3	2	2	.37062937
4	1	1	.63336056
4	1	2	.23641340
4	1	3	.14660746
4	1	4	.10949271
4	2	2	.29901023
4	2	3	.19209478
5	1	1	.62936272
5	1	2	.22919301
5	1	3	.14065072
5	1	4	.10203856
5	1	5	.08272514
5	2	2	.26850799
5	2	3	.16918993
5	2	4	.12456052
5	3	3	.22539837
6	1	1	.63065141
6	1	2	.22547484
6	1	3	.13745270
6	1	4	.09913176
6	1	5	.07794314
6	1	6	.06657130
6	2	2	.25202233
6	2	3	.15698973
6	2	4	.11453481
6	2	5	.09073032
6	3	3	.19371604
6	3	4	.14325463
7	1	1	.63451372
7	1	2	.22357317

N	R	S	COVARIANCE
7	1	3	.13559185
7	1	4	.09750813
7	1	5	.07626695
7	1	6	.06300999
7	1	7	.05581252
7	2	2	.24197835
7	2	3	.14949885
7	2	4	.10855339
7	2	5	.08542322
7	2	6	.07088112
7	3	3	.17616230
7	3	4	.12926770
7	3	5	.10242702
7	4	4	.16167695
8	1	1	.63972231
8	1	2	.22272092
8	1	3	.13449380
8	1	4	.09651636
8	1	5	.07534875
8	1	6	.06190644
8	1	7	.05290467
8	1	8	.04814767
8	2	2	.23542014
8	2	3	.14451349
8	2	4	.10458452
8	2	5	.08207320
8	2	6	.06767448
8	2	7	.05799291
8	3	3	.16508389
8	3	4	.12051553
8	3	5	.09510269
8	3	6	.07872906
8	4	4	.14388082
8	4	5	.11431353
9	1	1	.64565573

N	R	S	COVARIANCE
9	1	2	.22251942
9	1	3	.13387203
9	1	4	.09590036
9	1	5	.07478477
9	1	6	.06133781
9	1	7	.05209594
9	1	8	.04563395
9	1	9	.04241408
9	2	2	.23095436
9	2	3	.14102252
9	2	4	.10178703
9	2	5	.07974130
9	2	6	.06560854
9	2	7	.05585262
9	2	8	.04901560
9	3	3	.15751691
9	3	4	.11454567
9	3	5	.09016007
9	3	6	.07442490
9	3	7	.06351514
9	4	4	.13257338
9	4	5	.10492103
9	4	6	.08694786
9	5	5	.12598208
10	1	1	.65197428
10	1	2	.22274180
10	1	3	.13356571
10	1	4	.09552950
10	1	5	.07443080
10	1	6	.06099917
10	1	7	.05171348
10	1	8	.04498910
10	1	9	.04016122
10	1	10	.03796409
10	2	2	.22783806

N	R	S	COVARIANCE
10	2	3	.13849427
10	2	4	.09973668
10	2	5	.07803246
10	2	6	.06413064
10	2	7	.05447962
10	2	8	.04747095
10	2	9	.04243282
10	3	3	.15206902
10	3	4	.11023750
10	3	5	.08660338
10	3	6	.07137663
10	3	7	.06076265
10	3	8	.05303335
10	4	4	.12477366
10	4	5	.09847411
10	4	6	.08142266
10	4	7	.06948349
10	5	5	.11459910
10	5	6	.09512747
11	1	1	.65848196
11	1	2	.22325016
11	1	3	.13347725
11	1	4	.09532778
11	1	5	.07421602
11	1	6	.06079174
11	1	7	.05149986
11	1	8	.04471102
11	1	9	.03961321
11	1	10	.03589700
11	1	11	.03440946
11	2	2	.22563904
11	2	3	.13662176
11	2	4	.09819333
11	2	5	.07673902
11	2	6	.06301945

TABLE II

P = 3.5

N	R	S	COVARIANCE	N	R	S	COVARIANCE	N	R	S	COVARIANCE	N	R	S	COVARIANCE
11	2	7	.05348589	12	2	5	.07573826	13	1	9	.03929651	13	5	8	.05921553
11	2	8	.04650101	12	2	6	.06215900	13	1	10	.03522804	13	5	9	.05250237
11	2	9	.04124618	12	2	7	.05272720	13	1	11	.03203405	13	6	6	.08948348
11	2	10	.03741343	12	2	8	.04579956	13	1	12	.02968751	13	6	7	.07640781
11	3	3	.14799847	12	2	9	.04052471	13	1	13	.02908322	13	6	8	.06669787
11	3	4	.10700319	12	2	10	.03645493	13	2	2	.22301266	13	7	7	.08735649
11	3	5	.08393129	12	2	11	.03346687	13	2	3	.13415068	14	1	1	.67817409
11	3	6	.06909898	12	3	3	.14487330	13	2	4	.09609142	14	1	2	.22574619
11	3	7	.05875370	12	3	4	.10450367	13	2	5	.07495202	14	1	3	.13398790
11	3	8	.05115373	12	3	5	.08186059	13	2	6	.06147973	14	1	4	.09533227
11	3	9	.04542585	12	3	6	.06733575	13	2	7	.05213064	14	1	5	.07406443
11	4	4	.11908809	12	3	7	.05721260	13	2	8	.04526046	14	1	6	.06059459
11	4	5	.09378285	12	3	8	.04975845	13	2	9	.04000856	14	1	7	.05128717
11	4	6	.07742367	12	3	9	.04407227	13	2	10	.03589393	14	1	8	.04446579
11	4	7	.06596774	12	3	10	.03967988	13	2	11	.03266094	14	1	9	.03925380
11	4	8	.05752737	12	4	4	.11477666	13	2	12	.03028632	14	1	10	.03515388
11	5	5	.10672588	12	4	5	.09022518	13	3	3	.14242470	14	1	11	.03187473
11	5	6	.08839489	12	4	6	.07439707	13	3	4	.10252921	14	1	12	.02926575
11	5	7	.07549993	12	4	7	.06332590	13	3	5	.08021820	14	1	13	.02735757
11	6	6	.10317818	12	4	8	.05515193	13	3	6	.06593549	14	1	14	.02703486
12	1	1	.66506135	12	4	9	.04890440	13	3	7	.05599262	14	2	2	.22229102
12	1	2	.22395635	12	5	5	.10096318	13	3	8	.04866901	14	2	3	.13334417
12	1	3	.13354400	12	5	6	.08348214	13	3	9	.04306061	14	2	4	.09537294
12	1	4	.09524729	12	5	7	.07120606	13	3	10	.03866098	14	2	5	.07432785
12	1	5	.07409938	12	5	8	.06211543	13	3	11	.03520131	14	2	6	.06093638
12	1	6	.06067036	12	6	6	.09527551	13	4	4	.11140938	14	2	7	.05165286
12	1	7	.05137710	12	6	7	.08146694	13	4	5	.08744315	14	2	8	.04483270
12	1	8	.04456935	13	1	1	.67164035	13	4	6	.07203089	14	2	9	.03961223
12	1	9	.03939582	13	1	2	.22480176	13	4	7	.06126665	14	2	10	.03549996
12	1	10	.03540924	13	1	3	.13372393	13	4	8	.05331869	14	2	11	.03220763
12	1	11	.03248253	13	1	4	.09525615	13	4	9	.04722090	14	2	12	.02958666
12	1	12	.03150375	13	1	5	.07405476	13	4	10	.04243081	14	2	13	.02767065
12	2	2	.22408898	13	1	6	.06060990	13	5	5	.09657035	14	3	3	.14047652
12	2	3	.13521516	13	1	7	.05131201	13	5	6	.07974239	14	3	4	.10094296
12	2	4	.09701002	13	1	8	.04449667	13	5	7	.06794767	14	3	5	.07889194

TABLE II

P = 3.5

N	R	S	COVARIANCE	N	R	S	COVARIANCE	N	R	S	COVARIANCE	N	R	S	COVARIANCE
14	3	6	.06480182	15	1	9	.03924370	15	4	8	.05066582	16	1	15	.02369456
14	3	7	.05500519	15	1	10	.03512907	15	4	9	.04482454	16	1	16	.02375457
14	3	8	.04779237	15	1	11	.03181540	15	4	10	.04020216	16	2	2	.22160042
14	3	9	.04226214	15	1	12	.02911972	15	4	11	.03646859	16	2	3	.13228468
14	3	10	.03790032	15	1	13	.02695476	15	4	12	.03342483	16	2	4	.09436729
14	3	11	.03440492	15	1	14	.02538549	15	5	5	.09033978	16	2	5	.07342928
14	3	12	.03162086	15	1	15	.02527824	15	5	6	.07443928	16	2	6	.06014204
14	4	4	.10871897	15	2	2	.22184040	15	5	7	.06333449	16	2	7	.05094922
14	4	5	.08521575	15	2	3	.13273644	15	5	8	.05512920	16	2	8	.04420390
14	4	6	.07013488	15	2	4	.09480918	15	5	9	.04881606	16	2	9	.03904024
14	4	7	.05961803	15	2	5	.07382908	15	5	10	.04381353	16	2	10	.03496110
14	4	8	.05185759	15	2	6	.06049788	15	5	11	.03976864	16	2	11	.03166312
14	4	9	.04589724	15	2	7	.05126559	15	6	6	.08157252	16	2	12	.02895542
14	4	10	.04118993	15	2	8	.04448685	15	6	7	.06951498	16	2	13	.02672216
14	4	11	.03741401	15	2	9	.03929669	15	6	8	.06058413	16	2	14	.02491700
14	5	5	.09311800	15	2	10	.03519972	15	6	9	.05369970	16	2	15	.02362663
14	5	6	.07680437	15	2	11	.03189687	15	6	10	.04823627	16	3	3	.13763568
14	5	7	.06539143	15	2	12	.02920797	15	7	7	.07711349	16	3	4	.09858979
14	5	8	.05694936	15	2	13	.02704769	15	7	8	.06730309	16	3	5	.07690625
14	5	9	.05045335	15	2	14	.02548288	15	7	9	.05972439	16	3	6	.06309517
14	5	10	.04531556	15	3	3	.13890849	15	8	8	.07573832	16	3	7	.05351512
14	6	6	.08505908	15	3	4	.09965172	16	1	1	.69100470	16	3	8	.04647212
14	6	7	.07255106	15	3	5	.07780577	16	1	2	.22782683	16	3	9	.04107261
14	6	8	.06327400	15	3	6	.06387009	16	1	3	.13469044	16	3	10	.03680218
14	6	9	.05612046	15	3	7	.05419258	16	1	4	.09562731	16	3	11	.03334638
14	7	7	.08155129	15	3	8	.04707239	16	1	5	.07419997	16	3	12	.03050707
14	7	8	.07124245	15	3	9	.04161225	16	1	6	.06065989	16	3	13	.02816409
15	1	1	.68463461	15	3	10	.03729690	16	1	7	.05131976	16	3	14	.02626990
15	1	2	.22676133	15	3	11	.03381471	16	1	8	.04448119	16	4	4	.10472430
15	1	3	.13431513	15	3	12	.03097787	16	1	9	.03925479	16	4	5	.08189457
15	1	4	.09545985	15	3	13	.02869800	16	1	10	.03513140	16	4	6	.06730093
15	1	5	.07411588	15	4	4	.10653044	16	1	11	.03180105	16	4	7	.05715201
15	1	6	.06061378	15	4	5	.08339890	16	1	12	.02906891	16	4	8	.04967640
15	1	7	.05129207	15	4	6	.06858599	16	1	13	.02681674	16	4	9	.04393668
15	1	8	.04446320	15	4	7	.05827086	16	1	14	.02499668	16	4	10	.03939184

TABLE II

P = 3.5

N	R	S	COVARIANCE	N	R	S	COVARIANCE	N	R	S	COVARIANCE	N	R	S	COVARIANCE
16	4	11	.03571052	17	1	13	.02677050	17	4	6	.06622118	18	1	2	.23005321
16	4	12	.03268371	17	1	14	.02486338	17	4	7	.05621049	18	1	3	.13554288
16	4	13	.03018473	17	1	15	.02331651	17	4	8	.04884341	18	1	4	.09604929
16	5	5	.08806149	17	1	16	.02222845	17	4	9	.04319008	18	1	5	.07444036
16	5	6	.07249821	17	1	17	.02241985	17	4	10	.03871350	18	1	6	.06081191
16	5	7	.06164594	17	2	2	.22152648	17	4	11	.03508331	18	1	7	.05142527
16	5	8	.05363604	17	2	3	.13195717	17	4	12	.03208760	18	1	8	.04456070
16	5	9	.04747637	17	2	4	.09402292	17	4	13	.02958910	18	1	9	.03931806
16	5	10	.04259288	17	2	5	.07310885	17	4	14	.02750579	18	1	10	.03518133
16	5	11	.03863329	17	2	6	.05985253	17	5	5	.08616433	18	1	11	.03183430
16	5	12	.03537515	17	2	7	.05068956	17	5	6	.07087995	18	1	12	.02907362
16	6	6	.07875765	17	2	8	.04397071	17	5	7	.06023726	18	1	13	.02676474
16	6	7	.06706481	17	2	9	.03882923	17	5	8	.05239038	18	1	14	.02481923
16	6	8	.05841554	17	2	10	.03476696	17	5	9	.04636005	18	1	15	.02318586
16	6	9	.05175272	17	2	11	.03147858	17	5	10	.04157936	18	1	16	.02185900
16	6	10	.04646304	17	2	12	.02876881	17	5	11	.03769890	18	1	17	.02094492
16	6	11	.04216937	17	2	13	.02651127	17	5	12	.03449424	18	1	18	.02124057
16	7	7	.07361219	17	2	14	.02463027	17	5	13	.03181990	18	2	2	.22158498
16	7	8	.06419939	17	2	15	.02310445	17	6	6	.07644067	18	2	3	.13172991
16	7	9	.05693444	17	2	16	.02203226	17	6	7	.06504795	18	2	4	.09375764
16	7	10	.05115780	17	3	3	.13659640	17	6	8	.05663118	18	2	5	.07285296
16	8	8	.07129408	17	3	4	.09770950	17	6	9	.05015267	18	2	6	.05961703
16	8	9	.06330115	17	3	5	.07615467	17	6	10	.04501018	18	2	7	.05047601
17	1	1	.69727417	17	3	6	.06244449	17	6	11	.04083175	18	2	8	.04377774
17	1	2	.22892781	17	3	7	.05294452	17	6	12	.03737817	18	2	9	.03865433
17	1	3	.13510253	17	3	8	.04596587	17	7	7	.07078096	18	2	10	.03460685
17	1	4	.09582599	17	3	9	.04061810	17	7	8	.06169163	18	2	11	.03132889
17	1	5	.07430987	17	3	10	.03638817	17	7	9	.05468328	18	2	12	.02862309
17	1	6	.06072734	17	3	11	.03296104	17	7	10	.04911241	18	2	13	.02635871
17	1	7	.05136536	17	3	12	.03013493	17	7	11	.04458075	18	2	14	.02444979
17	1	8	.04451481	17	3	13	.02777917	17	8	8	.06778608	18	2	15	.02284665
17	1	9	.03928084	17	3	14	.02581563	17	8	9	.06014739	18	2	16	.02154434
17	1	10	.03515063	17	3	15	.02422272	17	8	10	.05406586	18	2	17	.02064820
17	1	11	.03181055	17	4	4	.10321618	17	9	9	.06684589	18	3	3	.13574478
17	1	12	.02906033	17	4	5	.08063369	18	1	1	.70343743	18	3	4	.09697543

TABLE II

P = 3.5

N	R	S	COVARIANCE	N	R	S	COVARIANCE	N	R	S	COVARIANCE	N	R	S	COVARIANCE
18	3	5	.07552228	18	6	7	.06336085	19	1	17	.02058259	19	4	4	.10086430
18	3	6	.06189393	18	6	8	.05513852	19	1	18	.01981167	19	4	5	.07865460
18	3	7	.05245993	18	6	9	.04881491	19	1	19	.02019068	19	4	6	.06451920
18	3	8	.04553501	18	6	10	.04379746	19	2	2	.22175018	19	4	7	.05472209
18	3	9	.04023119	18	6	11	.03971957	19	2	3	.13158448	19	4	8	.04752431
18	3	10	.03603677	18	6	12	.03634378	19	2	4	.09355724	19	4	9	.04200764
18	3	11	.03263692	18	6	13	.03351215	19	2	5	.07265021	19	4	10	.03764193
18	3	12	.02982859	18	7	7	.06844606	19	2	6	.05942606	19	4	11	.03410019
18	3	13	.02747711	18	7	8	.05962426	19	2	7	.05030047	19	4	12	.03117051
18	3	14	.02549391	18	7	9	.05282885	19	2	8	.04361782	19	4	13	.02871077
18	3	15	.02382794	18	7	10	.04743013	19	2	9	.03850880	19	4	14	.02662424
18	3	16	.02247461	18	7	11	.04303778	19	2	10	.03447375	19	4	15	.02484703
18	4	4	.10194473	18	7	12	.03939853	19	2	11	.03120554	19	4	16	.02334485
18	4	5	.07956612	18	8	8	.06494732	19	2	12	.02850568	19	5	5	.08320091
18	4	6	.06530442	18	8	9	.05759785	19	2	13	.02624136	19	5	6	.06834639
18	4	7	.05540958	18	8	10	.05175063	19	2	14	.02432227	19	5	7	.05802844
18	4	8	.04813407	18	8	11	.04698780	19	2	15	.02268875	19	5	8	.05043558
18	4	9	.04255439	18	9	9	.06333456	19	2	16	.02130856	19	5	9	.04460867
18	4	10	.03813720	18	9	10	.05695445	19	2	17	.02018779	19	5	10	.03999273
18	4	11	.03455376	19	1	1	.70949196	19	2	18	.01943553	19	5	11	.03624486
18	4	12	.03159174	19	1	2	.23119473	19	3	3	.13504600	19	5	12	.03314256
18	4	13	.02911020	19	1	3	.13600492	19	3	4	.09636068	19	5	13	.03053642
18	4	14	.02701641	19	1	4	.09629209	19	3	5	.07498728	19	5	14	.02832469
18	4	15	.02525709	19	1	5	.07458741	19	3	6	.06142523	19	5	15	.02644019
18	5	5	.08456444	19	1	6	.06091033	19	3	7	.05204563	19	6	6	.07286168
18	5	6	.06951323	19	1	7	.05149677	19	3	8	.04516564	19	6	7	.06193064
18	5	7	.05904639	19	1	8	.04461641	19	3	9	.03989908	19	6	8	.05387275
18	5	8	.05133683	19	1	9	.03936394	19	3	10	.03573542	19	6	9	.04768061
18	5	9	.04541622	19	1	10	.03522036	19	3	11	.03236029	19	6	10	.04277001
18	5	10	.04072400	19	1	11	.03186734	19	3	12	.02957028	19	6	11	.03877935
18	5	11	.03691408	19	1	12	.02909952	19	3	13	.02722906	19	6	12	.03547366
18	5	12	.03376258	19	1	13	.02677962	19	3	14	.02524394	19	6	13	.03269498
18	5	13	.03112076	19	1	14	.02481439	19	3	15	.02355365	19	6	14	.03033567
18	5	14	.02889074	19	1	15	.02314223	19	3	16	.02212522	19	7	7	.06648929
18	6	6	.07450318	19	1	16	.02172971	19	3	17	.02096537	19	7	8	.05789179

TABLE II

P = 3.5

N	R	S	COVARIANCE	N	R	S	COVARIANCE	N	R	S	COVARIANCE	N	R	S	COVARIANCE
19	7	9	.05127535	20	2	3	.13150644	20	4	6	.06384171	20	7	8	.05642013
19	7	10	.04602209	20	2	4	.09341060	20	4	7	.05412745	20	7	9	.04995579
19	7	11	.04174884	20	2	5	.07249162	20	4	8	.04699599	20	7	10	.04482658
19	7	12	.03820622	20	2	6	.05927219	20	4	9	.04153344	20	7	11	.04065566
19	7	13	.03522641	20	2	7	.05015665	20	4	10	.03721238	20	7	12	.03719746
19	8	8	.06260388	20	2	8	.04348541	20	4	11	.03370744	20	7	13	.03428588
19	8	9	.05549442	20	2	9	.03838759	20	4	12	.03080746	20	7	14	.03180595
19	8	10	.04984240	20	2	10	.03436269	20	4	13	.02837010	20	8	8	.06063755
19	8	11	.04523986	20	2	11	.03110301	20	4	14	.02629706	20	8	9	.05373003
19	8	12	.04142086	20	2	12	.02840936	20	4	15	.02452028	20	8	10	.04824265
19	9	9	.06049225	20	2	13	.02614780	20	4	16	.02299513	20	8	11	.04377610
19	9	10	.05437285	20	2	14	.02422594	20	4	17	.02170030	20	8	12	.04006975
19	9	11	.04938375	20	2	15	.02257988	20	5	5	.08202841	20	8	13	.03694711
19	10	10	.05982115	20	2	16	.02116767	20	5	6	.06734105	20	9	9	.05814472
20	1	1	.71543728	20	2	17	.01996903	20	5	7	.05715012	20	9	10	.05224225
20	1	2	.23234608	20	2	18	.01899763	20	5	8	.04965717	20	9	11	.04743260
20	1	3	.13648358	20	2	19	.01836431	20	5	9	.04391082	20	9	12	.04343790
20	1	4	.09655040	20	3	3	.13447318	20	5	10	.03936089	20	10	10	.05697705
20	1	5	.07474784	20	3	4	.09584449	20	5	11	.03566741	20	10	11	.05176552
20	1	6	.06102003	20	3	5	.07453283	20	5	12	.03260944	********************			
20	1	7	.05157773	20	3	6	.06102429	20	5	13	.03003791				
20	1	8	.04468009	20	3	7	.05168955	20	5	14	.02784979				
20	1	9	.03941670	20	3	8	.04484711	20	5	15	.02597370				
20	1	10	.03526570	20	3	9	.03961213	20	5	16	.02436288				
20	1	11	.03190689	20	3	10	.03547496	20	6	6	.07145556				
20	1	12	.02913336	20	3	11	.03212174	20	6	7	.06070455				
20	1	13	.02680613	20	3	12	.02934903	20	6	8	.05278707				
20	1	14	.02482945	20	3	13	.02701986	20	6	9	.04670748				
20	1	15	.02313711	20	3	14	.02503969	20	6	10	.04188880				
20	1	16	.02168562	20	3	15	.02334309	20	6	11	.03797391				
20	1	17	.02045384	20	3	16	.02188714	20	6	12	.03473041				
20	1	18	.01945543	20	3	17	.02065122	20	6	13	.03200129				
20	1	19	.01880359	20	3	18	.01964972	20	6	14	.02967796				
20	1	20	.01924969	20	4	4	.09994020	20	6	15	.02768519				
20	2	2	.22200204	20	4	5	.07787084	20	7	7	.06482733				

TABLE II

P = 4.0

N	R	S	COVARIANCE	N	R	S	COVARIANCE	N	R	S	COVARIANCE	N	R	S	COVARIANCE
2	1	1	.69963335	7	1	3	.13571433	9	1	2	.21571810	10	2	3	.14061415
2	1	2	.30036665	7	1	4	.09819085	9	1	3	.13239475	10	2	4	.10219468
3	1	1	.63206479	7	1	5	.07674182	9	1	4	.09568872	10	2	5	.08030138
3	1	2	.25657479	7	1	6	.06286726	9	1	5	.07488504	10	2	6	.06606437
3	1	3	.16267538	7	1	7	.05416715	9	1	6	.06139881	10	2	7	.05601416
3	2	2	.38422051	7	2	2	.24549538	9	1	7	.05191585	10	2	8	.04853741
4	1	1	.60547543	7	2	3	.15412682	9	1	8	.04498181	10	2	9	.04287844
4	1	2	.23882059	7	2	4	.11250431	9	1	9	.04058933	10	3	3	.15644905
4	1	3	.14912978	7	2	5	.08842059	9	2	2	.23129949	10	3	4	.11439562
4	1	4	.10920411	7	2	6	.07272407	9	2	3	.14389067	10	3	5	.09023015
4	2	2	.30973165	7	3	3	.18355913	9	2	4	.10471988	10	3	6	.07442760
4	2	3	.19968807	7	3	4	.13528883	9	2	5	.08230092	10	3	7	.06322791
5	1	1	.59313186	7	3	5	.10700335	9	2	6	.06767485	10	3	8	.05487229
5	1	2	.22929688	7	4	4	.16963841	9	2	7	.05734549	10	4	4	.13033794
5	1	3	.14253082	8	1	1	.58437579	9	2	8	.04977179	10	4	5	.10324150
5	1	4	.10317419	8	1	2	.21737996	9	3	3	.16276239	10	4	6	.08541475
5	1	5	.08168369	8	1	3	.13378763	9	3	4	.11927161	10	4	7	.07272511
5	2	2	.27636523	8	1	4	.09675683	9	3	5	.09414517	10	5	5	.12048252
5	2	3	.17595535	8	1	5	.07568218	9	3	6	.07764946	10	5	6	.10004055
5	2	4	.12908814	8	1	6	.06199251	9	3	7	.06594852	11	1	1	.58871581
5	3	3	.23563319	8	1	7	.05245517	9	4	4	.13884132	11	1	2	.21387320
6	1	1	.58727327	8	1	8	.04638525	9	4	5	.11014557	11	1	3	.13061481
6	1	2	.22354445	8	2	2	.23724596	9	4	6	.09117385	11	1	4	.09425228
6	1	3	.13845375	8	2	3	.14821254	9	5	5	.13245591	11	1	5	.07377752
6	1	4	.10018476	8	2	4	.10802099	10	1	1	.58666722	11	1	6	.06057341
6	1	5	.07828260	8	2	5	.08489833	10	1	2	.21460232	11	1	7	.05130564
6	1	6	.06513872	8	2	6	.06977293	10	1	3	.13137161	11	1	8	.04441953
6	2	2	.25751448	8	2	7	.05918843	10	1	4	.09487686	11	1	9	.03911288
6	2	3	.16265320	8	3	3	.17132299	10	1	5	.07426590	11	1	10	.03501219
6	2	4	.11894260	8	3	4	.12586681	10	1	6	.06094059	11	1	11	.03254574
6	2	5	.09357958	8	3	5	.09943225	10	1	7	.05157204	11	2	2	.22347058
6	3	3	.20246743	8	3	6	.08201629	10	1	8	.04462241	11	2	3	.13806185
6	3	4	.14990379	8	4	4	.15095227	10	1	9	.03937152	11	2	4	.10020839
7	1	1	.58483549	8	4	5	.11999673	10	1	10	.03610973	11	2	5	.07871653
7	1	2	.21984329	9	1	1	.58513195	10	2	2	.22686328	11	2	6	.06478173

TABLE II

P = 4.0

N	R	S	COVARIANCE	N	R	S	COVARIANCE	N	R	S	COVARIANCE	N	R	S	COVARIANCE
11	2	7	.05496423	12	2	5	.07743299	13	1	9	.03890819	13	5	8	.06217390
11	2	8	.04764962	12	2	6	.06373584	13	1	10	.03473971	13	5	9	.05500629
11	2	9	.04200162	12	2	7	.05410593	13	1	11	.03136457	13	6	6	.09412421
11	2	10	.03763236	12	2	8	.04693809	13	1	12	.02868968	13	6	7	.08045291
11	3	3	.15161322	12	2	9	.04138484	13	1	13	.02723451	13	6	8	.07018842
11	3	4	.11064736	12	2	10	.03697702	13	2	2	.21874337	13	7	7	.09204757
11	3	5	.08721161	12	2	11	.03351607	13	2	3	.13439272	14	1	1	.59651337
11	3	6	.07193977	12	3	3	.14780222	13	2	4	.09730894	14	1	2	.21311434
11	3	7	.06114159	12	3	4	.10768077	13	2	5	.07637598	14	1	3	.12935567
11	3	8	.05307499	12	3	5	.08481399	13	2	6	.06286828	14	1	4	.09310543
11	3	9	.04683437	12	3	6	.06995757	13	2	7	.05338976	14	1	5	.07283160
11	4	4	.12403491	12	3	7	.05947836	13	2	8	.04634494	14	1	6	.05983236
11	4	5	.09812373	12	3	8	.05165932	13	2	9	.04088750	14	1	7	.05075440
11	4	6	.08114870	12	3	9	.04559037	13	2	10	.03653309	14	1	8	.04403267
11	4	7	.06909990	12	3	10	.04076677	13	2	11	.03300385	14	1	9	.03883954
11	4	8	.06007311	12	4	4	.11917658	13	2	12	.03020578	14	1	10	.03469855
11	5	5	.11207427	12	4	5	.09417503	13	3	3	.14473192	14	1	11	.03132189
11	5	6	.09296422	12	4	6	.07785419	13	3	4	.10527922	14	1	12	.02854282
11	5	7	.07934045	12	4	7	.06630232	13	3	5	.08286551	14	1	13	.02632732
11	6	6	.10862121	12	4	8	.05766055	13	3	6	.06834087	14	1	14	.02520300
12	1	1	.59110835	12	4	9	.05093973	13	3	7	.05811797	14	2	2	.21708272
12	1	2	.21342620	12	5	5	.10583459	13	3	8	.05050277	14	2	3	.13305147
12	1	3	.13005582	12	5	6	.08771778	13	3	9	.04459327	14	2	4	.09623002
12	1	4	.09376882	12	5	7	.07484564	13	3	10	.03987192	14	2	5	.07549391
12	1	5	.07338910	12	5	8	.06518813	13	3	11	.03604163	14	2	6	.06213903
12	1	6	.06027511	12	6	6	.10029905	13	4	4	.11531978	14	2	7	.05278367
12	1	7	.05108750	12	6	7	.08577758	13	4	5	.09103545	14	2	8	.04584063
12	1	8	.04426650	13	1	1	.59373290	13	4	6	.07523057	14	2	9	.04046712
12	1	9	.03899217	13	1	2	.21318999	13	4	7	.06407191	14	2	10	.03617645
12	1	10	.03481158	13	1	3	.12964733	13	4	8	.05574010	14	2	11	.03267403
12	1	11	.03153126	13	1	4	.09339428	13	4	9	.04926273	14	2	12	.02978935
12	1	12	.02964345	13	1	5	.07307905	13	4	10	.04408040	14	2	13	.02748926
12	2	2	.22082840	13	1	6	.06003150	13	5	5	.10101600	14	3	3	.14221467
12	2	3	.13603237	13	1	7	.05090598	13	5	6	.08366608	14	3	4	.10329990
12	2	4	.09861252	13	1	8	.04413998	13	5	7	.07137500	14	3	5	.08125309

TABLE II

P = 4.0

N	R	S	COVARIANCE
14	3	6	.06699797
14	3	7	.05698397
14	3	8	.04953657
14	3	9	.04376341
14	3	10	.03914778
14	3	11	.03537642
14	3	12	.03226810
14	4	4	.11218686
14	4	5	.08848025
14	4	6	.07309119
14	4	7	.06224980
14	4	8	.05416974
14	4	9	.04789567
14	4	10	.04287296
14	4	11	.03876476
14	5	5	.09718124
14	5	6	.08043979
14	5	7	.06860967
14	5	8	.05977221
14	5	9	.05289752
14	5	10	.04738602
14	6	6	.08935385
14	6	7	.07634124
14	6	8	.06659517
14	6	9	.05899812
14	7	7	.08592924
14	7	8	.07507563
15	1	1	.59939679
15	1	2	.21316292
15	1	3	.12915618
15	1	4	.09288508
15	1	5	.07263489
15	1	6	.05966992
15	1	7	.05062801
15	1	8	.04394127
15	1	9	.03878054
15	1	10	.03466710
15	1	11	.03130747
15	1	12	.02851738
15	1	13	.02619187
15	1	14	.02433307
15	1	15	.02346654
15	2	2	.21575225
15	2	3	.13194362
15	2	4	.09532754
15	2	5	.07474987
15	2	6	.06151947
15	2	7	.05226522
15	2	8	.04540653
15	2	9	.04010427
15	2	10	.03587249
15	2	11	.03241263
15	2	12	.02953698
15	2	13	.02713887
15	2	14	.02522200
15	3	3	.14012144
15	3	4	.10164465
15	3	5	.07989904
15	3	6	.06586596
15	3	7	.05602452
15	3	8	.04871638
15	3	9	.04305812
15	3	10	.03853681
15	3	11	.03483671
15	3	12	.03175913
15	3	13	.02919135
15	4	4	.10959473
15	4	5	.08636156
15	4	6	.07131357
15	4	7	.06073261
15	4	8	.05285974
15	4	9	.04675484
15	4	10	.04187065
15	4	11	.03786966
15	4	12	.03453927
15	5	5	.09405692
15	5	6	.07780886
15	5	7	.06635236
15	5	8	.05781004
15	5	9	.05117503
15	5	10	.04585971
15	5	11	.04150090
15	6	6	.08555438
15	6	7	.07306614
15	6	8	.06373292
15	6	9	.05647029
15	6	10	.05064356
15	7	7	.08120363
15	7	8	.07092558
15	7	9	.06291108
15	8	8	.07985807
16	1	1	.60234561
16	1	2	.21330886
16	1	3	.12903044
16	1	4	.09272021
16	1	5	.07247984
16	1	6	.05953811
16	1	7	.05052313
16	1	8	.04386376
16	1	9	.03872940
16	1	10	.03464011
16	1	11	.03130015
16	1	12	.02851958
16	1	13	.02617624
16	1	14	.02420389
16	1	15	.02262749
16	1	16	.02196506
16	2	2	.21468324
16	2	3	.13102165
16	2	4	.09456619
16	2	5	.07411672
16	2	6	.06098851
16	2	7	.05181792
16	2	8	.04502954
16	2	9	.03978739
16	2	10	.03560703
16	2	11	.03218926
16	2	12	.02934163
16	2	13	.02694028
16	2	14	.02491831
16	2	15	.02330241
16	3	3	.13836050
16	3	4	.10024372
16	3	5	.07874809
16	3	6	.06490012
16	3	7	.05520288
16	3	8	.04801146
16	3	9	.04245019
16	3	10	.03801032
16	3	11	.03437711
16	3	12	.03134775
16	3	13	.02879172
16	3	14	.02663873
16	4	4	.10741762
16	4	5	.08457791
16	4	6	.06981378
16	4	7	.05944972
16	4	8	.05174964
16	4	9	.04578647
16	4	10	.04102031

TABLE II

P = 4.0

N	R	S	COVARIANCE
16	4	11	.03711647
16	4	12	.03385904
16	4	13	.03110898
16	5	5	.09146304
16	5	6	.07562221
16	5	7	.06447396
16	5	8	.05617523
16	5	9	.04973868
16	5	10	.04458789
16	5	11	.04036483
16	5	12	.03683821
16	6	6	.08245520
16	6	7	.07039363
16	6	8	.06139617
16	6	9	.05440610
16	6	10	.04880486
16	6	11	.04420746
16	7	7	.07743941
16	7	8	.06762055
16	7	9	.05997820
16	7	10	.05384505
16	8	8	.07517128
16	8	9	.06674896
17	1	1	.60533284
17	1	2	.21353189
17	1	3	.12896438
17	1	4	.09260082
17	1	5	.07235934
17	1	6	.05943208
17	1	7	.05043675
17	1	8	.04379853
17	1	9	.03868534
17	1	10	.03461629
17	1	11	.03129455
17	1	12	.02852789

N	R	S	COVARIANCE
17	1	13	.02618853
17	1	14	.02419340
17	1	15	.02250145
17	1	16	.02115235
17	1	17	.02065374
17	2	2	.21382447
17	2	3	.13025004
17	2	4	.09391946
17	2	5	.07357403
17	2	6	.06053017
17	2	7	.05142927
17	2	8	.04469985
17	2	9	.03950851
17	2	10	.03537233
17	2	11	.03199252
17	2	12	.02917530
17	2	13	.02679166
17	2	14	.02475779
17	2	15	.02303249
17	2	16	.02165706
17	3	3	.13686479
17	3	4	.09904609
17	3	5	.07775982
17	3	6	.06406767
17	3	7	.05449216
17	3	8	.04739949
17	3	9	.04192060
17	3	10	.03755064
17	3	11	.03397671
17	3	12	.03099559
17	3	13	.02847184
17	3	14	.02631747
17	3	15	.02448949
17	4	4	.10556613
17	4	5	.08305726

N	R	S	COVARIANCE
17	4	6	.06853228
17	4	7	.05835110
17	4	8	.05079683
17	4	9	.04495355
17	4	10	.04028796
17	4	11	.03646895
17	4	12	.03328111
17	4	13	.03058081
17	4	14	.02827470
17	5	5	.08927611
17	5	6	.07377633
17	5	7	.06288620
17	5	8	.05479144
17	5	9	.04852129
17	5	10	.04350924
17	5	11	.03940283
17	5	12	.03597252
17	5	13	.03306503
17	6	6	.07987845
17	6	7	.06817028
17	6	8	.05945077
17	6	9	.05268645
17	6	10	.04727277
17	6	11	.04283286
17	6	12	.03912087
17	7	7	.07436785
17	7	8	.06492339
17	7	9	.05758445
17	7	10	.05170290
17	7	11	.04687389
17	8	8	.07143886
17	8	9	.06342422
17	8	10	.05699131
17	9	9	.07051742
18	1	1	.60833881

N	R	S	COVARIANCE
18	1	2	.21381650
18	1	3	.12894712
18	1	4	.09251906
18	1	5	.07226774
18	1	6	.05934794
18	1	7	.05036636
18	1	8	.04374425
18	1	9	.03864785
18	1	10	.03459545
18	1	11	.03128947
18	1	12	.02853663
18	1	13	.02620686
18	1	14	.02421152
18	1	15	.02249311
18	1	16	.02102753
18	1	17	.01986403
18	1	18	.01949842
18	2	2	.21313691
18	2	3	.12960166
18	2	4	.09336707
18	2	5	.07310608
18	2	6	.06013212
18	2	7	.05108961
18	2	8	.04440989
18	2	9	.03926167
18	2	10	.03516335
18	2	11	.03181680
18	2	12	.02902806
18	2	13	.02666643
18	2	14	.02464277
18	2	15	.02289932
18	2	16	.02141207
18	2	17	.02023169
18	3	3	.13558414
18	3	4	.09801359

TABLE II

P = 4.0

N	R	S	COVARIANCE	N	R	S	COVARIANCE	N	R	S	COVARIANCE	N	R	S	COVARIANCE
18	3	5	.07690392	18	6	7	.06629116	19	1	17	.01973924	19	4	4	.10259529
18	3	6	.06334401	18	6	8	.05780517	19	1	18	.01872920	19	4	5	.08060774
18	3	7	.05387215	18	6	9	.05123053	19	1	19	.01847270	19	4	6	.06646106
18	3	8	.04686374	18	6	10	.04597469	19	2	2	.21259038	19	4	7	.05656970
18	3	9	.04145531	18	6	11	.04166824	19	2	3	.12905544	19	4	8	.04924676
18	3	10	.03714547	18	6	12	.03806941	19	2	4	.09289322	19	4	9	.04359397
18	3	11	.03362332	18	6	13	.03501462	19	2	5	.07270061	19	4	10	.03908893
18	3	12	.03068626	18	7	7	.07181254	19	2	6	.05978470	19	4	11	.03540717
18	3	13	.02819764	18	7	8	.06267884	19	2	7	.05079127	19	4	12	.03233673
18	3	14	.02606421	18	7	9	.05559163	19	2	8	.04415367	19	4	13	.02973376
18	3	15	.02422555	18	7	10	.04991904	19	2	9	.03904220	19	4	14	.02749829
18	3	16	.02265680	18	7	11	.04526638	19	2	10	.03497636	19	4	15	.02556097
18	4	4	.10397498	18	7	12	.04137492	19	2	11	.03165871	19	4	16	.02387771
18	4	5	.08174699	18	8	8	.06839330	19	2	12	.02889547	19	5	5	.08579616
18	4	6	.06742556	18	8	9	.06071154	19	2	13	.02655545	19	5	6	.07083318
18	4	7	.05740021	18	8	10	.05455462	19	2	14	.02454760	19	5	7	.06034945
18	4	8	.04997028	18	8	11	.04949894	19	2	15	.02280880	19	5	8	.05257584
18	4	9	.04422934	18	9	9	.06681272	19	2	16	.02129882	19	5	9	.04656783
18	4	10	.03964994	18	9	10	.06008565	19	2	17	.02000543	19	5	10	.04177501
18	4	11	.03590442	19	1	1	.61134910	19	2	18	.01898533	19	5	11	.03785491
18	4	12	.03277898	19	1	2	.21415061	19	3	3	.13448023	19	5	12	.03458351
18	4	13	.03012928	19	1	3	.12897010	19	3	4	.09711705	19	5	13	.03180864
18	4	14	.02785671	19	1	4	.09246872	19	3	5	.07615722	19	5	14	.02942443
18	4	15	.02589747	19	1	5	.07220054	19	3	6	.06271029	19	5	15	.02735742
18	5	5	.08740849	19	1	6	.05928249	19	3	7	.05332732	19	6	6	.07584016
18	5	6	.07219784	19	1	7	.05030986	19	3	8	.04639135	19	6	7	.06468195
18	5	7	.06152656	19	1	8	.04369971	19	3	9	.04104361	19	6	8	.05639465
18	5	8	.05360474	19	1	9	.03861650	19	3	10	.03678574	19	6	9	.04998135
18	5	9	.04747575	19	1	10	.03457761	19	3	11	.03330867	19	6	10	.04485987
18	5	10	.04258168	19	1	11	.03128490	19	3	12	.03041078	19	6	11	.04066737
18	5	11	.03857535	19	1	12	.02854445	19	3	13	.02795541	19	6	12	.03716613
18	5	12	.03522990	19	1	13	.02622514	19	3	14	.02584765	19	6	13	.03419453
18	5	13	.03239200	19	1	14	.02423607	19	3	15	.02402165	19	6	14	.03164000
18	5	14	.02995686	19	1	15	.02251423	19	3	16	.02243554	19	7	7	.06965275
18	6	6	.07770225	19	1	16	.02101945	19	3	17	.02107677	19	7	8	.06078085

TABLE II

P = 4.0

N	R	S	COVARIANCE
19	7	9	.05390558
19	7	10	.04840898
19	7	11	.04390518
19	7	12	.04014104
19	7	13	.03694419
19	8	8	.06585920
19	8	9	.05845412
19	8	10	.05252656
19	8	11	.04766462
19	8	12	.04359763
19	9	9	.06379044
19	9	10	.05736265
19	9	11	.05208429
19	10	10	.06313189
20	1	1	.61435308
20	1	2	.21452473
20	1	3	.12902651
20	1	4	.09244480
20	1	5	.07215406
20	1	6	.05923309
20	1	7	.05026549
20	1	8	.04366388
20	1	9	.03859083
20	1	10	.03456279
20	1	11	.03128107
20	1	12	.02855131
20	1	13	.02624181
20	1	14	.02426059
20	1	15	.02254243
20	1	16	.02104169
20	1	17	.01973023
20	1	18	.01860377
20	1	19	.01772198
20	1	20	.01755579
20	2	2	.21216115
20	2	3	.12859480
20	2	4	.09248541
20	2	5	.07234789
20	2	6	.05948019
20	2	7	.05052814
20	2	8	.04392637
20	2	9	.03884633
20	2	10	.03480843
20	2	11	.03151584
20	2	12	.02877509
20	2	13	.02645491
20	2	14	.02446355
20	2	15	.02273588
20	2	16	.02122633
20	2	17	.01990685
20	2	18	.01877340
20	2	19	.01788650
20	3	3	.13352332
20	3	4	.09633379
20	3	5	.07550167
20	3	6	.06215182
20	3	7	.05284555
20	3	8	.04597225
20	3	9	.04067714
20	3	10	.03646441
20	3	11	.03302668
20	3	12	.03016334
20	3	13	.02773813
20	3	14	.02565572
20	3	15	.02384840
20	3	16	.02226879
20	3	17	.02088778
20	3	18	.01970140
20	4	4	.10138977
20	4	5	.07960944
20	4	6	.06561391
20	4	7	.05583865
20	4	8	.04860851
20	4	9	.04303222
20	4	10	.03859180
20	4	11	.03496561
20	4	12	.03194346
20	4	13	.02938243
20	4	14	.02718246
20	4	15	.02527241
20	4	16	.02360254
20	4	17	.02214231
20	5	5	.08439128
20	5	6	.06964233
20	5	7	.05932078
20	5	8	.05167537
20	5	9	.04577203
20	5	10	.04106685
20	5	11	.03722154
20	5	12	.03401475
20	5	13	.03129582
20	5	14	.02895916
20	5	15	.02692969
20	5	16	.02515487
20	6	6	.07422916
20	6	7	.06328849
20	6	8	.05517206
20	6	9	.04889748
20	6	10	.04389155
20	6	11	.03979717
20	6	12	.03638040
20	6	13	.03348179
20	6	14	.03098954
20	6	15	.02882406
20	7	7	.06780294
20	7	8	.05915439
20	7	9	.05245984
20	7	10	.04711326
20	7	11	.04273649
20	7	12	.03908142
20	7	13	.03597876
20	7	14	.03330967
20	8	8	.06371664
20	8	9	.05654498
20	8	10	.05081084
20	8	11	.04611237
20	8	12	.04218551
20	8	13	.03884989
20	9	9	.06127588
20	9	10	.05509711
20	9	11	.05002900
20	9	12	.04578946
20	10	10	.06013006
20	10	11	.05463244

****

TABLE II

P = 4.5

N	R	S	COVARIANCE	N	R	S	COVARIANCE	N	R	S	COVARIANCE	N	R	S	COVARIANCE
2	1	1	.69644188	7	1	3	.13560781	9	1	2	.21063523	10	2	3	.14193606
2	1	2	.30355812	7	1	4	.09853969	9	1	3	.13113243	10	2	4	.10383885
3	1	1	.62009671	7	1	5	.07696778	9	1	4	.09538434	10	2	5	.08184772
3	1	2	.25980212	7	1	6	.06266339	9	1	5	.07483713	10	2	6	.06738637
3	1	3	.16340153	7	1	7	.05292559	9	1	6	.06134232	10	2	7	.05705354
3	2	2	.39379506	7	2	2	.24773322	9	1	7	.05169862	10	2	8	.04923692
4	1	1	.58667750	7	2	3	.15732271	9	1	8	.04443760	10	2	9	.04312415
4	1	2	.24029952	7	2	4	.11526945	9	1	9	.03925033	10	3	3	.15944540
4	1	3	.15077060	7	2	5	.09050729	9	2	2	.23121541	10	3	4	.11731533
4	1	4	.10885309	7	2	6	.07396302	9	2	3	.14576981	10	3	5	.09280077
4	2	2	.31727811	7	3	3	.18881653	9	2	4	.10672114	10	3	6	.07659299
4	2	3	.20505106	7	3	4	.13960268	9	2	5	.08406586	10	3	7	.06496740
5	1	1	.56881396	7	3	5	.11026685	9	2	6	.06909487	10	3	8	.05614766
5	1	2	.22910968	7	4	4	.17537242	9	2	7	.05835018	10	4	4	.13430964
5	1	3	.14370750	8	1	1	.54775844	9	2	8	.05023634	10	4	5	.10666879
5	1	4	.10385089	8	1	2	.21331677	9	3	3	.16640857	10	4	6	.08828689
5	1	5	.08082760	8	1	3	.13308047	9	3	4	.12261987	10	4	7	.07504530
5	2	2	.28180817	8	1	4	.09678206	9	3	5	.09698462	10	5	5	.12473573
5	2	3	.18074045	8	1	5	.07580492	9	3	6	.07994098	10	5	6	.10359356
5	2	4	.13225584	8	1	6	.06195446	9	3	7	.06765546	11	1	1	.54311601
5	3	3	.24295478	8	1	7	.05204473	9	4	4	.14333963	11	1	2	.20701417
6	1	1	.55830109	8	1	8	.04507685	9	4	5	.11391117	11	1	3	.12836554
6	1	2	.22189662	8	2	2	.23824221	9	4	6	.09421161	11	1	4	.09332292
6	1	3	.13898780	8	2	3	.15071273	9	5	5	.13713620	11	1	5	.07333696
6	1	4	.10080437	8	2	4	.11040099	10	1	1	.54382020	11	1	6	.06031271
6	1	5	.07840854	8	2	5	.08686131	10	1	2	.20859309	11	1	7	.05107714
6	1	6	.06402869	8	2	6	.07121263	10	1	3	.12959683	11	1	8	.04413181
6	2	2	.26120946	8	2	7	.05996487	10	1	4	.09425257	11	1	9	.03868419
6	2	3	.16662333	8	3	3	.17571492	10	1	5	.07402435	11	1	10	.03431995
6	2	4	.12204355	8	3	4	.12968449	10	1	6	.06079657	11	1	11	.03120061
6	2	5	.09554072	8	3	5	.10252551	10	1	7	.05138237	11	2	2	.22157046
6	3	3	.20872620	8	3	6	.08434381	10	1	8	.04428114	11	2	3	.13887980
6	3	4	.15466542	8	4	4	.15604471	10	1	9	.03873870	11	2	4	.10152017
7	1	1	.55181628	8	4	5	.12409206	10	1	10	.03476132	11	2	5	.08004336
7	1	2	.21691535	9	1	1	.54526222	10	2	2	.22582253	11	2	6	.06597513

TABLE II

P = 4.5

N	R	S	COVARIANCE	N	R	S	COVARIANCE	N	R	S	COVARIANCE	N	R	S	COVARIANCE
11	2	7	.05596288	12	2	5	.07854701	13	1	9	.03855843	13	5	8	.06430570
11	2	8	.04841319	12	2	6	.06478962	13	1	10	.03432569	13	5	9	.05680193
11	2	9	.04247962	12	2	7	.05503020	13	1	11	.03082847	13	6	6	.09748207
11	2	10	.03771969	12	2	8	.04769538	13	1	12	.02793102	13	6	7	.08338516
11	3	3	.15403579	12	2	9	.04194288	13	1	13	.02591609	13	6	8	.07271551
11	3	4	.11317851	12	2	10	.03728975	13	2	2	.21534529	13	7	7	.09545292
11	3	5	.08952039	12	2	11	.03348591	13	2	3	.13432901	14	1	1	.54367011
11	3	6	.07394810	12	3	3	.14971193	13	2	4	.09802524	14	1	2	.20405115
11	3	7	.06282681	12	3	4	.10985890	13	2	5	.07728668	14	1	3	.12586578
11	3	8	.05441898	12	3	5	.08687495	13	2	6	.06378026	14	1	4	.09135466
11	3	9	.04779798	12	3	6	.07180059	13	2	7	.05422413	14	1	5	.07181851
11	4	4	.12754168	12	3	7	.06107303	13	2	8	.04706216	14	1	6	.05917726
11	4	5	.10123115	12	3	8	.05299143	13	2	9	.04146065	14	1	7	.05027945
11	4	6	.08382400	12	3	9	.04664163	13	2	10	.03693431	14	1	8	.04364099
11	4	7	.07134568	12	3	10	.04149817	13	2	11	.03319049	14	1	9	.03847052
11	4	8	.06188510	12	4	4	.12227034	13	2	12	.03008673	14	1	10	.03430750
11	5	5	.11593216	12	4	5	.09698714	13	3	3	.14617824	14	1	11	.03086761
11	5	6	.09626901	12	4	6	.08032875	13	3	4	.10713505	14	1	12	.02797422
11	5	7	.08211324	12	4	7	.06843441	13	3	5	.08469432	14	1	13	.02555154
11	6	6	.11256581	12	4	8	.05945109	13	3	6	.07001978	14	1	14	.02390305
12	1	1	.54294286	12	4	9	.05237888	13	3	7	.05960696	14	2	2	.21302139
12	1	2	.20578115	12	5	5	.10933590	13	3	8	.05178587	14	2	3	.13259727
12	1	3	.12736525	12	5	6	.09077591	13	3	9	.04565849	14	2	4	.09667880
12	1	4	.09255041	12	5	7	.07747465	13	3	10	.04070062	14	2	5	.07621149
12	1	5	.07275172	12	5	8	.06740011	13	3	11	.03659569	14	2	6	.06291120
12	1	6	.05988561	12	6	6	.10393951	13	4	4	.11804350	14	2	7	.05352147
12	1	7	.05078930	12	6	7	.08890210	13	4	5	.09357655	14	2	8	.04650046
12	1	8	.04396999	13	1	1	.54316034	13	4	6	.07751062	14	2	9	.04102271
12	1	9	.03863202	13	1	2	.20481281	13	4	7	.06607619	14	2	10	.03660634
12	1	10	.03432046	13	1	3	.12654444	13	4	8	.05746815	14	2	11	.03295322
12	1	11	.03079912	13	1	4	.09190241	13	4	9	.05071219	14	2	12	.02987798
12	1	12	.02830933	13	1	5	.07225038	13	4	10	.04523799	14	2	13	.02730188
12	2	2	.21814718	13	1	6	.05950918	13	5	5	.10419681	14	3	3	.14323839
12	2	3	.13639081	13	1	7	.05052300	13	5	6	.08649068	14	3	4	.10485977
12	2	4	.09961572	13	1	8	.04380400	13	5	7	.07384759	14	3	5	.08286501

TABLE II

P = 4.5

N	R	S	COVARIANCE	N	R	S	COVARIANCE	N	R	S	COVARIANCE	N	R	S	COVARIANCE
14	3	6	.06851799	15	1	9	.03837582	15	4	8	.05441781	16	1	15	.02183571
14	3	7	.05836196	15	1	10	.03427070	15	4	9	.04812452	16	1	16	.02070604
14	3	8	.05075260	15	1	11	.03088452	15	4	10	.04304965	16	2	2	.20942513
14	3	9	.04480648	15	1	12	.02803293	15	4	11	.03885120	16	2	3	.12986408
14	3	10	.04000650	15	1	13	.02559983	15	4	12	.03530717	16	2	4	.09452953
14	3	11	.03603212	15	1	14	.02354686	15	5	5	.09668641	16	2	5	.07447771
14	3	12	.03268389	15	1	15	.02218666	15	5	6	.08021461	16	2	6	.06149374
14	4	4	.11457618	15	2	2	.21107306	15	5	7	.06851743	16	2	7	.05235871
14	4	5	.09077251	15	2	3	.13112602	15	5	8	.05973677	16	2	8	.04555234
14	4	6	.07518653	15	2	4	.09552599	15	5	9	.05286902	16	2	9	.04026229
14	4	7	.06412318	15	2	5	.07528440	15	5	10	.04732372	16	2	10	.03601457
14	4	8	.05581675	15	2	6	.06215578	15	5	11	.04273120	16	2	11	.03251387
14	4	9	.04931535	15	2	7	.05290434	15	6	6	.08841872	16	2	12	.02956719
14	4	10	.04406029	15	2	8	.04600004	15	6	7	.07563157	16	2	13	.02704544
14	4	11	.03970455	15	2	9	.04062452	15	6	8	.06601113	16	2	14	.02486771
14	5	5	.10007297	15	2	10	.03630001	15	6	9	.05847312	16	2	15	.02301999
14	5	6	.08304696	15	2	11	.03272918	15	6	10	.05237779	16	3	3	.13863606
14	5	7	.07092589	15	2	12	.02971962	15	7	7	.08416929	16	3	4	.10127650
14	5	8	.06180482	15	2	13	.02715013	15	7	8	.07355567	16	3	5	.07996740
14	5	9	.05465309	15	2	14	.02498143	15	7	9	.06522256	16	3	6	.06612319
14	5	10	.04886411	15	3	3	.14075665	15	8	8	.08285236	16	3	7	.05635984
14	6	6	.09245347	15	3	4	.10293120	16	1	1	.54530274	16	3	8	.04907242
14	6	7	.07908526	15	3	5	.08130822	16	1	2	.20298835	16	3	9	.04340069
14	6	8	.06900037	15	3	6	.06723389	16	1	3	.12483072	16	3	10	.03884148
14	6	9	.06107707	15	3	7	.05729103	16	1	4	.09049065	16	3	11	.03508072
14	7	7	.08910721	15	3	8	.04985670	16	1	5	.07111974	16	3	12	.03191282
14	7	8	.07785856	15	3	9	.04406003	16	1	6	.05862497	16	3	13	.02920016
15	1	1	.54440153	15	3	10	.03939125	16	1	7	.04985781	16	3	14	.02685652
15	1	2	.20345363	15	3	11	.03553251	16	1	8	.04333857	16	4	4	.10922274
15	1	3	.12530154	15	3	12	.03227782	16	1	9	.03827947	16	4	5	.08642895
15	1	4	.09088881	15	3	13	.02949745	16	1	10	.03422214	16	4	6	.07157168
15	1	5	.07144468	15	4	4	.11167941	16	1	11	.03088163	16	4	7	.06106925
15	1	6	.05888420	15	4	5	.08842460	16	1	12	.02807205	16	4	8	.05321626
15	1	7	.05005816	15	4	6	.07323492	16	1	13	.02566920	16	4	9	.04709587
15	1	8	.04348514	15	4	7	.06247698	16	1	14	.02359514	16	4	10	.04217053

TABLE II

P = 4.5

N	R	S	COVARIANCE	N	R	S	COVARIANCE	N	R	S	COVARIANCE	N	R	S	COVARIANCE
16	4	11	.03810405	17	1	13	.02572215	17	4	6	.07013660	18	1	2	.20236239
16	4	12	.03467606	17	1	14	.02366952	17	4	7	.05985090	18	1	3	.12410781
16	4	13	.03173890	17	1	15	.02188102	17	4	8	.05217232	18	1	4	.08985523
16	5	5	.09385316	17	1	16	.02035856	17	4	9	.04619769	18	1	5	.07058833
16	5	6	.07784138	17	1	17	.01941585	17	4	10	.04139802	18	1	6	.05819127
16	5	7	.06649508	17	2	2	.20802117	17	4	11	.03744255	18	1	7	.04951323
16	5	8	.05799525	17	2	3	.12877290	17	4	12	.03411393	18	1	8	.04307590
16	5	9	.05136097	17	2	4	.09366121	17	4	13	.03126426	18	1	9	.03809335
16	5	10	.04601573	17	2	5	.07377031	17	4	14	.02879223	18	1	10	.03410885
16	5	11	.04159827	17	2	6	.06090942	17	5	5	.09144641	18	1	11	.03083857
16	5	12	.03787140	17	2	7	.05187342	17	5	6	.07582232	18	1	12	.02809677
16	6	6	.08510477	17	2	8	.04515018	17	5	7	.06477109	18	1	13	.02575696
16	6	7	.07279165	17	2	9	.03993245	17	5	8	.05650677	18	1	14	.02373075
16	6	8	.06354884	17	2	10	.03574950	17	5	9	.05006766	18	1	15	.02195682
16	6	9	.05632304	17	2	11	.03230812	17	5	10	.04488917	18	1	16	.02039923
16	6	10	.05049360	17	2	12	.02941618	17	5	11	.04061767	18	1	17	.01907084
16	6	11	.04567077	17	2	13	.02694323	17	5	12	.03702041	18	1	18	.01828161
16	7	7	.08020910	17	2	14	.02480002	17	5	13	.03393882	18	2	2	.20681798
16	7	8	.07010210	17	2	15	.02293188	17	6	6	.08233118	18	2	3	.12782286
16	7	9	.06218647	17	2	16	.02134148	17	6	7	.07041205	18	2	4	.09289927
16	7	10	.05579109	17	3	3	.13680540	17	6	8	.06148220	18	2	5	.07314580
16	8	8	.07798924	17	3	4	.09984209	17	6	9	.05451435	18	2	6	.06039046
16	8	9	.06925509	17	3	5	.07880082	17	6	10	.04890397	18	2	7	.05143948
17	1	1	.54633496	17	3	6	.06515305	17	6	11	.04427166	18	2	8	.04478752
17	1	2	.20263092	17	3	7	.05554270	17	6	12	.04036733	18	2	9	.03963161
17	1	3	.12443705	17	3	8	.04838015	17	7	7	.07695713	18	2	10	.03550386
17	1	4	.09014907	17	3	9	.04281426	17	7	8	.06726365	18	2	11	.03211295
17	1	5	.07083629	17	3	10	.03834758	17	7	9	.05968772	18	2	12	.02926789
17	1	6	.05839523	17	3	11	.03466967	17	7	10	.05357966	18	2	13	.02683845
17	1	7	.04967674	17	3	12	.03157680	17	7	11	.04853090	18	2	14	.02473352
17	1	8	.04320211	17	3	13	.02893050	17	8	8	.07409138	18	2	15	.02288992
17	1	9	.03818469	17	3	14	.02663599	17	8	9	.06580615	18	2	16	.02127067
17	1	10	.03416701	17	3	15	.02463529	17	8	10	.05911624	18	2	17	.01988964
17	1	11	.03086471	17	4	4	.10711307	17	9	9	.07318844	18	3	3	.13521117
17	1	12	.02809180	17	4	5	.08471138	18	1	1	.54746875	18	3	4	.09858766

TABLE II

P = 4.5

N	R	S	COVARIANCE	N	R	S	COVARIANCE	N	R	S	COVARIANCE	N	R	S	COVARIANCE
18	3	5	.07777700	18	6	7	.06838715	19	1	17	.01910600	19	4	4	.10367851
18	3	6	.06429852	18	6	8	.05972082	19	1	18	.01793854	19	4	5	.08190615
18	3	7	.05481995	18	6	9	.05296958	19	1	19	.01727668	19	4	6	.06778466
18	3	8	.04776472	18	6	10	.04754262	19	2	2	.20578190	19	4	7	.05784618
18	3	9	.04228951	18	6	11	.04306950	19	2	3	.12699086	19	4	8	.05044614
18	3	10	.03790183	18	6	12	.03930599	19	2	4	.09222664	19	4	9	.04470338
18	3	11	.03429450	18	6	13	.03608474	19	2	5	.07259122	19	4	10	.04010268
18	3	12	.03126585	18	7	7	.07423552	19	2	6	.05992703	19	4	11	.03632233
18	3	13	.02867820	18	7	8	.06488596	19	2	7	.05104959	19	4	12	.03315093
18	3	14	.02643517	18	7	9	.05759175	19	2	8	.04445927	19	4	13	.03044380
18	3	15	.02446987	18	7	10	.05172131	19	2	9	.03935670	19	4	14	.02809888
18	3	16	.02274327	18	7	11	.04687784	19	2	10	.03527645	19	4	15	.02604302
18	4	4	.10528210	18	7	12	.04279932	19	2	11	.03192882	19	4	16	.02422521
18	4	5	.08321743	18	8	8	.07089188	19	2	12	.02912396	19	5	5	.08757492
18	4	6	.06888538	18	8	9	.06297318	19	2	13	.02673226	19	5	6	.07256737
18	4	7	.05878568	18	8	10	.05659162	19	2	14	.02466241	19	5	7	.06198432
18	4	8	.05125644	18	8	11	.05132058	19	2	15	.02284907	19	5	8	.05409252
18	4	9	.04540626	18	9	9	.06934305	19	2	16	.02124667	19	5	9	.04796091
18	4	10	.04071352	18	9	10	.06236373	19	2	17	.01983049	19	5	10	.04304402
18	4	11	.03685231	19	1	1	.54868144	19	2	18	.01862209	19	5	11	.03900071
18	4	12	.03360835	19	1	2	.20216782	19	3	3	.13381234	19	5	12	.03560648
18	4	13	.03083520	19	1	3	.12383284	19	3	4	.09748227	19	5	13	.03270756
18	4	14	.02843025	19	1	4	.08960199	19	3	5	.07687170	19	5	14	.03019533
18	4	15	.02632229	19	1	5	.07037092	19	3	6	.06354033	19	5	15	.02799192
18	5	5	.08937574	19	1	6	.05800990	19	3	7	.05417624	19	6	6	.07794400
18	5	6	.07408261	19	1	7	.04936560	19	3	8	.04721412	19	6	7	.06664174
18	5	7	.06328284	19	1	8	.04295964	19	3	9	.04181739	19	6	8	.05820030
18	5	8	.05521876	19	1	9	.03800652	19	3	10	.03749787	19	6	9	.05163347
18	5	9	.04894513	19	1	10	.03405000	19	3	11	.03395123	19	6	10	.04636226
18	5	10	.04390760	19	1	11	.03080672	19	3	12	.03097776	19	6	11	.04202396
18	5	11	.03975921	19	1	12	.02809125	19	3	13	.02844093	19	6	12	.03837[illegible]56
18	5	12	.03627156	19	1	13	.02577715	19	3	14	.02624450	19	6	13	.03526511
18	5	13	.03328833	19	1	14	.02377548	19	3	15	.02431955	19	6	14	.03256475
18	5	14	.03069992	19	1	15	.02202261	19	3	16	.02261800	19	7	7	.07192201
18	6	6	.07997374	19	1	16	.02047417	19	3	17	.02111389	19	7	8	.06286291

TABLE II

P = 4.5

N	R	S	COVARIANCE
19	7	9	.05580610
19	7	10	.05013537
19	7	11	.04546401
19	7	12	.04153682
19	7	13	.03817852
19	8	8	.06821487
19	8	9	.06060107
19	8	10	.05447536
19	8	11	.04942413
19	8	12	.04517395
19	9	9	.06618817
19	9	10	.05953791
19	9	11	.05404792
19	10	10	.06554223
20	1	1	.54995537
20	1	2	.20203529
20	1	3	.12360397
20	1	4	.08938357
20	1	5	.07017999
20	1	6	.05784839
20	1	7	.04923231
20	1	8	.04285284
20	1	9	.03792469
20	1	10	.03399196
20	1	11	.03077160
20	1	12	.02807849
20	1	13	.02578636
20	1	14	.02380619
20	1	15	.02207367
20	1	16	.02054202
20	1	17	.01917865
20	1	18	.01796783
20	1	19	.01693531
20	1	20	.01638014
20	2	2	.20488632
20	2	3	.12625855
20	2	4	.09162975
20	2	5	.07209619
20	2	6	.05951118
20	2	7	.05069778
20	2	8	.04416113
20	2	9	.03910497
20	2	10	.03506592
20	2	11	.03175573
20	2	12	.02898557
20	2	13	.02662650
20	2	14	.02458749
20	2	15	.02280275
20	2	16	.02122437
20	2	17	.01981902
20	2	18	.01857070
20	2	19	.01750625
20	3	3	.13257690
20	3	4	.09650174
20	3	5	.07606594
20	3	6	.06286332
20	3	7	.05359945
20	3	8	.04671877
20	3	9	.04139054
20	3	10	.03713036
20	3	11	.03363640
20	3	12	.03071069
20	3	13	.02821788
20	3	14	.02606235
20	3	15	.02417493
20	3	16	.02250523
20	3	17	.02101821
20	3	18	.01969712
20	4	4	.10226300
20	4	5	.08074614
20	4	6	.06680882
20	4	7	.05701125
20	4	8	.04972403
20	4	9	.04407486
20	4	10	.03955414
20	4	11	.03584387
20	4	12	.03273518
20	4	13	.03008514
20	4	14	.02779267
20	4	15	.02578462
20	4	16	.02400766
20	4	17	.02242472
20	5	5	.08599421
20	5	6	.07123541
20	5	7	.06084096
20	5	8	.05309886
20	5	9	.04709042
20	5	10	.04227790
20	5	11	.03832525
20	5	12	.03501145
20	5	13	.03218510
20	5	14	.02973902
20	5	15	.02759561
20	5	16	.02569826
20	6	6	.07617721
20	6	7	.06512075
20	6	8	.05687342
20	6	9	.05046545
20	6	10	.04532809
20	6	11	.04110537
20	6	12	.03756285
20	6	13	.03453974
20	6	14	.03192215
20	6	15	.02962751
20	7	7	.06992962
20	7	8	.06111908
20	7	9	.05426500
20	7	10	.04876443
20	7	11	.04423938
20	7	12	.04044054
20	7	13	.03719675
20	7	14	.03438664
20	8	8	.06593954
20	8	9	.05858335
20	8	10	.05267332
20	8	11	.04780695
20	8	12	.04371840
20	8	13	.04022491
20	9	9	.06354942
20	9	10	.05717294
20	9	11	.05191712
20	9	12	.04749754
20	10	10	.06242555
20	10	11	.05671977

****

TABLE II

P = 5.0

N	R	S	COVARIANCE	N	R	S	COVARIANCE	N	R	S	COVARIANCE	N	R	S	COVARIANCE
2	1	1	.69419594	7	1	3	.13542790	9	1	2	.20672413	10	2	3	.14282333
2	1	2	.30580406	7	1	4	.09872845	9	1	3	.13008177	10	2	4	.10501017
3	1	1	.61149074	7	1	5	.07707764	9	1	4	.09507701	10	2	5	.08296574
3	1	2	.26210614	7	1	6	.06245765	9	1	5	.07473889	10	2	6	.06834441
3	1	3	.16384685	7	1	7	.05196408	9	1	6	.06124771	10	2	7	.05780114
3	2	2	.40090025	7	2	2	.24926563	9	1	7	.05148979	10	2	8	.04972695
4	1	1	.57317151	7	2	3	.15965833	9	1	8	.04398991	10	2	9	.04326906
4	1	2	.24128447	7	2	4	.11731054	9	1	9	.03823239	10	3	3	.16161890
4	1	3	.15191542	7	2	5	.09204114	9	2	2	.23098024	10	3	4	.11947580
4	1	4	.10851606	7	2	6	.07484904	9	2	3	.14708657	10	3	5	.09471533
4	2	2	.32287516	7	3	3	.19274326	9	2	4	.10816973	10	3	6	.07820812
4	2	3	.20903748	7	3	4	.14284351	9	2	5	.08535366	10	3	7	.06625904
5	1	1	.55139871	7	3	5	.11270990	9	2	6	.07012817	10	3	8	.05708215
5	1	2	.22883715	7	4	4	.17969712	9	2	7	.05906913	10	4	4	.13728521
5	1	3	.14450220	8	1	1	.52181853	9	2	8	.05054339	10	4	5	.10925014
5	1	4	.10428714	8	1	2	.21015459	9	3	3	.16908641	10	4	6	.09045129
5	1	5	.08012994	8	1	3	.13244690	9	3	4	.12511434	10	4	7	.07678692
5	2	2	.28579782	8	1	4	.09672470	9	3	5	.09910889	10	5	5	.12795250
5	2	3	.18430122	8	1	5	.07583646	9	3	6	.08165183	10	5	6	.10628143
5	2	4	.13459344	8	1	6	.06187390	9	3	7	.06891726	11	1	1	.51110999
5	3	3	.24844933	8	1	7	.05169120	9	4	4	.14672342	11	1	2	.20181110
6	1	1	.53762942	8	1	8	.04407432	9	4	5	.11675275	11	1	3	.12657922
6	1	2	.22052798	8	2	2	.23882734	9	4	6	.09649928	11	1	4	.09254445
6	1	3	.13929388	8	2	3	.15250958	9	5	5	.14067612	11	1	5	.07294111
6	1	4	.10119900	8	2	4	.11214377	10	1	1	.51365962	11	1	6	.06006300
6	1	5	.07844186	8	2	5	.08830211	10	1	2	.20400614	11	1	7	.05085928
6	1	6	.06315462	8	2	6	.07225886	10	1	3	.12816321	11	1	8	.04387505
6	2	2	.26385796	8	2	7	.06050440	10	1	4	.09370356	11	1	9	.03832644
6	2	3	.16955807	8	3	3	.17897157	10	1	5	.07377913	11	1	10	.03377141
6	2	4	.12434149	8	3	4	.13254352	10	1	6	.06063543	11	1	11	.03018965
6	2	5	.09696973	8	3	5	.10484437	10	1	7	.05119411	11	2	2	.21996683
6	3	3	.21342232	8	3	6	.08607671	10	1	8	.04398445	11	2	3	.13938003
6	3	4	.15824141	8	4	4	.15988524	10	1	9	.03822997	11	2	4	.10243388
7	1	1	.52834339	8	4	5	.12718197	10	1	10	.03374260	11	2	5	.08099074
7	1	2	.21459289	9	1	1	.51711096	10	2	2	.22486470	11	2	6	.06683346

TABLE II

P = 5.0

N	R	S	COVARIANCE	N	R	S	COVARIANCE	N	R	S	COVARIANCE	N	R	S	COVARIANCE
11	2	7	.05667966	12	2	5	.07932910	13	1	9	.03825905	13	5	8	.06591405
11	2	8	.04895408	12	2	6	.06553947	13	1	10	.03398211	13	5	9	.05815173
11	2	9	.04280413	12	2	7	.05568950	13	1	11	.03039764	13	6	6	.10002338
11	2	10	.03774803	12	2	8	.04823207	13	1	12	.02734059	13	6	7	.08560737
11	3	3	.15576460	12	2	9	.04233012	13	1	13	.02493354	13	6	8	.07462884
11	3	4	.11503587	12	2	10	.03749131	13	2	2	.21262436	13	7	7	.09803642
11	3	5	.09123167	12	2	11	.03342844	13	2	3	.13415598	14	1	1	.50686178
11	3	6	.07544163	12	3	3	.15104413	13	2	4	.09847925	14	1	2	.19726743
11	3	7	.06407834	12	3	4	.11144103	13	2	5	.07791113	14	1	3	.12316633
11	3	8	.05541032	12	3	5	.08839255	13	2	6	.06441993	14	1	4	.08996355
11	3	9	.04849647	12	3	6	.07316525	13	2	7	.05481367	14	1	5	.07099240
11	4	4	.13015534	12	3	7	.06225486	13	2	8	.04756803	14	1	6	.05862984
11	4	5	.10356413	12	3	8	.05397537	13	2	9	.04186010	14	1	7	.04987505
11	4	6	.08583742	12	3	9	.04741093	13	2	10	.03720469	14	1	8	.04330511
11	4	7	.07303368	12	3	10	.04202132	13	2	11	.03329872	14	1	9	.03815638
11	4	8	.06323949	12	4	4	.12456192	13	2	12	.02996487	14	1	10	.03398074
11	5	5	.11884515	12	4	5	.09908974	13	3	3	.14715303	14	1	11	.03049706
11	5	6	.09876932	12	4	6	.08218640	13	3	4	.10846623	14	1	12	.02752216
11	5	7	.08420825	12	4	7	.07003580	13	3	5	.08603043	14	1	13	.02495158
11	6	6	.11555468	12	4	8	.06079226	13	3	6	.07125616	14	1	14	.02293773
12	1	1	.50922716	12	4	9	.05344915	13	3	7	.06070658	14	2	2	.20980938
12	1	2	.20001152	12	5	5	.11197259	13	3	8	.05273247	14	2	3	.13212783
12	1	3	.12525261	12	5	6	.09308661	13	3	9	.04644021	14	2	4	.09692725
12	1	4	.09155546	12	5	7	.07946176	13	3	10	.04130142	14	2	5	.07668658
12	1	5	.07220743	12	5	8	.06906770	13	3	11	.03698534	14	2	6	.06344236
12	1	6	.05953838	12	6	6	.10669787	13	4	4	.12004641	14	2	7	.05403602
12	1	7	.05051853	12	6	7	.09126991	13	4	5	.09546723	14	2	8	.04696190
12	1	8	.04370631	13	1	1	.50784938	13	4	6	.07921646	14	2	9	.04140880
12	1	9	.03832516	13	1	2	.19851835	13	4	7	.06757867	14	2	10	.03689927
12	1	10	.03391984	13	1	3	.12412841	13	4	8	.05876239	14	2	11	.03313300
12	1	11	.03022485	13	1	4	.09070353	13	4	9	.05179337	14	2	12	.02991406
12	1	12	.02731114	13	1	5	.07156233	13	4	10	.04609384	14	2	13	.02713099
12	2	2	.21595956	13	1	6	.05906167	13	5	5	.10658397	14	3	3	.14388818
12	2	3	.13653950	13	1	7	.05018804	13	5	6	.08862025	14	3	4	.10596095
12	2	4	.10029086	13	1	8	.04351030	13	5	7	.07571473	14	3	5	.08403171

TABLE II

P = 5.0

N	R	S	COVARIANCE	N	R	S	COVARIANCE	N	R	S	COVARIANCE	N	R	S	COVARIANCE
14	3	6	.06963005	15	1	9	.03803459	15	4	8	.05558041	16	1	15	.02122940
14	3	7	.05937489	15	1	10	.03393963	15	4	9	.04914585	16	1	16	.01977688
14	3	8	.05164721	15	1	11	.03053695	15	4	10	.04392601	16	2	2	.20533286
14	3	9	.04557176	15	1	12	.02764263	15	4	11	.03957592	16	2	3	.12886379
14	3	10	.04063195	15	1	13	.02513305	15	4	12	.03586685	16	2	4	.09440678
14	3	11	.03650222	15	1	14	.02294206	15	5	5	.09864287	16	2	5	.07467805
14	3	12	.03296981	15	1	15	.02123930	15	5	6	.08201713	16	2	6	.06181833
14	4	4	.11631834	15	2	2	.20740565	15	5	7	.07014551	16	2	7	.05272046
14	4	5	.09246844	15	2	3	.13038182	15	5	8	.06118754	16	2	8	.04590747
14	4	6	.07674780	15	2	4	.09558272	15	5	9	.05414371	16	2	9	.04058595
14	4	7	.06552358	15	2	5	.07561826	15	5	10	.04842231	16	2	10	.03629047
14	4	8	.05704855	15	2	6	.06258168	15	5	11	.04364922	16	2	11	.03272910
14	4	9	.05037490	15	2	7	.05334230	15	6	6	.09057705	16	2	12	.02970878
14	4	10	.04494172	15	2	8	.04641002	15	6	7	.07757081	16	2	13	.02709682
14	4	11	.04039476	15	2	9	.04098331	15	6	8	.06773522	16	2	14	.02480223
14	5	5	.10223445	15	2	10	.03659128	15	6	9	.05998779	16	2	15	.02278192
14	5	6	.08500701	15	2	11	.03293804	15	6	10	.05368590	16	3	3	.13871310
14	5	7	.07267183	15	2	12	.02982801	15	7	7	.08641727	16	3	4	.10196947
14	5	8	.06333748	15	2	13	.02712961	15	7	8	.07555126	16	3	5	.08082811
14	5	9	.05597435	15	2	14	.02477275	15	7	9	.06697515	16	3	6	.06700308
14	5	10	.04997137	15	3	3	.14110857	15	8	8	.08512610	16	3	7	.05719982
14	6	6	.09479485	15	3	4	.10382046	16	1	1	.50574344	16	3	8	.04984596
14	6	7	.08116287	15	3	5	.08231679	16	1	2	.19531413	16	3	9	.04409427
14	6	8	.07082184	15	3	6	.06822698	16	1	3	.12161484	16	3	10	.03944657
14	6	9	.06264855	15	3	7	.05821673	16	1	4	.08874618	16	3	11	.03558979
14	7	7	.09151818	15	3	8	.05069229	16	1	5	.07003523	16	3	12	.03231656
14	7	8	.07997004	15	3	9	.04479359	16	1	6	.05788428	16	3	13	.02948417
15	1	1	.50618061	15	3	10	.04001409	16	1	7	.04931028	16	3	14	.02699473
15	1	2	.19621131	15	3	11	.03603484	16	1	8	.04290202	16	4	4	.11051020
15	1	3	.12233618	15	3	12	.03264466	16	1	9	.03790411	16	4	5	.08777971
15	1	4	.08931616	15	3	13	.02970139	16	1	10	.03387465	16	4	6	.07286842
15	1	5	.07048645	15	4	4	.11318428	16	1	11	.03053706	16	4	7	.06227081
15	1	6	.05823878	15	4	5	.08994115	16	1	12	.02770879	16	4	8	.05430736
15	1	7	.04958231	15	4	6	.07465983	16	1	13	.02526453	16	4	9	.04807046
15	1	8	.04310060	15	4	7	.06377642	16	1	14	.02311838	16	4	10	.04302519

TABLE II

P = 5.0

N	R	S	COVARIANCE	N	R	S	COVARIANCE	N	R	S	COVARIANCE	N	R	S	COVARIANCE
16	4	11	.03883474	17	1	13	.02534719	17	4	6	.07131296	18	1	2	.19389207
16	4	12	.03527566	17	1	14	.02325455	17	4	7	.06095852	18	1	3	.12043039
16	4	13	.03219402	17	1	15	.02139782	17	4	8	.05319200	18	1	4	.08779269
16	5	5	.09562238	17	1	16	.01975313	17	4	9	.04712136	18	1	5	.06926775
16	5	6	.07949773	17	1	17	.01850478	17	4	10	.04222170	18	1	6	.05726936
16	5	7	.06801050	17	2	2	.20353047	17	4	11	.03816306	18	1	7	.04882530
16	5	8	.05936288	17	2	3	.12753278	17	4	12	.03472694	18	1	8	.04253197
16	5	9	.05258042	17	2	4	.09336986	17	4	13	.03176243	18	1	9	.03763936
16	5	10	.04708743	17	2	5	.07384429	17	4	14	.02916236	18	1	10	.03370958
16	5	11	.04252072	17	2	6	.06113683	17	5	5	.09304369	18	1	11	.03046931
16	5	12	.03863896	17	2	7	.05216038	17	5	6	.07734290	18	1	12	.02773864
16	6	6	.08709546	17	2	8	.04544915	17	5	7	.06617970	18	1	13	.02539392
16	6	7	.07460052	17	2	9	.04021659	17	5	8	.05779251	18	1	14	.02334720
16	6	8	.06517573	17	2	10	.03600190	17	5	9	.05122807	18	1	15	.02153466
16	6	9	.05777211	17	2	11	.03251645	17	5	10	.04592418	18	1	16	.01991225
16	6	10	.05176837	17	2	12	.02956969	17	5	11	.04152681	18	1	17	.01846802
16	6	11	.04677172	17	2	13	.02703033	17	5	12	.03780116	18	1	18	.01738834
16	7	7	.08230552	17	2	14	.02480532	17	5	13	.03458483	18	2	2	.20195192
16	7	8	.07198364	17	2	15	.02283030	17	6	6	.08416791	18	2	3	.12635718
16	7	9	.06386105	17	2	16	.02108040	17	6	7	.07209880	18	2	4	.09244899
16	7	10	.05726478	17	3	3	.13662725	17	6	8	.06301466	18	2	5	.07309999
16	8	8	.08012906	17	3	4	.10035237	17	6	9	.05589459	18	2	6	.06052484
16	8	9	.07115825	17	3	5	.07952297	17	6	10	.05013507	18	2	7	.05165363
17	1	1	.50550284	17	3	6	.06592547	17	6	11	.04535533	18	2	8	.04503021
17	1	2	.19454825	17	3	7	.05629946	17	6	12	.04130241	18	2	9	.03987393
17	1	3	.12098433	17	3	8	.04909089	17	7	7	.07891330	18	2	10	.03572787
17	1	4	.08824158	17	3	9	.04346348	17	7	8	.06903586	18	2	11	.03230624
17	1	5	.06963111	17	3	10	.03892615	17	7	9	.06128180	18	2	12	.02942062
17	1	6	.05756234	17	3	11	.03517078	17	7	10	.05500132	18	2	13	.02694133
17	1	7	.04905833	17	3	12	.03199364	17	7	11	.04978363	18	2	14	.02477601
17	1	8	.04271207	17	3	13	.02925416	17	8	8	.07610424	18	2	15	.02285760
17	1	9	.03777113	17	3	14	.02685264	17	8	9	.06761503	18	2	16	.02113985
17	1	10	.03379584	17	3	15	.02472014	17	8	10	.06072907	18	2	17	.01961049
17	1	11	.03051141	17	4	4	.10820008	17	9	9	.07521813	18	3	3	.13479484
17	1	12	.02773673	17	4	5	.08590830	18	1	1	.50542219	18	3	4	.09892717

TABLE II

P = 5.0

N	R	S	COVARIANCE	N	R	S	COVARIANCE	N	R	S	COVARIANCE	N	R	S	COVARIANCE
18	3	5	.07836893	18	6	7	.06995970	19	1	17	.01861724	19	4	4	.10440655
18	3	6	.06496898	18	6	8	.06116276	19	1	18	.01733960	19	4	5	.08282576
18	3	7	.05549643	18	6	9	.05428068	19	1	19	.01640078	19	4	6	.06874129
18	3	8	.04841314	18	6	10	.04872497	19	2	2	.20056081	19	4	7	.05877859
18	3	9	.04289225	18	6	11	.04412499	19	2	3	.12531221	19	4	8	.05132761
18	3	10	.03844878	18	6	12	.04023497	19	2	4	.09162609	19	4	9	.04552146
18	3	11	.03477883	18	6	13	.03688496	19	2	5	.07243165	19	4	10	.04085121
18	3	12	.03168176	18	7	7	.07606197	19	2	6	.05997238	19	4	11	.03699776
18	3	13	.02901934	18	7	8	.06655458	19	2	7	.05119320	19	4	12	.03375047
18	3	14	.02669299	18	7	9	.05910606	19	2	8	.04464630	19	4	13	.03096426
18	3	15	.02463112	18	7	10	.05308604	19	2	9	.03955614	19	4	14	.02853562
18	3	16	.02278434	18	7	11	.04809676	19	2	10	.03546916	19	4	15	.02638866
18	4	4	.10618327	18	7	12	.04387405	19	2	11	.03210197	19	4	16	.02446724
18	4	5	.08427104	18	8	8	.07278580	19	2	12	.02926794	19	5	5	.08886626
18	4	6	.06994857	18	8	9	.06468976	19	2	13	.02683885	19	5	6	.07384347
18	4	7	.05980359	18	8	10	.05813787	19	2	14	.02472338	19	5	7	.06319662
18	4	8	.05220601	18	8	11	.05270185	19	2	15	.02285470	19	5	8	.05522236
18	4	9	.04627728	18	9	9	.07126581	19	2	16	.02118339	19	5	9	.04900134
18	4	10	.04150102	18	9	10	.06409490	19	2	17	.01967584	19	5	10	.04399274
18	4	11	.03755313	19	1	1	.50547282	19	2	18	.01832869	19	5	11	.03985697
18	4	12	.03421934	19	1	2	.19332838	19	3	3	.13317269	19	5	12	.03636954
18	4	13	.03135183	19	1	3	.11994163	19	3	4	.09766149	19	5	13	.03337564
18	4	14	.02884511	19	1	4	.08739166	19	3	5	.07734091	19	5	14	.03076476
18	4	15	.02662249	19	1	5	.06893985	19	3	6	.06411398	19	5	15	.02845578
18	5	5	.09081431	19	1	6	.05700214	19	3	7	.05477556	19	6	6	.07950745
18	5	6	.07547676	19	1	7	.04860985	19	3	8	.04780142	19	6	7	.06810747
18	5	7	.06459048	19	1	8	.04236214	19	3	9	.04237289	19	6	8	.05955602
18	5	8	.05642505	19	1	9	.03751095	19	3	10	.03801028	19	6	9	.05287663
18	5	9	.05004538	19	1	10	.03361993	19	3	11	.03441332	19	6	10	.04749370
18	5	10	.04490074	19	1	11	.03041689	19	3	12	.03138403	19	6	11	.04304521
18	5	11	.04064487	19	1	12	.02772293	19	3	13	.02878622	19	6	12	.03929150
18	5	12	.03704854	19	1	13	.02541525	19	3	14	.02652280	19	6	13	.03606712
18	5	13	.03395340	19	1	14	.02340654	19	3	15	.02452267	19	6	14	.03325383
18	5	14	.03124635	19	1	15	.02163291	19	3	16	.02273323	19	7	7	.07362827
18	6	6	.08166854	19	1	16	.02004720	19	3	17	.02111872	19	7	8	.06443390

TABLE II

P = 5.0

N	R	S	COVARIANCE	N	R	S	COVARIANCE	N	R	S	COVARIANCE	N	R	S	COVARIANCE
19	7	9	.05724292	20	2	3	.12437813	20	4	6	.06766488	20	7	8	.06259819
19	7	10	.05144151	20	2	4	.09088668	20	4	7	.05786226	20	7	9	.05562735
19	7	11	.04664292	20	2	5	.07182838	20	4	8	.05053968	20	7	10	.05001212
19	7	12	.04259074	20	2	6	.05947129	20	4	9	.04484054	20	7	11	.04537519
19	7	13	.03910775	20	2	7	.05077320	20	4	10	.04026239	20	7	12	.04146680
19	8	8	.06999780	20	2	8	.04429355	20	4	11	.03649051	20	7	13	.03811457
19	8	9	.06222899	20	2	9	.03926124	20	4	12	.03331738	20	7	14	.03519537
19	8	10	.05595400	20	2	10	.03522568	20	4	13	.03060025	20	8	8	.06761899
19	8	11	.05075858	20	2	11	.03190550	20	4	14	.02823747	20	8	9	.06012689
19	8	12	.04636763	20	2	12	.02911563	20	4	15	.02615443	20	8	10	.05408526
19	9	9	.06800922	20	2	13	.02672905	20	4	16	.02429520	20	8	11	.04909176
19	9	10	.06119097	20	2	14	.02465547	20	4	17	.02261791	20	8	12	.04487961
19	9	11	.05553949	20	2	15	.02282873	20	5	5	.08714838	20	8	13	.04126448
19	10	10	.06737485	20	2	16	.02119929	20	5	6	.07240095	20	9	9	.06527458
20	1	1	.50563197	20	2	17	.01973007	20	5	7	.06196319	20	9	10	.05874973
20	1	2	.19284335	20	2	18	.01839649	20	5	8	.05415546	20	9	11	.05335144
20	1	3	.11950879	20	2	19	.01720164	20	5	9	.04807221	20	9	12	.04879396
20	1	4	.08703204	20	3	3	.13172700	20	5	10	.04318124	20	10	10	.06417084
20	1	5	.06864301	20	3	4	.09652997	20	5	11	.03914876	20	10	11	.05830684
20	1	6	.05675786	20	3	5	.07641918	20	5	12	.03575436	*********************			
20	1	7	.04841058	20	3	6	.06334494	20	5	13	.03284628				
20	1	8	.04220248	20	3	7	.05412474	20	5	14	.03031633				
20	1	9	.03738717	20	3	8	.04724652	20	5	15	.02808508				
20	1	10	.03352950	20	3	9	.04189886	20	5	16	.02609292				
20	1	11	.03035826	20	3	10	.03760669	20	6	6	.07761859				
20	1	12	.02769536	20	3	11	.03407288	20	6	7	.06648639				
20	1	13	.02541869	20	3	12	.03110175	20	6	8	.05814727				
20	1	14	.02344157	20	3	13	.02855886	20	6	9	.05164266				
20	1	15	.02170053	20	3	14	.02634851	20	6	10	.04640813				
20	1	16	.02014809	20	3	15	.02440057	20	6	11	.04208913				
20	1	17	.01874871	20	3	16	.02266247	20	6	12	.03845122				
20	1	18	.01747883	20	3	17	.02109486	20	6	13	.03533282				
20	1	19	.01634113	20	3	18	.01967170	20	6	14	.03261864				
20	1	20	.01552109	20	4	4	.10282901	20	6	15	.03022393				
20	2	2	.19932825	20	4	5	.08154001	20	7	7	.07152422				

TABLE II

P = 5.5

N	R	S	COVARIANCE	N	R	S	COVARIANCE	N	R	S	COVARIANCE	N	R	S	COVARIANCE
2	1	1	.69253398	7	1	3	.13523146	9	1	2	.20363370	10	2	3	.14345227
2	1	2	.30746602	7	1	4	.09883351	9	1	3	.12920717	10	2	4	.10588427
3	1	1	.60501126	7	1	5	.07712887	9	1	4	.09479333	10	2	5	.08381013
3	1	2	.26383093	7	1	6	.06226766	9	1	5	.07462701	10	2	6	.06906928
3	1	3	.16413668	7	1	7	.05120081	9	1	6	.06114434	10	2	7	.05836331
3	2	2	.40638040	7	2	2	.25037288	9	1	7	.05130134	10	2	8	.05008736
4	1	1	.56300888	7	2	3	.16143809	9	1	8	.04361979	10	2	9	.04335862
4	1	2	.24198016	7	2	4	.11887795	9	1	9	.03743497	10	3	3	.16326511
4	1	3	.15275603	7	2	5	.09321502	9	2	2	.23069927	10	3	4	.12113801
4	1	4	.10821266	7	2	6	.07551224	9	2	3	.14805600	10	3	5	.09619819
4	2	2	.32719043	7	3	3	.19578677	9	2	4	.10926492	10	3	6	.07945845
4	2	3	.21211565	7	3	4	.14536662	9	2	5	.08633349	10	3	7	.06725538
5	1	1	.53832754	7	3	5	.11460657	9	2	6	.07091257	10	3	8	.05779539
5	1	2	.22855165	7	4	4	.18307434	9	2	7	.05960748	10	4	4	.13959687
5	1	3	.14506973	8	1	1	.50251383	9	2	8	.05075775	10	4	5	.11126376
5	1	4	.10458529	8	1	2	.20763607	9	3	3	.17113477	10	4	6	.09214033
5	1	5	.07955690	8	1	3	.13189645	9	3	4	.12704379	10	4	7	.07814186
5	2	2	.28884610	8	1	4	.09663668	9	3	5	.10075729	10	5	5	.13046992
5	2	3	.18705311	8	1	5	.07582635	9	3	6	.08297724	10	5	6	.10838532
5	2	4	.13638811	8	1	6	.06178155	9	3	7	.06988716	11	1	1	.48746307
5	3	3	.25272362	8	1	7	.05139056	9	4	4	.14936052	11	1	2	.19774170
6	1	1	.52215886	8	1	8	.04328451	9	4	5	.11897265	11	1	3	.12513652
6	1	2	.21939104	8	2	2	.23918928	9	4	6	.09828359	11	1	4	.09189372
6	1	3	.13947805	8	2	3	.15386052	9	5	5	.14344651	11	1	5	.07259641
6	1	4	.10146554	8	2	4	.11347365	10	1	1	.49132566	11	1	6	.05983811
6	1	5	.07843275	8	2	5	.08940350	10	1	2	.20040237	11	1	7	.05066351
6	1	6	.06245273	8	2	6	.07305223	10	1	3	.12699228	11	1	8	.04365223
6	2	2	.26584611	8	2	7	.06089855	10	1	4	.09323130	11	1	9	.03802797
6	2	3	.17181450	8	3	3	.18148154	10	1	5	.07355213	11	1	10	.03332897
6	2	4	.12611150	8	3	4	.13476394	10	1	6	.06047989	11	1	11	.02940437
6	2	5	.09805582	8	3	5	.10664660	10	1	7	.05102158	11	2	2	.21861979
6	3	3	.21707503	8	3	6	.08741631	10	1	8	.04373077	11	2	3	.13970302
6	3	4	.16102466	8	4	4	.16288419	10	1	9	.03781563	11	2	4	.10310288
7	1	1	.51082676	8	4	5	.12959556	10	1	10	.03294822	11	2	5	.08169895
7	1	2	.21271923	9	1	1	.49621448	10	2	2	.22401967	11	2	6	.06747892

TABLE II

P = 5.5

N	R	S	COVARIANCE	N	R	S	COVARIANCE	N	R	S	COVARIANCE	N	R	S	COVARIANCE
11	2	7	.05721772	12	2	5	.07990550	13	1	9	.03800540	13	5	8	.06717028
11	2	8	.04935568	12	2	6	.06609837	13	1	10	.03369646	13	5	9	.05920300
11	2	9	.04303627	12	2	7	.05618188	13	1	11	.03004683	13	6	6	.10201320
11	2	10	.03774838	12	2	8	.04863069	13	1	12	.02687029	13	6	7	.08734917
11	3	3	.15705668	12	2	9	.04261258	13	1	13	.02417516	13	6	8	.07612741
11	3	4	.11645544	12	2	10	.03762854	13	2	2	.21041398	13	7	7	.10006310
11	3	5	.09254995	12	2	11	.03336352	13	2	3	.13394918	14	1	1	.47982960
11	3	6	.07659515	12	3	3	.15202092	13	2	4	.09878344	14	1	2	.19201397
11	3	7	.06504386	12	3	4	.11264040	13	2	5	.07836202	14	1	3	.12102554
11	3	8	.05617097	12	3	5	.08955556	13	2	6	.06489095	14	1	4	.08883959
11	3	9	.04902494	12	3	6	.07421563	13	2	7	.05525046	14	1	5	.07031337
11	4	4	.13217762	12	3	7	.06316512	13	2	8	.04794226	14	1	6	.05817283
11	4	5	.10537950	12	3	8	.05473117	13	2	9	.04215257	14	1	7	.04953355
11	4	6	.08740704	12	3	9	.04799748	13	2	10	.03739680	14	1	8	.04302025
11	4	7	.07434826	12	3	10	.04241281	13	2	11	.03336419	14	1	9	.03789110
11	4	8	.06428966	12	4	4	.12632639	13	2	12	.02985132	14	1	10	.03370791
11	5	5	.12112193	12	4	5	.10072058	13	3	3	.14784632	14	1	11	.03019236
11	5	6	.10072653	12	4	6	.08363174	13	3	4	.10946507	14	1	12	.02715672
11	5	7	.08584646	12	4	7	.07128225	13	3	5	.08704795	14	1	13	.02447583
11	6	6	.11789707	12	4	8	.06183390	13	3	6	.07220367	14	1	14	.02219463
12	1	1	.48436992	12	4	9	.05427571	13	3	7	.06155116	14	2	2	.20722176
12	1	2	.19551597	12	5	5	.11402914	13	3	8	.05345891	14	2	3	.13168889
12	1	3	.12355951	12	5	6	.09489358	13	3	9	.04703759	14	2	4	.09706893
12	1	4	.09073695	12	5	7	.08101606	13	3	10	.04175602	14	2	5	.07701876
12	1	5	.07174705	12	5	8	.07036943	13	3	11	.03727280	14	2	6	.06382692
12	1	6	.05923728	12	6	6	.10885958	13	4	4	.12157976	14	2	7	.05441301
12	1	7	.05028133	12	6	7	.09312575	13	4	5	.09692808	14	2	8	.04730076
12	1	8	.04347792	13	1	1	.48186798	13	4	6	.08054017	14	2	9	.04169078
12	1	9	.03806589	13	1	2	.19362977	13	4	7	.06874634	14	2	10	.03710949
12	1	10	.03359041	13	1	3	.12220341	13	4	8	.05976751	14	2	11	.03325527
12	1	11	.02976490	13	1	4	.08972753	13	4	9	.05263034	14	2	12	.02992432
12	1	12	.02653838	13	1	5	.07099026	13	4	10	.04675174	14	2	13	.02698068
12	2	2	.21416042	13	1	6	.05868239	13	5	5	.10844094	14	3	3	.14432428
12	2	3	.13658543	13	1	7	.04990069	13	5	6	.09028273	14	3	4	.10677630
12	2	4	.10077032	13	1	8	.04325839	13	5	7	.07717414	14	3	5	.08491352

TABLE II

P = 5.5

N	R	S	COVARIANCE	N	R	S	COVARIANCE	N	R	S	COVARIANCE	N	R	S	COVARIANCE
14	3	6	.07047783	15	1	9	.03774814	15	4	8	.05648067	16	1	15	.02075200
14	3	7	.06014999	15	1	10	.03366329	15	4	9	.04993629	16	1	16	.01906498
14	3	8	.05233223	15	1	11	.03024978	15	4	10	.04460251	16	2	2	.20207264
14	3	9	.04615646	15	1	12	.02732427	15	4	11	.04013242	16	2	3	.12801100
14	3	10	.04110699	15	1	13	.02475784	15	4	12	.03629217	16	2	4	.09425612
14	3	11	.03685466	15	1	14	.02246428	15	5	5	.10015453	16	2	5	.07479257
14	3	12	.03317666	15	1	15	.02051182	15	5	6	.08341746	16	2	6	.06203866
14	4	4	.11764304	15	2	2	.20446947	15	5	7	.07141389	16	2	7	.05297576
14	4	5	.09377298	15	2	3	.12972835	15	5	8	.06231895	16	2	8	.04616166
14	4	6	.07795545	15	2	4	.09557321	15	5	9	.05513728	16	2	9	.04081829
14	4	7	.06660955	15	2	5	.07583844	15	5	10	.04927674	16	2	10	.03648732
14	4	8	.05800412	15	2	6	.06288250	15	5	11	.04436005	16	2	11	.03287971
14	4	9	.05119547	15	2	7	.05365827	15	6	6	.09226126	16	2	12	.02980258
14	4	10	.04562147	15	2	8	.04670789	15	6	7	.07908781	16	2	13	.02712061
14	4	11	.04092249	15	2	9	.04124361	15	6	8	.06908508	16	2	14	.02473530
14	5	5	.10391056	15	2	10	.03680031	15	6	9	.06117306	16	2	15	.02258191
14	5	6	.08653375	15	2	11	.03308352	15	6	10	.05470744	16	3	3	.13869760
14	5	7	.07403459	15	2	12	.02989544	15	7	7	.08817956	16	3	4	.10246002
14	5	8	.06453408	15	2	13	.02709676	15	7	8	.07711687	16	3	5	.08146526
14	5	9	.05700435	15	2	14	.02459436	15	7	9	.06834935	16	3	6	.06766476
14	5	10	.05083141	15	3	3	.14130953	15	8	8	.08691104	16	3	7	.05783621
14	6	6	.09662541	15	3	4	.10446703	16	1	1	.47679149	16	3	8	.05043402
14	6	7	.08279014	15	3	5	.08307207	16	1	2	.18940033	16	3	9	.04462187
14	6	8	.07224872	15	3	6	.06897936	16	1	3	.11908291	16	3	10	.03990597
14	6	9	.06387783	15	3	7	.05892188	16	1	4	.08735154	16	3	11	.03597433
14	7	7	.09340949	15	3	8	.05133006	16	1	5	.06915689	16	3	12	.03261838
14	7	8	.08162654	15	3	9	.04535313	16	1	6	.05727753	16	3	13	.02969162
15	1	1	.47816095	15	3	10	.04048748	16	1	7	.04885753	16	3	14	.02708729
15	1	2	.19061718	15	3	11	.03641363	16	1	8	.04253876	16	4	4	.11147160
15	1	3	.11993364	15	3	12	.03291658	16	1	9	.03759104	16	4	5	.08880734
15	1	4	.08805293	15	3	13	.02984471	16	1	10	.03358544	16	4	6	.07386351
15	1	5	.06970536	15	4	4	.11431928	16	1	11	.03025204	16	4	7	.06319706
15	1	6	.05770576	15	4	5	.09110175	16	1	12	.02741105	16	4	8	.05515027
15	1	7	.04918521	15	4	6	.07575791	16	1	13	.02493653	16	4	9	.04882363
15	1	8	.04277779	15	4	7	.06478134	16	1	14	.02273694	16	4	10	.04368478

TABLE II

P = 5.5

N	R	S	COVARIANCE	N	R	S	COVARIANCE	N	R	S	COVARIANCE	N	R	S	COVARIANCE
16	4	11	.03939673	17	1	13	.02504179	17	4	6	.07221135	18	1	2	.18739201
16	4	12	.03573385	17	1	14	.02291993	17	4	7	.06180923	18	1	3	.11755118
16	4	13	.03253740	17	1	15	.02101275	17	4	8	.05397751	18	1	4	.08615592
16	5	5	.09698382	17	1	16	.01927782	17	4	9	.04783374	18	1	5	.06820834
16	5	6	.08078064	17	1	17	.01780822	17	4	10	.04285667	18	1	6	.05652292
16	5	7	.06918845	17	2	2	.19996737	17	4	11	.03871733	18	1	7	.04826396
16	5	8	.06042776	17	2	3	.12649386	17	4	12	.03519648	18	1	8	.04208544
16	5	9	.05353014	17	2	4	.09308694	17	4	13	.03214100	18	1	9	.03726518
16	5	10	.04792110	17	2	5	.07385904	17	4	14	.02943927	18	1	10	.03338004
16	5	11	.04323626	17	2	6	.06128012	17	5	5	.09426720	18	1	11	.03016491
16	5	12	.03923123	17	2	7	.05235624	17	5	6	.07851667	18	1	12	.02744451
16	6	6	.08864532	17	2	8	.04565864	17	5	7	.06727176	18	1	13	.02509752
16	6	7	.07601318	17	2	9	.04041749	17	5	8	.05879165	18	1	14	.02303642
16	6	8	.06644815	17	2	10	.03618006	17	5	9	.05213063	18	1	15	.02119569
16	6	9	.05890564	17	2	11	.03266161	17	5	10	.04672887	18	1	16	.01952548
16	6	10	.05276446	17	2	12	.02967299	17	5	11	.04223240	18	1	17	.01799602
16	6	11	.04762976	17	2	13	.02708240	17	5	12	.03840498	18	1	18	.01670672
16	7	7	.08394714	17	2	14	.02479399	17	5	13	.03508138	18	2	2	.19810440
16	7	8	.07345897	17	2	15	.02273619	17	6	6	.08559422	18	2	3	.12514370
16	7	9	.06517425	17	2	16	.02086365	17	6	7	.07341350	18	2	4	.09204172
16	7	10	.05841917	17	3	3	.13641130	17	6	8	.06421153	18	2	5	.07302038
16	8	8	.08180886	17	3	4	.10069802	17	6	9	.05697338	18	2	6	.06059452
16	8	9	.07265236	17	3	5	.08004957	17	6	10	.05109690	18	2	7	.05179172
17	1	1	.47566721	17	3	6	.06650114	17	6	11	.04620062	18	2	8	.04519497
17	1	2	.18833320	17	3	7	.05686912	17	6	12	.04202952	18	2	9	.04004155
17	1	3	.11827390	17	3	8	.04962858	17	7	7	.08044286	18	2	10	.03588346
17	1	4	.08672266	17	3	9	.04395556	17	7	8	.07042412	18	2	11	.03243947
17	1	5	.06866010	17	3	10	.03936442	17	7	9	.06253133	18	2	12	.02952332
17	1	6	.05688443	17	3	11	.03554918	17	7	10	.05611525	18	2	13	.02700591
17	1	7	.04855069	17	3	12	.03230624	17	7	11	.05076368	18	2	14	.02479402
17	1	8	.04230727	17	3	13	.02949359	17	8	8	.07768358	18	2	15	.02281773
17	1	9	.03742835	17	3	14	.02700781	17	8	9	.06903515	18	2	16	.02102387
17	1	10	.03348817	17	3	15	.02477163	17	8	10	.06199470	18	2	17	.01938076
17	1	11	.03021949	17	4	4	.10900167	17	9	9	.07681236	18	3	3	.13439229
17	1	12	.02744513	17	4	5	.08681268	18	1	1	.47474601	18	3	4	.09913764

TABLE II

P = 5.5

N	R	S	COVARIANCE	N	R	S	COVARIANCE	N	R	S	COVARIANCE	N	R	S	COVARIANCE
18	3	5	.07879183	18	6	7	.07118267	19	1	17	.01823024	19	4	4	.10492263
18	3	6	.06546313	18	6	8	.06228699	19	1	18	.01687175	19	4	5	.08350836
18	3	7	.05600178	18	6	9	.05530418	19	1	19	.01573362	19	4	6	.06946335
18	3	8	.04890085	18	6	10	.04964812	19	2	2	.19644509	19	4	7	.05948846
18	3	9	.04334708	18	6	11	.04494836	19	2	3	.12393438	19	4	8	.05200197
18	3	10	.03886177	18	6	12	.04095815	19	2	4	.09110152	19	4	9	.04614901
18	3	11	.03514389	18	6	13	.03750561	19	2	5	.07226262	19	4	10	.04142601
18	3	12	.03199381	18	7	7	.07748762	19	2	6	.05997175	19	4	11	.03751624
18	3	13	.02927298	18	7	8	.06786002	19	2	7	.05127540	19	4	12	.03420985
18	3	14	.02688123	18	7	9	.06029208	19	2	8	.04476688	19	4	13	.03136156
18	3	15	.02474341	18	7	10	.05415504	19	2	9	.03968967	19	4	14	.02886686
18	3	16	.02280227	18	7	11	.04905074	19	2	10	.03559987	19	4	15	.02664778
18	4	4	.10683694	18	7	12	.04471354	19	2	11	.03221919	19	4	16	.02464426
18	4	5	.08506073	18	8	8	.07427050	19	2	12	.02936370	19	5	5	.08984457
18	4	6	.07075611	18	8	9	.06603682	19	2	13	.02690637	19	5	6	.07482078
18	4	7	.06058216	18	8	10	.05935136	19	2	14	.02475600	19	5	7	.06413075
18	4	8	.05293516	18	8	11	.05378495	19	2	15	.02284466	19	5	8	.05609611
18	4	9	.04694739	18	9	9	.07277605	19	2	16	.02112013	19	5	9	.04980756
18	4	10	.04210707	18	9	10	.06545471	19	2	17	.01954214	19	5	10	.04472847
18	4	11	.03809185	19	1	1	.47399466	19	2	18	.01808889	19	5	11	.04052079
18	4	12	.03468765	19	1	2	.18655777	19	3	3	.13259560	19	5	12	.03696021
18	4	13	.03174571	19	1	3	.11690229	19	3	4	.09774536	19	5	13	.03389131
18	4	14	.02915839	19	1	4	.08564286	19	3	5	.07766627	19	5	14	.03120213
18	4	15	.02684482	19	1	5	.06779605	19	3	6	.06453086	19	5	15	.02880910
18	5	5	.09191057	19	1	6	.05618977	19	3	7	.05521918	19	6	6	.08071424
18	5	6	.07654891	19	1	7	.04799622	19	3	8	.04824015	19	6	7	.06924452
18	5	7	.06560132	19	1	8	.04187417	19	3	9	.04278985	19	6	8	.06061095
18	5	8	.05736032	19	1	9	.03710457	19	3	10	.03839562	19	6	9	.05384563
18	5	9	.05089968	19	1	10	.03326647	19	3	11	.03476064	19	6	10	.04837619
18	5	10	.04567204	19	1	11	.03009649	19	3	12	.03168848	19	6	11	.04384150
18	5	11	.04133204	19	1	12	.02742081	19	3	13	.02904334	19	6	12	.04000161
18	5	12	.03764998	19	1	13	.02511957	19	3	14	.02672758	19	6	13	.03669006
18	5	13	.03446608	19	1	14	.02310680	19	3	15	.02466847	19	6	14	.03378680
18	5	14	.03166456	19	1	15	.02131853	19	3	16	.02280998	19	7	7	.07495747
18	6	6	.08298087	19	1	16	.01970564	19	3	17	.02110896	19	7	8	.06566106

TABLE II

P = 5.5

N	R	S	COVARIANCE	N	R	S	COVARIANCE	N	R	S	COVARIANCE	N	R	S	COVARIANCE
19	7	9	.05836700	20	2	3	.12284503	20	4	6	.06830637	20	7	8	.06375156
19	7	10	.05246393	20	2	4	.09025115	20	4	7	.05850675	20	7	9	.05669171
19	7	11	.04756544	20	2	5	.07157445	20	4	8	.05116112	20	7	10	.05098782
19	7	12	.04341443	20	2	6	.05940349	20	4	9	.04542595	20	7	11	.04626352
19	7	13	.03983228	20	2	7	.05080147	20	4	10	.04080483	20	7	12	.04226889
19	8	8	.07139388	20	2	8	.04437083	20	4	11	.03698593	20	7	13	.03883076
19	8	9	.06350549	20	2	9	.03936051	20	4	12	.03376297	20	7	14	.03582472
19	8	10	.05711404	20	2	10	.03533021	20	4	13	.03099350	20	8	8	.06893223
19	8	11	.05180515	20	2	11	.03200414	20	4	14	.02857548	20	8	9	.06133598
19	8	12	.04730266	20	2	12	.02920032	20	4	15	.02643326	20	8	10	.05519224
19	9	9	.06943898	20	2	13	.02679337	20	4	16	.02450895	20	8	11	.05009915
19	9	10	.06248944	20	2	14	.02469363	20	4	17	.02275696	20	8	12	.04578946
19	9	11	.05671070	20	2	15	.02283475	20	5	5	.08801690	20	8	13	.04207779
19	10	10	.06881491	20	2	16	.02116600	20	5	6	.07328947	20	9	9	.06662811
20	1	1	.47338653	20	2	17	.01964752	20	5	7	.06282485	20	9	10	.05998785
20	1	2	.18581514	20	2	18	.01824837	20	5	8	.05497024	20	9	11	.05447776
20	1	3	.11631712	20	2	19	.01695407	20	5	9	.04883124	20	9	12	.04981130
20	1	4	.08517650	20	3	3	.13098598	20	5	10	.04388053	20	10	10	.06554227
20	1	5	.06741851	20	3	4	.09649483	20	5	11	.03978639	20	10	11	.05955399
20	1	6	.05588208	20	3	5	.07665253	20	5	12	.03632910	*****	*****	*****	**********
20	1	7	.04774618	20	3	6	.06368849	20	5	13	.03335678				
20	1	8	.04167373	20	3	7	.05450918	20	5	14	.03076053				
20	1	9	.03694834	20	3	8	.04763758	20	5	15	.02845955				
20	1	10	.03315095	20	3	9	.04227802	20	5	16	.02639194				
20	1	11	.03001961	20	3	10	.03796313	20	6	6	.07872732				
20	1	12	.02738169	20	3	11	.03439972	20	6	7	.06754291				
20	1	13	.02511843	20	3	12	.03139409	20	6	8	.05913632				
20	1	14	.02314499	20	3	13	.02881263	20	6	9	.05255863				
20	1	15	.02139865	20	3	14	.02655970	20	6	10	.04724941				
20	1	16	.01983149	20	3	15	.02456447	20	6	11	.04285553				
20	1	17	.01840590	20	3	16	.02277275	20	6	12	.03914280				
20	1	18	.01709267	20	3	17	.02114193	20	6	13	.03594916				
20	1	19	.01587804	20	3	18	.01963893	20	6	14	.03315830				
20	1	20	.01486790	20	4	4	.10321655	20	6	15	.03068384				
20	2	2	.19495870	20	4	5	.08212224	20	7	7	.07276372				

TABLE II

P = 6.0

N	R	S	COVARIANCE	N	R	S	COVARIANCE	N	R	S	COVARIANCE	N	R	S	COVARIANCE
2	1	1	.69125659	7	1	3	.13504010	9	1	2	.20113564	10	2	3	.14391722
2	1	2	.30874341	7	1	4	.09889197	9	1	3	.12847337	10	2	4	.10656018
3	1	1	.59995989	7	1	5	.07714885	9	1	4	.09453935	10	2	5	.08446958
3	1	2	.26516930	7	1	6	.06209744	9	1	5	.07451587	10	2	6	.06963621
3	1	3	.16433406	7	1	7	.05058167	9	1	6	.06104347	10	2	7	.05880072
3	2	2	.41073488	7	2	2	.25120631	9	1	7	.05113476	10	2	8	.05036254
4	1	1	.55508977	7	2	3	.16283847	9	1	8	.04331063	10	2	9	.04341582
4	1	2	.24249394	7	2	4	.12011886	9	1	9	.03679460	10	3	3	.16455393
4	1	3	.15339768	7	2	5	.09414177	9	2	2	.23041468	10	3	4	.12245597
4	1	4	.10794513	7	2	6	.07602631	9	2	3	.14879717	10	3	5	.09737830
4	2	2	.33061834	7	3	3	.19821442	9	2	4	.11012102	10	3	6	.08045471
4	2	3	.21456353	7	3	4	.14738625	9	2	5	.08710336	10	3	7	.06804697
5	1	1	.52816244	7	3	5	.11612134	9	2	6	.07152773	10	3	8	.05835717
5	1	2	.22827921	7	4	4	.18578425	9	2	7	.06002488	10	4	4	.14144411
5	1	3	.14549263	8	1	1	.48760309	9	2	8	.05091387	10	4	5	.11287809
5	1	4	.10479843	8	1	2	.20558831	9	3	3	.17275143	10	4	6	.09349485
5	1	5	.07908052	8	1	3	.13142147	9	3	4	.12858023	10	4	7	.07922577
5	2	2	.29125025	8	1	4	.09653983	9	3	5	.10207329	10	5	5	.13249337
5	2	3	.18924306	8	1	5	.07579707	9	3	6	.08403397	10	5	6	.11007663
5	2	4	.13780868	8	1	6	.06168952	9	3	7	.07065553	11	1	1	.46930493
5	3	3	.25614291	8	1	7	.05113460	9	4	4	.15147314	11	1	2	.19447770
6	1	1	.51015560	8	1	8	.04264749	9	4	5	.12075447	11	1	3	.12395146
6	1	2	.21843887	8	2	2	.23942160	9	4	6	.09971392	11	1	4	.09134613
6	1	3	.13959249	8	2	3	.15491185	9	5	5	.14567336	11	1	5	.07229843
6	1	4	.10165391	8	2	4	.11452113	10	1	1	.47414440	11	1	6	.05963962
6	1	5	.07840401	8	2	5	.09027222	10	1	2	.19750188	11	1	7	.05049091
6	1	6	.06187854	8	2	6	.07367387	10	1	3	.12602270	11	1	8	.04345998
6	2	2	.26739188	8	2	7	.06119777	10	1	4	.09282626	11	1	9	.03777708
6	2	3	.17360283	8	3	3	.18347462	10	1	5	.07334852	11	1	10	.03296583
6	2	4	.12751625	8	3	4	.13653783	10	1	6	.06033709	11	1	11	.02877784
6	2	5	.09890845	8	3	5	.10808722	10	1	7	.05086768	11	2	2	.21748227
6	3	3	.21999679	8	3	6	.08848246	10	1	8	.04351399	11	2	3	.13992032
6	3	4	.16325202	8	4	4	.16529048	10	1	9	.03747316	11	2	4	.10361182
7	1	1	.49726729	8	4	5	.13153264	10	1	10	.03231253	11	2	5	.08224732
7	1	2	.21118155	9	1	1	.48010773	10	2	2	.22328306	11	2	6	.06798116

TABLE II

P = 6.0

N	R	S	COVARIANCE	N	R	S	COVARIANCE	N	R	S	COVARIANCE	N	R	S	COVARIANCE
11	2	7	.05763576	12	2	5	.08034645	13	1	9	.03778996	13	5	8	.06817842
11	2	8	.04966479	12	2	6	.06653004	13	1	10	.03345692	13	5	9	.06004472
11	2	9	.04320920	12	2	7	.05656279	13	1	11	.02975691	13	6	6	.10361324
11	2	10	.03773518	12	2	8	.04893761	13	1	12	.02648783	13	6	7	.08875094
11	3	3	.15805710	12	2	9	.04282669	13	1	13	.02357307	13	6	8	.07733270
11	3	4	.11757497	12	2	10	.03772604	13	2	2	.20859030	13	7	7	.10169521
11	3	5	.09359624	12	2	11	.03329907	13	2	3	.13373925	14	1	1	.45917239
11	3	6	.07751255	12	3	3	.15276498	13	2	4	.09899621	14	1	2	.18783230
11	3	7	.06581104	12	3	4	.11357995	13	2	5	.07870080	14	1	3	.11929047
11	3	8	.05677269	12	3	5	.09047471	13	2	6	.06525098	14	1	4	.08791606
11	3	9	.04943821	12	3	6	.07504870	13	2	7	.05558609	14	1	5	.06974853
11	4	4	.13378839	12	3	7	.06388742	13	2	8	.04822942	14	1	6	.05778857
11	4	5	.10683201	12	3	8	.05532959	13	2	9	.04237500	14	1	7	.04924416
11	4	6	.08866477	12	3	9	.04845906	13	2	10	.03753899	14	1	8	.04277815
11	4	7	.07540076	12	3	10	.04271615	13	2	11	.03340485	14	1	9	.03766628
11	4	8	.06512751	12	4	4	.12772629	13	2	12	.02974892	14	1	10	.03347847
11	5	5	.12295011	12	4	5	.10202208	13	3	3	.14836026	14	1	11	.02993880
11	5	6	.10230001	12	4	6	.08478806	13	3	4	.11024088	14	1	12	.02685630
11	5	7	.08716237	12	4	7	.07227975	13	3	5	.08784801	14	1	13	.02409025
11	6	6	.11978198	12	4	8	.06266604	13	3	6	.07295250	14	1	14	.02160592
12	1	1	.46531510	12	4	9	.05493302	13	3	7	.06221982	14	2	2	.20509969
12	1	2	.19192056	12	5	5	.11567775	13	3	8	.05403364	14	2	3	.13129250
12	1	3	.12217656	12	5	6	.09634511	13	3	9	.04750858	14	2	4	.09715017
12	1	4	.09005572	12	5	7	.08226486	13	3	10	.04211153	14	2	5	.07726106
12	1	5	.07135653	12	5	8	.07141360	13	3	11	.03749282	14	2	6	.06411653
12	1	6	.05897768	12	6	6	.11059907	13	4	4	.12279063	14	2	7	.05469991
12	1	7	.05007548	12	6	7	.09461924	13	4	5	.09809032	14	2	8	.04755914
12	1	8	.04328113	13	1	1	.46198351	13	4	6	.08159694	14	2	9	.04190476
12	1	9	.03784609	13	1	2	.18972984	13	4	7	.06967967	14	2	10	.03726654
12	1	10	.03331626	13	1	3	.12063780	13	4	8	.06057042	14	2	11	.03334203
12	1	11	.02938934	13	1	4	.08892120	13	4	9	.05329719	14	2	12	.02992141
12	1	12	.02592344	13	1	5	.07051060	13	4	10	.04727294	14	2	13	.02684979
12	2	2	.21266303	13	1	6	.05836020	13	5	5	.10992638	14	3	3	.14462986
12	2	3	.13657905	13	1	7	.04965463	13	5	6	.09161637	14	3	4	.10740248
12	2	4	.10112525	13	1	8	.04304270	13	5	7	.07834601	14	3	5	.08560255

TABLE II

P = 6.0

N	R	S	COVARIANCE	N	R	S	COVARIANCE	N	R	S	COVARIANCE	N	R	S	COVARIANCE
14	3	6	.07114498	15	1	9	.03750635	15	4	8	.05719812	16	1	15	.02036718
14	3	7	.06076180	15	1	10	.03343095	15	4	9	.05056594	16	1	16	.01850309
14	3	8	.05287321	15	1	11	.03001002	15	4	10	.04514029	16	2	2	.19942091
14	3	9	.04661738	15	1	12	.02706087	15	4	11	.04057290	16	2	3	.12728341
14	3	10	.04147964	15	1	13	.02445067	15	4	12	.03662593	16	2	4	.09410148
14	3	11	.03712816	15	1	14	.02207817	15	5	5	.10135718	16	2	5	.07485841
14	3	12	.03333227	15	1	15	.01993660	15	5	6	.08453643	16	2	6	.06219476
14	4	4	.11868333	15	2	2	.20207254	15	5	7	.07242970	16	2	7	.05316359
14	4	5	.09480711	15	2	3	.12916008	15	5	8	.06322580	16	2	8	.04635118
14	4	6	.07891707	15	2	4	.09553234	15	5	9	.05593331	16	2	9	.04099190
14	4	7	.06747602	15	2	5	.07598985	15	5	10	.04996009	16	2	10	.03663358
14	4	8	.05876677	15	2	6	.06310398	15	5	11	.04492652	16	2	11	.03298960
14	4	9	.05184950	15	2	7	.05389546	15	6	6	.09361186	16	2	12	.02986735
14	4	10	.04616140	15	2	8	.04693289	15	6	7	.08030672	16	2	13	.02712953
14	4	11	.04133875	15	2	9	.04143996	15	6	8	.07017045	16	2	14	.02467187
14	5	5	.10524793	15	2	10	.03695645	15	6	9	.06212569	16	2	15	.02241291
14	5	6	.08775632	15	2	11	.03318920	15	6	10	.05552712	16	3	3	.13863843
14	5	7	.07512764	15	2	12	.02993880	15	7	7	.08959799	16	3	4	.10282196
14	5	8	.06549403	15	2	13	.02705997	15	7	8	.07837778	16	3	5	.08195443
14	5	9	.05782966	15	2	14	.02444186	15	7	9	.06945560	16	3	6	.06817957
14	5	10	.05151850	15	3	3	.14142500	15	8	8	.08834931	16	3	7	.05833442
14	6	6	.09809565	15	3	4	.10495579	16	1	1	.45472919	16	3	8	.05089567
14	6	7	.08409898	15	3	5	.08365768	16	1	2	.18471116	16	3	9	.04503627
14	6	8	.07339651	15	3	6	.06956838	16	1	3	.11704204	16	3	10	.04026623
14	6	9	.06486553	15	3	7	.05947640	16	1	4	.08621445	16	3	11	.03627461
14	7	7	.09493258	15	3	8	.05183242	16	1	5	.06843393	16	3	12	.03285202
14	7	8	.08296061	15	3	9	.04579360	16	1	6	.05677402	16	3	13	.02984902
15	1	1	.45677993	15	3	10	.04085904	16	1	7	.04847935	16	3	14	.02715192
15	1	2	.18617329	15	3	11	.03670898	16	1	8	.04223398	16	4	4	.11221529
15	1	3	.11810063	15	3	12	.03312554	16	1	9	.03732796	16	4	5	.08961465
15	1	4	.08701936	15	3	13	.02994970	16	1	10	.03334276	16	4	6	.07465078
15	1	5	.06905946	15	4	4	.11520462	16	1	11	.03001388	16	4	7	.06393255
15	1	6	.05726089	15	4	5	.09201794	16	1	12	.02716384	16	4	8	.05582075
15	1	7	.04885140	15	4	6	.07662966	16	1	13	.02466639	16	4	9	.04942289
15	1	8	.04250534	15	4	7	.06558138	16	1	14	.02242575	16	4	10	.04420900

TABLE II

P = 6.0

N	R	S	COVARIANCE	N	R	S	COVARIANCE	N	R	S	COVARIANCE	N	R	S	COVARIANCE
16	4	11	.03984214	17	1	13	.02478937	17	4	6	.07291931	18	1	2	.18225487
16	4	12	.03609506	17	1	14	.02264536	17	4	7	.06248273	18	1	3	.11524027
16	4	13	.03280527	17	1	15	.02069952	17	4	8	.05460092	18	1	4	.08482874
16	5	5	.09806338	17	1	16	.01889555	17	4	9	.04839961	18	1	5	.06734238
16	5	6	.08180328	17	1	17	.01725933	17	4	10	.04336088	18	1	6	.05590864
16	5	7	.07013012	17	2	2	.19707814	17	4	11	.03915668	18	1	7	.04779940
16	5	8	.06128021	17	2	3	.12561787	17	4	12	.03556736	18	1	8	.04171430
16	5	9	.05429054	17	2	4	.09282592	17	4	13	.03243810	18	1	9	.03695332
16	5	10	.04858795	17	2	5	.07384422	17	4	14	.02965376	18	1	10	.03310511
16	5	11	.04380730	17	2	6	.06137373	17	5	5	.09523377	18	1	11	.02991118
16	5	12	.03970189	17	2	7	.05249576	17	5	6	.07944976	18	1	12	.02719998
16	6	6	.08988591	17	2	8	.04581171	17	5	7	.06814295	18	1	13	.02485211
16	6	7	.07714677	17	2	9	.04056551	17	5	8	.05959021	18	1	14	.02278052
16	6	8	.06747040	17	2	10	.03631114	17	5	9	.05285249	18	1	15	.02091844
16	6	9	.05981643	17	2	11	.03276711	17	5	10	.04737225	18	1	16	.01921168
16	6	10	.05356410	17	2	12	.02974549	17	5	11	.04279573	18	1	17	.01761717
16	6	11	.04831714	17	2	13	.02711421	17	5	12	.03888571	18	1	18	.01617043
16	7	7	.08526728	17	2	14	.02477530	17	5	13	.03547473	18	2	2	.19499272
16	7	8	.07464663	17	2	15	.02265159	17	6	6	.08673354	18	2	3	.12412886
16	7	9	.06623149	17	2	16	.02068202	17	6	7	.07446681	18	2	4	.09168090
16	7	10	.05934775	17	3	3	.13619074	17	6	8	.06517198	18	2	5	.07292941
16	8	8	.08316241	17	3	4	.10094236	17	6	9	.05783958	18	2	6	.06062852
16	8	9	.07385632	17	3	5	.08044863	17	6	10	.05186895	18	2	7	.05188435
17	1	1	.45296096	17	3	6	.06694563	17	6	11	.04687823	18	2	8	.04531169
17	1	2	.18341353	17	3	7	.05731270	17	6	12	.04261093	18	2	9	.04016251
17	1	3	.11609407	17	3	8	.05004899	17	7	7	.08167143	18	2	10	.03599618
17	1	4	.08548788	17	3	9	.04434091	17	7	8	.07154082	18	2	11	.03253527
17	1	5	.06786385	17	3	10	.03970747	17	7	9	.06353697	18	2	12	.02959536
17	1	6	.05632444	17	3	11	.03584457	17	7	10	.05701145	18	2	13	.02704789
17	1	7	.04812881	17	3	12	.03254887	17	7	11	.05155120	18	2	14	.02479902
17	1	8	.04196937	17	3	13	.02967727	17	8	8	.07895565	18	2	15	.02277668
17	1	9	.03714153	17	3	14	.02712348	17	8	9	.07017951	18	2	16	.02092234
17	1	10	.03323070	17	3	15	.02480371	17	8	10	.06301421	18	2	17	.01918943
17	1	11	.02997576	17	4	4	.10961511	17	9	9	.07809752	18	3	3	.13402179
17	1	12	.02720269	17	4	5	.08751913	18	1	1	.45142886	18	3	4	.09927248

TABLE II

P = 6.0

N	R	S	COVARIANCE
18	3	5	.07910647
18	3	6	.06584108
18	3	7	.05639276
18	3	8	.04928036
18	3	9	.04370195
18	3	10	.03918416
18	3	11	.03542846
18	3	12	.03223611
18	3	13	.02946841
18	3	14	.02702399
18	3	15	.02482492
18	3	16	.02280782
18	4	4	.10732999
18	4	5	.08567344
18	4	6	.07138962
18	4	7	.06119648
18	4	8	.05351231
18	4	9	.04747861
18	4	10	.04258764
18	4	11	.03851864
18	4	12	.03505778
18	4	13	.03205566
18	4	14	.02940296
18	4	15	.02701554
18	5	5	.09277295
18	5	6	.07739864
18	5	7	.06640581
18	5	8	.05810644
18	5	9	.05158201
18	5	10	.04628819
18	5	11	.04188055
18	5	12	.03812916
18	5	13	.03487315
18	5	14	.03199469
18	6	6	.08402672

N	R	S	COVARIANCE
18	6	7	.07216071
18	6	8	.06318790
18	6	9	.05612518
18	6	10	.05038872
18	6	11	.04560844
18	6	12	.04153693
18	6	13	.03800088
18	7	7	.07863113
18	7	8	.06890902
18	7	9	.06124598
18	7	10	.05501491
18	7	11	.04981755
18	7	12	.04538726
18	8	8	.07546551
18	8	9	.06712193
18	8	10	.06032894
18	8	11	.05465688
18	9	9	.07399349
18	9	10	.06655092
19	1	1	.45009608
19	1	2	.18121472
19	1	3	.11446735
19	1	4	.08422805
19	1	5	.06686366
19	1	6	.05552337
19	1	7	.04749020
19	1	8	.04147012
19	1	9	.03676708
19	1	10	.03297248
19	1	11	.02982996
19	1	12	.02716986
19	1	13	.02487465
19	1	14	.02285950
19	1	15	.02106049
19	1	16	.01942782

N	R	S	CCVARIANCE
19	1	17	.01791695
19	1	18	.01649693
19	1	19	.01520949
19	2	2	.19312418
19	2	3	.12278911
19	2	4	.09064677
19	2	5	.07209962
19	2	6	.05994890
19	2	7	.05132273
19	2	8	.04484772
19	2	9	.03978293
19	2	10	.03569237
19	2	11	.03230198
19	2	12	.02943009
19	2	13	.02695073
19	2	14	.02477282
19	2	15	.02282766
19	2	16	.02106084
19	2	17	.01942697
19	2	18	.01789017
19	3	3	.13208514
19	3	4	.09777782
19	3	5	.07790161
19	3	6	.06484578
19	3	7	.05555970
19	3	8	.04857958
19	3	9	.04311373
19	3	10	.03869540
19	3	11	.03503072
19	3	12	.03192463
19	3	13	.02924171
19	3	14	.02688395
19	3	15	.02477736
19	3	16	.02286327
19	3	17	.02109271

N	R	S	COVARIANCE
19	4	4	.10530385
19	4	5	.08403361
19	4	6	.07002687
19	4	7	.06004643
19	4	8	.05253416
19	4	9	.04664532
19	4	10	.04188098
19	4	11	.03792651
19	4	12	.03457280
19	4	13	.03167452
19	4	14	.02912639
19	4	15	.02684883
19	4	16	.02477871
19	5	5	.09061045
19	5	6	.07559270
19	5	7	.06487224
19	5	8	.05679170
19	5	9	.05045045
19	5	10	.04531551
19	5	11	.04105030
19	5	12	.03743082
19	5	13	.03430119
19	5	14	.03154839
19	5	15	.02908691
19	6	6	.08167353
19	6	7	.07015205
19	6	8	.06145499
19	6	9	.05462197
19	6	10	.04908359
19	6	11	.04447964
19	6	12	.04057009
19	6	13	.03718772
19	6	14	.03421113
19	7	7	.07602193
19	7	8	.06664595

TABLE II

P = 6.0

N	R	S	COVARIANCE
19	7	9	.05927026
19	7	10	.05328587
19	7	11	.04830689
19	7	12	.04407578
19	7	13	.04041292
19	8	8	.07251652
19	8	9	.06453312
19	8	10	.05804829
19	8	11	.05264780
19	8	12	.04805477
19	9	9	.07059118
19	9	10	.06353622
19	9	11	.05765461
19	10	10	.06997619
20	1	1	.44893298
20	1	2	.18027651
20	1	3	.11376446
20	1	4	.08367839
20	1	5	.06642273
20	1	6	.05516565
20	1	7	.04720000
20	1	8	.04123731
20	1	9	.03658505
20	1	10	.03283698
20	1	11	.02973853
20	1	12	.02712150
20	1	13	.02486980
20	1	14	.02290011
20	1	15	.02115035
20	1	16	.01957234
20	1	17	.01812693
20	1	18	.01678071
20	1	19	.01550766
20	1	20	.01435544
20	2	2	.19144025

N	R	S	COVARIANCE
20	2	3	.12157678
20	2	4	.08970770
20	2	5	.07134319
20	2	6	.05932641
20	2	7	.05080515
20	2	8	.04441656
20	2	9	.03942596
20	2	10	.03540144
20	2	11	.03207184
20	2	12	.02925773
20	2	13	.02683513
20	2	14	.02471493
20	2	15	.02283069
20	2	16	.02113080
20	2	17	.01957324
20	2	18	.01812219
20	2	19	.01674973
20	3	3	.13034432
20	3	4	.09643120
20	3	5	.07681319
20	3	6	.06394368
20	3	7	.05480139
20	3	8	.04793803
20	3	9	.04257097
20	3	10	.03823927
20	3	11	.03465308
20	3	12	.03162038
20	3	13	.02900833
20	3	14	.02672138
20	3	15	.02468820
20	3	16	.02285336
20	3	17	.02117169
20	3	18	.01960463
20	4	4	.10349345
20	4	5	.08256564

N	R	S	COVARIANCE
20	4	6	.06880397
20	4	7	.05901113
20	4	8	.05164989
20	4	9	.04588768
20	4	10	.04123328
20	4	11	.03737734
20	4	12	.03411471
20	4	13	.03130327
20	4	14	.02884073
20	4	15	.02665066
20	4	16	.02467360
20	4	17	.02286108
20	5	5	.08869301
20	5	6	.07398861
20	5	7	.06350684
20	5	8	.05561739
20	5	9	.04943536
20	5	10	.04443768
20	5	11	.04029450
20	5	12	.03678679
20	5	13	.03376266
20	5	14	.03111270
20	5	15	.02875505
20	5	16	.02662601
20	6	6	.07960619
20	6	7	.06838431
20	6	8	.05992624
20	6	9	.05329147
20	6	10	.04792306
20	6	11	.04346930
20	6	12	.03969631
20	6	13	.03644176
20	6	14	.03358857
20	6	15	.03104910
20	7	7	.07375461

N	R	S	COVARIANCE
20	7	8	.06467592
20	7	9	.05754606
20	7	10	.05177160
20	7	11	.04697718
20	7	12	.04291289
20	7	13	.03940505
20	7	14	.03632826
20	8	8	.06998708
20	8	9	.06230854
20	8	10	.05608328
20	8	11	.05091010
20	8	12	.04652149
20	8	13	.04273132
20	9	9	.06771826
20	9	10	.06098569
20	9	11	.05538553
20	9	12	.05063079
20	10	10	.06664821
20	10	11	.06055972

****

TABLE II

P = 6.5

N	R	S	COVARIANCE	N	R	S	COVARIANCE	N	R	S	COVARIANCE	N	R	S	COVARIANCE
2	1	1	.69024526	7	1	3	.13486164	9	1	2	.19907724	10	2	3	.14427255
2	1	2	.30975474	7	1	4	.09892329	9	1	3	.12785152	10	2	4	.10709768
3	1	1	.59591297	7	1	5	.07715161	9	1	4	.09431428	10	2	5	.08499840
3	1	2	.26623740	7	1	6	.06194649	9	1	5	.07441104	10	2	6	.07009138
3	1	3	.16447337	7	1	7	.05007011	9	1	6	.06094924	10	2	7	.05915036
3	2	2	.41427772	7	2	2	.25185411	9	1	7	.05098834	10	2	8	.05057895
4	1	1	.54874759	7	2	3	.16396862	9	1	8	.04304943	10	2	9	.04345312
4	1	2	.24288687	7	2	4	.12112539	9	1	9	.03626963	10	3	3	.16558967
4	1	3	.15390256	7	2	5	.09489171	9	2	2	.23014331	10	3	4	.12352629
4	1	4	.10771046	7	2	6	.07643594	9	2	3	.14938094	10	3	5	.09834010
4	2	2	.33340673	7	3	3	.20019570	9	2	4	.11080812	10	3	6	.08126703
4	2	3	.21655636	7	3	4	.14903922	9	2	5	.08772389	10	3	7	.06869085
5	1	1	.52003486	7	3	5	.11735879	9	2	6	.07202274	10	3	8	.05881087
5	1	2	.22802842	7	4	4	.18800662	9	2	7	.06035753	10	4	4	.14295392
5	1	3	.14581848	8	1	1	.47574758	9	2	8	.05103147	10	4	5	.11420103
5	1	4	.10495633	8	1	2	.20389329	9	3	3	.17405940	10	4	6	.09460514
5	1	5	.07867954	8	1	3	.13101087	9	3	4	.12983240	10	4	7	.08011244
5	2	2	.29319452	8	1	4	.09644359	9	3	5	.10314803	10	5	5	.13415509
5	2	3	.19102694	8	1	5	.07575944	9	3	6	.08489600	10	5	6	.11146573
5	2	4	.13896072	8	1	6	.06160255	9	3	7	.07127905	11	1	1	.45493670
5	3	3	.25894007	8	1	7	.05091534	9	4	4	.15320340	11	1	2	.19180445
6	1	1	.50057647	8	1	8	.04212348	9	4	5	.12221606	11	1	3	.12296283
6	1	2	.21763320	8	2	2	.23957461	9	4	6	.10088594	11	1	4	.09088105
6	1	3	.13966493	8	2	3	.15575254	9	5	5	.14750213	11	1	5	.07204047
6	1	4	.10179191	8	2	4	.11536717	10	1	1	.46052870	11	1	6	.05946532
6	1	5	.07836664	8	2	5	.09097465	10	1	2	.19511991	11	1	7	.05033947
6	1	6	.06140101	8	2	6	.07417373	10	1	3	.12520895	11	1	8	.04329375
6	2	2	.26862728	8	2	7	.06143192	10	1	4	.09247757	11	1	9	.03756414
6	2	3	.17505469	8	3	3	.18509526	10	1	5	.07316784	11	1	10	.03266304
6	2	4	.12865793	8	3	4	.13798738	10	1	6	.06020845	11	1	11	.02826687
6	2	5	.09959518	8	3	5	.10926494	10	1	7	.05073159	11	2	2	.21651353
6	3	3	.22238678	8	3	6	.08935095	10	1	8	.04332782	11	2	3	.14007126
6	3	4	.16507467	8	4	4	.16726378	10	1	9	.03718610	11	2	4	.10401087
7	1	1	.48646665	8	4	5	.13312147	10	1	10	.03179286	11	2	5	.08268388
7	1	2	.20989973	9	1	1	.46732310	10	2	2	.22264163	11	2	6	.06838267

TABLE II

P = 6.5

N	R	S	COVARIANCE	N	R	S	COVARIANCE	N	R	S	COVARIANCE	N	R	S	COVARIANCE
11	2	7	.05796952	12	2	5	.08069385	13	1	9	.03760574	13	5	8	.06900526
11	2	8	.04990961	12	2	6	.06687294	13	1	10	.03325398	13	5	9	.06073375
11	2	9	.04334223	12	2	7	.05686579	13	1	11	.02951393	13	6	6	.10492768
11	2	10	.03771556	12	2	8	.04918076	13	1	12	.02617122	13	6	7	.08990328
11	3	3	.15885359	12	2	9	.04299400	13	1	13	.02308399	13	6	8	.07832303
11	3	4	.11848011	12	2	10	.03779771	13	2	2	.20706363	13	7	7	.10303764
11	3	5	.09444660	12	2	11	.03323821	13	2	3	.13353866	14	1	1	.44289188
11	3	6	.07825941	12	3	3	.15334905	13	2	4	.09915020	14	1	2	.18442824
11	3	7	.06643512	12	3	4	.11433533	13	2	5	.07896348	14	1	3	.11785788
11	3	8	.05726039	12	3	5	.09121914	13	2	6	.06553441	14	1	4	.08714542
11	3	9	.04976993	12	3	6	.07572535	13	2	7	.05585150	14	1	5	.06927284
11	4	4	.13510141	12	3	7	.06447436	13	2	8	.04845624	14	1	6	.05746237
11	4	5	.10802040	12	3	8	.05581498	13	2	9	.04254932	14	1	7	.04899711
11	4	6	.08969504	12	3	9	.04883155	13	2	10	.03764773	14	1	8	.04257104
11	4	7	.07626232	12	3	10	.04295774	13	2	11	.03343041	14	1	9	.03747435
11	4	8	.06581137	12	4	4	.12886373	13	2	12	.02965765	14	1	10	.03328366
11	5	5	.12445022	12	4	5	.10308468	13	3	3	.14875396	14	1	11	.02972518
11	5	6	.10359240	12	4	6	.08573404	13	3	4	.11086014	14	1	12	.02660552
11	5	7	.08824244	12	4	7	.07309600	13	3	5	.08849321	14	1	13	.02377189
11	6	6	.12133135	12	4	8	.06334599	13	3	6	.07355897	14	1	14	.02112853
12	1	1	.45025861	12	4	9	.05546807	13	3	7	.06276216	14	2	2	.20333137
12	1	2	.18898264	12	5	5	.11702869	13	3	8	.05449951	14	2	3	.13093897
12	1	3	.12102780	12	5	6	.09753656	13	3	9	.04788924	14	2	4	.09719561
12	1	4	.08948178	12	5	7	.08329004	13	3	10	.04239690	14	2	5	.07744384
12	1	5	.07102292	12	5	8	.07226966	13	3	11	.03766620	14	2	6	.06434154
12	1	6	.05875333	12	6	6	.11202889	13	4	4	.12377067	14	2	7	.05492488
12	1	7	.04989678	12	6	7	.09584695	13	4	5	.09903679	14	2	8	.04776211
12	1	8	.04311114	13	1	1	.44629235	13	4	6	.08245997	14	2	9	.04207212
12	1	9	.03765837	13	1	2	.18654945	13	4	7	.07044263	14	2	10	.03738766
12	1	10	.03308530	13	1	3	.11934162	13	4	8	.06122645	14	2	11	.03340572
12	1	11	.02907745	13	1	4	.08824559	13	4	9	.05384089	14	2	12	.02991174
12	1	12	.02542299	13	1	5	.07010425	13	4	10	.04769588	14	2	13	.02673582
12	2	2	.21140130	13	1	6	.05808469	13	5	5	.11114148	14	3	3	.14485147
12	2	3	.13654574	13	1	7	.04944300	13	5	6	.09270983	14	3	4	.10789748
12	2	4	.10139676	13	1	8	.04285721	13	5	7	.07930759	14	3	5	.08615530

TABLE II

P = 6.5

N	R	S	COVARIANCE	N	R	S	COVARIANCE	N	R	S	COVARIANCE	N	R	S	COVARIANCE
14	3	6	.07168336	15	1	9	.03730052	15	4	8	.05778320	16	1	15	.02005080
14	3	7	.06125678	15	1	10	.03323372	15	4	9	.05107921	16	1	16	.01804880
14	3	8	.05331105	15	1	11	.02980753	15	4	10	.04557793	16	2	2	.19722520
14	3	9	.04698986	15	1	12	.02683991	15	4	11	.04093006	16	2	3	.12665908
14	3	10	.04177955	15	1	13	.02419507	15	4	12	.03689464	16	2	4	.09395277
14	3	11	.03734627	15	1	14	.02176010	15	5	5	.10233655	16	2	5	.07489539
14	3	12	.03345303	15	1	15	.01947086	15	5	6	.08545093	16	2	6	.06230919
14	4	4	.11952141	15	2	2	.20008218	15	5	7	.07326144	16	2	7	.05330643
14	4	5	.09564676	15	2	3	.12866588	15	5	8	.06396881	16	2	8	.04649689
14	4	6	.07970072	15	2	4	.09547719	15	5	9	.05658530	16	2	9	.04112582
14	4	7	.06818330	15	2	5	.07609749	15	5	10	.05051897	16	2	10	.03674583
14	4	8	.05938945	15	2	6	.06327253	15	5	11	.04538844	16	2	11	.03307252
14	4	9	.05238289	15	2	7	.05407921	15	6	6	.09471891	16	2	12	.02991357
14	4	10	.04660048	15	2	8	.04710817	15	6	7	.08130743	16	2	13	.02713020
14	4	11	.04167529	15	2	9	.04159274	15	6	8	.07106204	16	2	14	.02461352
14	5	5	.10633963	15	2	10	.03707686	15	6	9	.06290794	16	2	15	.02226890
14	5	6	.08875724	15	2	11	.03326859	15	6	10	.05619930	16	3	3	.13855952
14	5	7	.07602373	15	2	12	.02996730	15	7	7	.09076416	16	3	4	.10309795
14	5	8	.06628112	15	2	13	.02702301	15	7	8	.07941495	16	3	5	.08234099
14	5	9	.05850569	15	2	14	.02431080	15	7	9	.07036521	16	3	6	.06859105
14	5	10	.05207994	15	3	3	.14148985	15	8	8	.08953285	16	3	7	.05873471
14	6	6	.09930227	15	3	4	.10533683	16	1	1	.43738145	16	3	8	.05126746
14	6	7	.08517441	15	3	5	.08412439	16	1	2	.18090580	16	3	9	.04537015
14	6	8	.07433972	15	3	6	.07004167	16	1	3	.11536418	16	3	10	.04055610
14	6	9	.06567639	15	3	7	.05992362	16	1	4	.08527123	16	3	11	.03651535
14	7	7	.09618533	15	3	8	.05223812	16	1	5	.06782986	16	3	12	.03303794
14	7	8	.08405794	15	3	9	.04614916	16	1	6	.05635073	16	3	13	.02997210
15	1	1	.43994895	15	3	10	.04115822	16	1	7	.04815986	16	3	14	.02719857
15	1	2	.18256164	15	3	11	.03694548	16	1	8	.04197566	16	4	4	.11280681
15	1	3	.11654121	15	3	12	.03329079	16	1	9	.03710472	16	4	5	.09026521
15	1	4	.08615972	15	3	13	.03002918	16	1	10	.03313705	16	4	6	.07528892
15	1	5	.06851789	15	4	4	.11591384	16	1	11	.02981262	16	4	7	.06453054
15	1	6	.05688531	15	4	5	.09275924	16	1	12	.02695594	16	4	8	.05636666
15	1	7	.04856808	15	4	6	.07733830	16	1	13	.02444055	16	4	9	.04991091
15	1	8	.04227341	15	4	7	.06622324	16	1	14	.02216748	16	4	10	.04463552

TABLE II

P = 6.5

N	R	S	COVARIANCE	N	R	S	COVARIANCE	N	R	S	COVARIANCE	N	R	S	COVARIANCE
16	4	11	.04020370	17	1	13	.02457780	17	4	6	.07349130	18	1	2	.17809702
16	4	12	.03638697	17	1	14	.02241647	17	4	7	.06302897	18	1	3	.11334682
16	4	13	.03301984	17	1	15	.02044015	17	4	8	.05510753	18	1	4	.08373256
16	5	5	.09894011	17	1	16	.01858182	17	4	9	.04885983	18	1	5	.06662266
16	5	6	.08263739	17	1	17	.01681615	17	4	10	.04377082	18	1	6	.05539546
16	5	7	.07089999	17	2	2	.19469147	17	4	11	.03951338	18	1	7	.04740967
16	5	8	.06197791	17	2	3	.12487257	17	4	12	.03586758	18	1	8	.04140193
16	5	9	.05491300	17	2	4	.09258989	17	4	13	.03267727	18	1	9	.03669030
16	5	10	.04913339	17	2	5	.07381469	17	4	14	.02982453	18	1	10	.03287306
16	5	11	.04427349	17	2	6	.06143651	17	5	5	.09601632	18	1	11	.02969715
16	5	12	.04008477	17	2	7	.05259858	17	5	6	.08020912	18	1	12	.02699411
16	6	6	.09090127	17	2	8	.04592736	17	5	7	.06885396	18	1	13	.02464615
16	6	7	.07807642	17	2	9	.04067823	17	5	8	.06024296	18	1	14	.02256663
16	6	8	.06830955	17	2	10	.03641081	17	5	9	.05344288	18	1	15	.02068788
16	6	9	.06056416	17	2	11	.03284639	17	5	10	.04789830	18	1	16	.01895236
16	6	10	.05422010	17	2	12	.02979813	17	5	11	.04325580	18	1	17	.01730675
16	6	11	.04888008	17	2	13	.02713378	17	5	12	.03927739	18	1	18	.01573797
16	7	7	.08635183	17	2	14	.02475370	17	5	13	.03579388	18	2	2	.19242759
16	7	8	.07562320	17	2	15	.02257633	17	6	6	.08766442	18	2	3	.12327061
16	7	9	.06710087	17	2	16	.02052817	17	6	7	.07532952	18	2	4	.09136308
16	7	10	.06011079	17	3	3	.13597920	17	6	8	.06595966	18	2	5	.07283720
16	8	8	.08427622	17	3	4	.10112109	17	6	9	.05855031	18	2	6	.06064207
16	8	9	.07484707	17	3	5	.08076038	17	6	10	.05250225	18	2	7	.05194839
17	1	1	.43512597	17	3	6	.06729861	17	6	11	.04743348	18	2	8	.04539723
17	1	2	.17942660	17	3	7	.05766749	17	6	12	.04308635	18	2	9	.04025280
17	1	3	.11430514	17	3	8	.05038642	17	7	7	.08267980	18	2	10	.03608065
17	1	4	.08446598	17	3	9	.04465061	17	7	8	.07245846	18	2	11	.03260653
17	1	5	.06720047	17	3	10	.03998306	17	7	9	.06436370	18	2	12	.02964761
17	1	6	.05585527	17	3	11	.03608133	17	7	10	.05774800	18	2	13	.02707584
17	1	7	.04777375	17	3	12	.03274239	17	7	11	.05219776	18	2	14	.02479694
17	1	8	.04168405	17	3	13	.02982230	17	8	8	.08000207	18	2	15	.02273712
17	1	9	.03689892	17	3	14	.02721250	17	8	9	.07112123	18	2	16	.02083361
17	1	10	.03301291	17	3	15	.02482395	17	8	10	.06385294	18	2	17	.01902813
17	1	11	.02976991	17	4	4	.11009852	17	9	9	.07915544	18	3	3	.13368725
17	1	12	.02699858	17	4	5	.08808571	18	1	1	.43313269	18	3	4	.09936077

TABLE II

P = 6.5

N	R	S	COVARIANCE	N	R	S	COVARIANCE	N	R	S	COVARIANCE	N	R	S	COVARIANCE
18	3	5	.07934821	18	6	7	.07296058	19	1	17	.01765853	19	4	4	.10559482
18	3	6	.06613873	18	6	8	.06392593	19	1	18	.01619027	19	4	5	.08444948
18	3	7	.05670375	18	6	9	.05679829	19	1	19	.01478733	19	4	6	.07047844
18	3	8	.04958371	18	6	10	.05099597	19	2	2	.19039154	19	4	7	.06049623
18	3	9	.04398625	18	6	11	.04614935	19	2	3	.12182494	19	4	8	.05296462
18	3	10	.03944257	18	6	12	.04201057	19	2	4	.09025209	19	4	9	.04704749
18	3	11	.03565626	18	6	13	.03840518	19	2	5	.07194821	19	4	10	.04224992
18	3	12	.03242942	18	7	7	.07956860	19	2	6	.05991558	19	4	11	.03825910
18	3	13	.02962330	18	7	8	.06977030	19	2	7	.05134934	19	4	12	.03486666
18	3	14	.02713558	18	7	9	.06202975	19	2	8	.04490359	19	4	13	.03192726
18	3	15	.02488610	18	7	10	.05572146	19	2	9	.03985029	19	4	14	.02933503
18	3	16	.02280660	18	7	11	.05044729	19	2	10	.03576008	19	4	15	.02700913
18	4	4	.10771357	18	7	12	.04593983	19	2	11	.03236246	19	4	16	.02488393
18	4	5	.08616201	18	8	8	.07644797	19	2	12	.02947768	19	5	5	.09122581
18	4	6	.07189952	18	8	9	.06801464	19	2	13	.02698066	19	5	6	.07621754
18	4	7	.06169330	18	8	10	.06113322	19	2	14	.02478049	19	5	7	.06547491
18	4	8	.05398030	18	8	11	.05537384	19	2	15	.02280787	19	5	8	.05735844
18	4	9	.04790992	18	9	9	.07499567	19	2	16	.02100667	19	5	9	.05097494
18	4	10	.04297791	18	9	10	.06745332	19	2	17	.01932746	19	5	10	.04579469
18	4	11	.03886494	19	1	1	.43136191	19	2	18	.01772324	19	5	11	.04148243
18	4	12	.03535752	19	1	2	.17689540	19	3	3	.13163597	19	5	12	.03781448
18	4	13	.03230574	19	1	3	.11247519	19	3	4	.09778166	19	5	13	.03463469
18	4	14	.02959897	19	1	4	.08306157	19	3	5	.07807768	19	5	14	.03182920
18	4	15	.02715044	19	1	5	.06609037	19	3	6	.06509110	19	5	15	.02931093
18	5	5	.09346868	19	1	6	.05496801	19	3	7	.05582872	19	6	6	.08245417
18	5	6	.07808840	19	1	7	.04706682	19	3	8	.04884956	19	6	7	.07089302
18	5	7	.06706111	19	1	8	.04113099	19	3	9	.04337223	19	6	8	.06214551
18	5	8	.05871540	19	1	9	.03648319	19	3	10	.03893499	19	6	9	.05525783
18	5	9	.05213944	19	1	10	.03272490	19	3	11	.03524649	19	6	10	.04966322
18	5	10	.04679163	19	1	11	.02960550	19	3	12	.03211288	19	6	11	.04500241
18	5	11	.04232843	19	1	12	.02695872	19	3	13	.02939911	19	6	12	.04103538
18	5	12	.03851981	19	1	13	.02466902	19	3	14	.02700691	19	6	13	.03759436
18	5	13	.03520409	19	1	14	.02265251	19	3	15	.02486130	19	6	14	.03455689
18	5	14	.03226176	19	1	15	.02084533	19	3	16	.02290150	19	7	7	.07689346
18	6	6	.08487959	19	1	16	.01919582	19	3	17	.02107389	19	7	8	.06745377

TABLE II

P = 6.5

N	R	S	COVARIANCE
19	7	9	.06001187
19	7	10	.05396096
19	7	11	.04891573
19	7	12	.04461841
19	7	13	.04088858
19	8	8	.07343878
19	8	9	.06537812
19	8	10	.05881676
19	8	11	.05334076
19	8	12	.04867278
19	9	9	.07153940
19	9	10	.06439793
19	9	11	.05843146
19	10	10	.07093241
20	1	1	.42978159
20	1	2	.17580413
20	1	3	.11167880
20	1	4	.08244521
20	1	5	.06559842
20	1	6	.05456987
20	1	7	.04674407
20	1	8	.04087191
20	1	9	.03628019
20	1	10	.03257314
20	1	11	.02950220
20	1	12	.02690282
20	1	13	.02466111
20	1	14	.02269501
20	1	15	.02094299
20	1	16	.01935673
20	1	17	.01789587
20	1	18	.01652379
20	1	19	.01520504
20	1	20	.01394314
20	2	2	.18854979

N	R	S	COVARIANCE
20	2	3	.12051295
20	2	4	.08924059
20	2	5	.07113576
20	2	6	.05924845
20	2	7	.05079574
20	2	8	.04444341
20	2	9	.03947031
20	2	10	.03545155
20	2	11	.03211983
20	2	12	.02929789
20	2	13	.02686291
20	2	14	.02472630
20	2	15	.02282156
20	2	16	.02109640
20	2	17	.01950712
20	2	18	.01801401
20	2	19	.01657861
20	3	3	.12978787
20	3	4	.09635592
20	3	5	.07692748
20	3	6	.06413945
20	3	7	.05503027
20	3	8	.04817559
20	3	9	.04280372
20	3	10	.03845917
20	3	11	.03485493
20	3	12	.03180044
20	3	13	.02916355
20	3	14	.02684881
20	3	15	.02478452
20	3	16	.02291426
20	3	17	.02119083
20	3	18	.01957128
20	4	4	.10369813
20	4	5	.08291353

N	R	S	COVARIANCE
20	4	6	.06920066
20	4	7	.05941625
20	4	8	.05204410
20	4	9	.04626098
20	4	10	.04158008
20	4	11	.03769423
20	4	12	.03439928
20	4	13	.03155344
20	4	14	.02905426
20	4	15	.02682470
20	4	16	.02480406
20	4	17	.02294153
20	5	5	.08923367
20	5	6	.07455275
20	5	7	.06405982
20	5	8	.05614366
20	5	9	.04992747
20	5	10	.04489192
20	5	11	.04070881
20	5	12	.03715978
20	5	13	.03409299
20	5	14	.03139863
20	5	15	.02899404
20	5	16	.02681406
20	6	6	.08031971
20	6	7	.06907005
20	6	8	.06057156
20	6	9	.05389099
20	6	10	.04847456
20	6	11	.04397183
20	6	12	.04014928
20	6	13	.03684442
20	6	14	.03393958
20	6	15	.03134610
20	7	7	.07456472

N	R	S	COVARIANCE
20	7	8	.06543323
20	7	9	.05824687
20	7	10	.05241494
20	7	11	.04756301
20	7	12	.04344130
20	7	13	.03987576
20	7	14	.03674023
20	8	8	.07085286
20	8	9	.06310772
20	8	10	.05681589
20	8	11	.05157689
20	8	12	.04712311
20	8	13	.04326789
20	9	9	.06861500
20	9	10	.06180694
20	9	11	.05613268
20	9	12	.05130497
20	10	10	.06755885
20	10	11	.06138786

****

TABLE II

P = 7.0

N	R	S	COVARIANCE	N	R	S	COVARIANCE	N	R	S	COVARIANCE	N	R	S	COVARIANCE
2	1	1	.68942533	7	1	3	.13469829	9	1	2	.19735326	10	2	3	.14455153
2	1	2	.31057467	7	1	4	.09893827	9	1	3	.12731916	10	2	4	.10753491
3	1	1	.59259887	7	1	5	.07714475	9	1	4	.09411517	10	2	5	.08543166
3	1	2	.26710923	7	1	6	.06181287	9	1	5	.07431435	10	2	6	.07046468
3	1	3	.16457455	7	1	7	.04964071	9	1	6	.06086286	10	2	7	.05943603
3	2	2	.41721622	7	2	2	.25237078	9	1	7	.05085953	10	2	8	.05075326
4	1	1	.54355535	7	2	3	.16489963	9	1	8	.04282632	10	2	9	.04347772
4	1	2	.24319592	7	2	4	.12195803	9	1	9	.03583176	10	3	3	.16643983
4	1	3	.15430964	7	2	5	.09551084	9	2	2	.22989129	10	3	4	.12441261
4	1	4	.10750441	7	2	6	.07676972	9	2	3	.14985189	10	3	5	.09913894
4	2	2	.33571910	7	3	3	.20184320	9	2	4	.11137147	10	3	6	.08194195
4	2	3	.21821002	7	3	4	.15041696	9	2	5	.08823449	10	3	7	.06922474
5	1	1	.51338998	7	3	5	.11838858	9	2	6	.07242950	10	3	8	.05918478
5	1	2	.22780091	7	4	4	.18986201	9	2	7	.06062860	10	4	4	.14421088
5	1	3	.14607638	8	1	1	.46610000	9	2	8	.05112253	10	4	5	.11530485
5	1	4	.10507676	8	1	2	.20246852	9	3	3	.17513913	10	4	6	.09553171
5	1	5	.07833803	8	1	3	.13065411	9	3	4	.13087239	10	4	7	.08085113
5	2	2	.29479911	8	1	4	.09635198	9	3	5	.10404219	10	5	5	.13554402
5	2	3	.19250795	8	1	5	.07571888	9	3	6	.08561252	10	5	6	.11262689
5	2	4	.13991360	8	1	6	.06152222	9	3	7	.07179503	11	1	1	.44329133
5	3	3	.26127056	8	1	7	.05072607	9	4	4	.15464638	11	1	2	.18957648
6	1	1	.49275720	8	1	8	.04168520	9	4	5	.12343655	11	1	3	.12212669
6	1	2	.21694437	8	2	2	.23967707	9	4	6	.10186375	11	1	4	.09048220
6	1	3	.13971104	8	2	3	.15643970	9	5	5	.14903072	11	1	5	.07181607
6	1	4	.10189601	8	2	4	.11606457	10	1	1	.44947935	11	1	6	.05931212
6	1	5	.07832617	8	2	5	.09155420	10	1	2	.19313033	11	1	7	.05020642
6	1	6	.06099810	8	2	6	.07458420	10	1	3	.12451746	11	1	8	.04314926
6	2	2	.26963678	8	2	7	.06161971	10	1	4	.09217549	11	1	9	.03738160
6	2	3	.17625670	8	3	3	.18643874	10	1	5	.07300785	11	1	10	.03240706
6	2	4	.12960397	8	3	4	.13919396	10	1	6	.06009335	11	1	11	.02784249
6	2	5	.10015991	8	3	5	.11024562	10	1	7	.05061136	11	2	2	.21568097
6	3	3	.22437790	8	3	6	.09007195	10	1	8	.04316680	11	2	3	.14017876
6	3	4	.16659359	8	4	4	.16891119	10	1	9	.03694238	11	2	4	.10433150
7	1	1	.47766418	8	4	5	.13444810	10	1	10	.03136040	11	2	5	.08303933
7	1	2	.20881628	9	1	1	.45693425	10	2	2	.22208121	11	2	6	.06871074

TABLE II

P = 7.0

N	R	S	COVARIANCE	N	R	S	COVARIANCE	N	R	S	COVARIANCE	N	R	S	COVARIANCE
11	2	7	.05824191	12	2	5	.08097413	13	1	9	.03744693	13	5	8	.06969561
11	2	8	.05010804	12	2	6	.06715159	13	1	10	.03308025	13	5	9	.06130813
11	2	9	.04344727	12	2	7	.05711230	13	1	11	.02930767	13	6	6	.10602664
11	2	10	.03769321	12	2	8	.04937789	13	1	12	.02590506	13	6	7	.09086725
11	3	3	.15950215	12	2	9	.04312799	13	1	13	.02267911	13	6	8	.07915113
11	3	4	.11922683	12	2	10	.03785187	13	2	2	.20576879	13	7	7	.10416115
11	3	5	.09515123	12	2	11	.03318197	13	2	3	.13335199	14	1	1	.42974075
11	3	6	.07887914	12	3	3	.15381883	13	2	4	.09926482	14	1	2	.18160525
11	3	7	.06695262	12	3	4	.11495555	13	2	5	.07917242	14	1	3	.11665614
11	3	8	.05766354	12	3	5	.09183417	13	2	6	.06576292	14	1	4	.08649352
11	3	9	.05004192	12	3	6	.07628575	13	2	7	.05606634	14	1	5	.06886753
11	4	4	.13619212	12	3	7	.06496062	13	2	8	.04863964	14	1	6	.05718274
11	4	5	.10901062	12	3	8	.05621648	13	2	9	.04268929	14	1	7	.04878441
11	4	6	.09055436	12	3	9	.04913834	13	2	10	.03773312	14	1	8	.04239245
11	4	7	.07698050	12	3	10	.04315449	13	2	11	.03344645	14	1	9	.03730909
11	4	8	.06638003	12	4	4	.12980602	13	2	12	.02957653	14	1	10	.03311663
11	5	5	.12570321	12	4	5	.10396851	13	3	3	.14906371	14	1	11	.02954311
11	5	6	.10467276	12	4	6	.08652220	13	3	4	.11136547	14	1	12	.02639329
11	5	7	.08914479	12	4	7	.07377620	13	3	5	.08902433	14	1	13	.02350483
11	6	6	.12262738	12	4	8	.06391191	13	3	6	.07405999	14	1	14	.02073387
12	1	1	.43806984	12	4	9	.05591200	13	3	7	.06321075	14	2	2	.20183699
12	1	2	.18653865	12	5	5	.11815580	13	3	8	.05488466	14	2	3	.13062467
12	1	3	.12005950	12	5	6	.09853200	13	3	9	.04820317	14	2	4	.09721917
12	1	4	.08899261	12	5	7	.08414667	13	3	10	.04263087	14	2	5	.07758557
12	1	5	.07073556	12	5	8	.07298417	13	3	11	.03780610	14	2	6	.06452084
12	1	6	.05855842	12	6	6	.11322491	13	4	4	.12457993	14	2	7	.05510564
12	1	7	.04974100	12	6	7	.09687395	13	4	5	.09982235	14	2	8	.04792543
12	1	8	.04296348	13	1	1	.43360392	13	4	6	.08317797	14	2	9	.04220627
12	1	9	.03749669	13	1	2	.18390809	13	4	7	.07107790	14	2	10	.03748351
12	1	10	.03288843	13	1	3	.11825195	13	4	8	.06177245	14	2	11	.03345379
12	1	11	.02881459	13	1	4	.08767223	13	4	9	.05429258	14	2	12	.02989872
12	1	12	.02500805	13	1	5	.06975646	13	4	10	.04804585	14	2	13	.02663623
12	2	2	.21032572	13	1	6	.05784718	13	5	5	.11215378	14	3	3	.14501674
12	2	3	.13649893	13	1	7	.04925980	13	5	6	.09362255	14	3	4	.10829803
12	2	4	.10161010	13	1	8	.04269663	13	5	7	.08011076	14	3	5	.08660829

TABLE II

P = 7.0

N	R	S	COVARIANCE	N	R	S	COVARIANCE	N	R	S	COVARIANCE	N	R	S	COVARIANCE
14	3	6	.07212679	15	1	9	.03712368	15	4	8	.05826936	16	1	15	.01978631
14	3	7	.06166533	15	1	10	.03306464	15	4	9	.05150556	16	1	16	.01767417
14	3	8	.05367257	15	1	11	.02963462	15	4	10	.04594093	16	2	2	.19537899
14	3	9	.04729701	15	1	12	.02665222	15	4	11	.04122541	16	2	3	.12611939
14	3	10	.04202600	15	1	13	.02397931	15	4	12	.03711549	16	2	4	.09381382
14	3	11	.03752408	15	1	14	.02149377	15	5	5	.10314941	16	2	5	.07491467
14	3	12	.03354913	15	1	15	.01908635	15	5	6	.08621224	16	2	6	.06239545
14	4	4	.12021072	15	2	2	.19840482	15	5	7	.07395491	16	2	7	.05341800
14	4	5	.09634190	15	2	3	.12823439	15	5	8	.06458865	16	2	8	.04661205
14	4	6	.08035152	15	2	4	.09541648	15	5	9	.05712903	16	2	9	.04123184
14	4	7	.06877149	15	2	5	.07617607	15	5	10	.05098449	16	2	10	.03683426
14	4	8	.05990738	15	2	6	.06340430	15	5	11	.04577224	16	2	11	.03313680
14	4	9	.05282613	15	2	7	.05422523	15	6	6	.09564276	16	2	12	.02994745
14	4	10	.04696448	15	2	8	.04724817	15	6	7	.08214367	16	2	13	.02712622
14	4	11	.04195291	15	2	9	.04171463	15	6	8	.07180743	16	2	14	.02456053
14	5	5	.10724754	15	2	10	.03717216	15	6	9	.06356174	16	2	15	.02214505
14	5	6	.08959171	15	2	11	.03332988	15	6	10	.05676046	16	3	3	.13847304
14	5	7	.07677162	15	2	12	.02998628	15	7	7	.09173980	16	3	4	.10331410
14	5	8	.06693813	15	2	13	.02698755	15	7	8	.08028302	16	3	5	.08265368
14	5	9	.05906951	15	2	14	.02419736	15	7	9	.07112628	16	3	6	.06892720
14	5	10	.05254725	15	3	3	.14152371	15	8	8	.09052377	16	3	7	.05906318
14	6	6	.10031028	15	3	4	.10564137	16	1	1	.42339538	16	3	8	.05157315
14	6	7	.08607370	15	3	5	.08450471	16	1	2	.17775806	16	3	9	.04564477
14	6	8	.07512849	15	3	6	.07043006	16	1	3	.11396156	16	3	10	.04079424
14	6	9	.06635396	15	3	7	.06029179	16	1	4	.08447709	16	3	11	.03671252
14	7	7	.09723377	15	3	8	.05257249	16	1	5	.06731833	16	3	12	.03318923
14	7	8	.08497635	15	3	9	.04644207	16	1	6	.05599057	16	3	13	.03007072
15	1	1	.42636668	15	3	10	.04140416	16	1	7	.04788698	16	3	14	.02723317
15	1	2	.17957047	15	3	11	.03713898	16	1	8	.04175449	16	4	4	.11328801
15	1	3	.11523550	15	3	12	.03342452	16	1	9	.03691340	16	4	5	.09080041
15	1	4	.08543440	15	3	13	.03009099	16	1	10	.03296089	16	4	6	.07581649
15	1	5	.06805803	15	4	4	.11649434	16	1	11	.02964069	16	4	7	.06502618
15	1	6	.05656468	15	4	5	.09337118	16	1	12	.02677897	16	4	8	.05681967
15	1	7	.04832523	15	4	6	.07792558	16	1	13	.02424923	16	4	9	.05031597
15	1	8	.04207416	15	4	7	.06677452	16	1	14	.02194994	16	4	10	.04498925

TABLE II

P = 7.0

N	R	S	COVARIANCE
16	4	11	.04050298
16	4	12	.03662768
16	4	13	.03319544
16	5	5	.09966611
16	5	6	.08333061
16	5	7	.07154107
16	5	8	.06255944
16	5	9	.05543187
16	5	10	.04958777
16	5	11	.04466123
16	5	12	.04040227
16	6	6	.09174754
16	6	7	.07885257
16	6	8	.06901070
16	6	9	.06118898
16	6	10	.05476793
16	6	11	.04934951
16	7	7	.08725864
16	7	8	.07644031
16	7	9	.06782833
16	7	10	.06074888
16	8	8	.08520875
16	8	9	.07567659
17	1	1	.42075997
17	1	2	.17613238
17	1	3	.11281184
17	1	4	.08360717
17	1	5	.06663999
17	1	6	.05545713
17	1	7	.04747138
17	1	8	.04144046
17	1	9	.03669151
17	1	10	.03282670
17	1	11	.02959414
17	1	12	.02682471
17	1	13	.02439819
17	1	14	.02222301
17	1	15	.02022209
17	1	16	.01831993
17	1	17	.01645110
17	2	2	.19268853
17	2	3	.12423246
17	2	4	.09237801
17	2	5	.07377824
17	2	6	.06147938
17	2	7	.05267651
17	2	8	.04601715
17	2	9	.04076639
17	2	10	.03648866
17	2	11	.03290762
17	2	12	.02983740
17	2	13	.02714565
17	2	14	.02473141
17	2	15	.02250954
17	2	16	.02039648
17	3	3	.13578186
17	3	4	.10125551
17	3	5	.08100999
17	3	6	.06758535
17	3	7	.05795749
17	3	8	.05066305
17	3	9	.04490478
17	3	10	.04020916
17	3	11	.03627519
17	3	12	.03290017
17	3	13	.02993952
17	3	14	.02728279
17	3	15	.02483670
17	4	4	.11048859
17	4	5	.08854991
17	4	6	.07396287
17	4	7	.06348077
17	4	8	.05552728
17	4	9	.04924137
17	4	10	.04411059
17	4	11	.03980865
17	4	12	.03611548
17	4	13	.03287386
17	4	14	.02996355
17	5	5	.09666263
17	5	6	.08083902
17	5	7	.06944518
17	5	8	.06078642
17	5	9	.05393467
17	5	10	.04833640
17	5	11	.04363855
17	5	12	.03960261
17	5	13	.03605795
17	6	6	.08843917
17	6	7	.07604899
17	6	8	.06661729
17	6	9	.05914394
17	6	10	.05303109
17	6	11	.04789672
17	6	12	.04348231
17	7	7	.08352222
17	7	8	.07322585
17	7	9	.06505530
17	7	10	.05836403
17	7	11	.05273806
17	8	8	.08087794
17	8	9	.07190971
17	8	10	.06455502
17	9	9	.08004145
18	1	1	.41840781
18	1	2	.17466521
18	1	3	.11176833
18	1	4	.08281280
18	1	5	.06601576
18	1	6	.05496096
18	1	7	.04707858
18	1	8	.04113591
18	1	9	.03646595
18	1	10	.03267501
18	1	11	.02951457
18	1	12	.02681875
18	1	13	.02447113
18	1	14	.02238547
18	1	15	.02049337
18	1	16	.01873469
18	1	17	.01704797
18	1	18	.01538212
18	2	2	.19027845
18	2	3	.12253687
18	2	4	.09108300
18	2	5	.07274827
18	2	6	.06064364
18	2	7	.05199371
18	2	8	.04546169
18	2	9	.04032210
18	2	10	.03614570
18	2	11	.03266102
18	2	12	.02968657
18	2	13	.02709477
18	2	14	.02479100
18	2	15	.02270009
18	2	16	.02075586
18	2	17	.01889055
18	3	3	.13338723
18	3	4	.09941927

TABLE II

P = 7.0

N	R	S	COVARIANCE	N	R	S	COVARIANCE	N	R	S	COVARIANCE	N	R	S	COVARIANCE
18	3	5	.07953887	18	6	7	.07362683	19	1	17	.01744191	19	4	4	.10582291
18	3	6	.06637877	18	6	8	.06454151	19	1	18	.01593493	19	4	5	.08478643
18	3	7	.05695675	18	6	9	.05736012	19	1	19	.01444029	19	4	6	.07084816
18	3	8	.04983154	18	6	10	.05150287	19	2	2	.18810544	19	4	7	.06086638
18	3	9	.04421897	18	6	11	.04660064	19	2	3	.12100357	19	4	8	.05331985
18	3	10	.03965416	18	6	12	.04240528	19	2	4	.08990801	19	4	9	.04737989
18	3	11	.03584258	18	6	13	.03874142	19	2	5	.07180980	19	4	10	.04255502
18	3	12	.03258709	18	7	7	.08035105	19	2	6	.05987787	19	4	11	.03853409
18	3	13	.02974889	18	7	8	.07049006	19	2	7	.05136324	19	4	12	.03510936
18	3	14	.02722495	18	7	9	.06268512	19	2	8	.04494310	19	4	13	.03213554
18	3	15	.02493329	18	7	10	.05631230	19	2	9	.03990029	19	4	14	.02950632
18	3	16	.02280163	18	7	11	.05097365	19	2	10	.03581105	19	4	15	.02713980
18	4	4	.10801958	18	7	12	.04640119	19	2	11	.03240790	19	4	16	.02496831
18	4	5	.08656032	18	8	8	.07726992	19	2	12	.02951273	19	5	5	.09173076
18	4	6	.07231856	18	8	9	.06876191	19	2	13	.02700125	19	5	6	.07673353
18	4	7	.06210326	18	8	10	.06180649	19	2	14	.02478272	19	5	7	.06597431
18	4	8	.05436733	18	8	11	.05597373	19	2	15	.02278737	19	5	8	.05782902
18	4	9	.04826700	18	9	9	.07583499	19	2	16	.02095770	19	5	9	.05141093
18	4	10	.04330106	18	9	10	.06820909	19	2	17	.01924099	19	5	10	.04619318
18	4	11	.03915150	19	1	1	.41629683	19	2	18	.01758128	19	5	11	.04184172
18	4	12	.03560512	19	1	2	.17333384	19	3	3	.13124041	19	5	12	.03813321
18	4	13	.03251168	19	1	3	.11081640	19	3	4	.09776937	19	5	13	.03491130
18	4	14	.02975947	19	1	4	.08208423	19	3	5	.07821308	19	5	14	.03206144
18	4	15	.02725954	19	1	5	.06543940	19	3	6	.06528701	19	5	15	.02949530
18	5	5	.09404157	19	1	6	.05449869	19	3	7	.05604627	19	6	6	.08310169
18	5	6	.07865935	19	1	7	.04670791	19	3	8	.04906919	19	6	7	.07150936
18	5	7	.06760512	19	1	8	.04084279	19	3	9	.04358313	19	6	8	.06272086
18	5	8	.05922176	19	1	9	.03624153	19	3	10	.03913067	19	6	9	.05578812
18	5	9	.05260333	19	1	10	.03251395	19	3	11	.03542267	19	6	10	.05014679
18	5	10	.04721064	19	1	11	.02941424	19	3	12	.03226629	19	6	11	.04543846
18	5	11	.04270099	19	1	12	.02677897	19	3	13	.02952686	19	6	12	.04142319
18	5	12	.03884434	19	1	13	.02449424	19	3	14	.02710592	19	6	13	.03793282
18	5	13	.03547836	19	1	14	.02247698	19	3	15	.02492770	19	6	14	.03484398
18	5	14	.03248219	19	1	15	.02066344	19	3	16	.02292966	19	7	7	.07762008
18	6	6	.08558829	19	1	16	.01900110	19	3	17	.02105442	19	7	8	.06812828

TABLE II

P = 7.0

N	R	S	COVARIANCE	N	R	S	COVARIANCE	N	R	S	COVARIANCE	N	R	S	COVARIANCE
19	7	9	.06063162	20	2	3	.11960924	20	4	6	.06952399	20	7	8	.06606495
19	7	10	.05452529	20	2	4	.08883627	20	4	7	.05974857	20	7	9	.05883209
19	7	11	.04942459	20	2	5	.07095058	20	4	8	.05236864	20	7	10	.05295243
19	7	12	.04507161	20	2	6	.05917336	20	4	9	.04656892	20	7	11	.04805248
19	7	13	.04128533	20	2	7	.05077943	20	4	10	.04186644	20	7	12	.04388262
19	8	8	.07420987	20	2	8	.04445855	20	4	11	.03795594	20	7	13	.04026854
19	8	9	.06608515	20	2	9	.03950097	20	4	12	.03463415	20	7	14	.03708348
19	8	10	.05945992	20	2	10	.03548772	20	4	13	.03175961	20	8	8	.07157618
19	8	11	.05392063	20	2	11	.03215476	20	4	14	.02922977	20	8	9	.06377603
19	8	12	.04918959	20	2	12	.02932668	20	4	15	.02696707	20	8	10	.05742883
19	9	9	.07233336	20	2	13	.02688168	20	4	16	.02490982	20	8	11	.05213478
19	9	10	.06511963	20	2	14	.02473162	20	4	17	.02300528	20	8	12	.04762629
19	9	11	.05908197	20	2	15	.02280998	20	5	5	.08967552	20	8	13	.04371628
19	10	10	.07173343	20	2	16	.02106393	20	5	6	.07501735	20	9	9	.06936555
20	1	1	.41439302	20	2	17	.01944840	20	5	7	.06451712	20	9	10	.06249461
20	1	2	.17211983	20	2	18	.01792055	20	5	8	.05657992	20	9	11	.05675833
20	1	3	.10994407	20	2	19	.01643344	20	5	9	.05033601	20	9	12	.05186930
20	1	4	.08141331	20	3	3	.12930303	20	5	10	.04526928	20	10	10	.06832169
20	1	5	.06490553	20	3	4	.09627738	20	5	11	.04105305	20	10	11	.06208161
20	1	6	.05406725	20	3	5	.07701096	20	5	12	.03746954				
20	1	7	.04635828	20	3	6	.06429361	20	5	13	.03436700				
20	1	8	.04056199	20	3	7	.05521394	20	5	14	.03163535				
20	1	9	.03602116	20	3	8	.04836782	20	5	15	.02919125				
20	1	10	.03234871	20	3	9	.04299286	20	5	16	.02696833				
20	1	11	.02930110	20	3	10	.03863823	20	6	6	.08091040				
20	1	12	.02671680	20	3	11	.03501935	20	6	7	.06963958				
20	1	13	.02448376	20	3	12	.03194696	20	6	8	.06110858				
20	1	14	.02252101	20	3	13	.02928949	20	6	9	.05439049				
20	1	15	.02076748	20	3	14	.02695162	20	6	10	.04893432				
20	1	16	.01917475	20	3	15	.02486138	20	6	11	.04439080				
20	1	17	.01770157	20	3	16	.02296152	20	6	12	.04052678				
20	1	18	.01630873	20	3	17	.02120320	20	6	13	.03717966				
20	1	19	.01495335	20	3	18	.01953988	20	6	14	.03423132				
20	1	20	.01360453	20	4	4	.10385369	20	6	15	.03159230				
20	2	2	.18613486	20	4	5	.08319315	20	7	7	.07523931				

TABLE II

P = 7.5

N	R	S	COVARIANCE	N	R	S	COVARIANCE	N	R	S	COVARIANCE	N	R	S	COVARIANCE
2	1	1	.68874753	7	1	3	.13454993	9	1	2	.19588915	10	2	3	.14477551
2	1	2	.31125247	7	1	4	.09894317	9	1	3	.12685901	10	2	4	.10789728
3	1	1	.58983559	7	1	5	.07713255	9	1	4	.09393869	10	2	5	.08579297
3	1	2	.26783414	7	1	6	.06169436	9	1	5	.07422605	10	2	6	.07077623
3	1	3	.16464979	7	1	7	.04927538	9	1	6	.06078430	10	2	7	.05967367
3	2	2	.41969270	7	2	2	.25279171	9	1	7	.05074584	10	2	8	.05089647
4	1	1	.53922707	7	2	3	.16567971	9	1	8	.04263380	10	2	9	.04349397
4	1	2	.24344463	7	2	4	.12265816	9	1	9	.03546117	10	3	3	.16714997
4	1	3	.15464451	7	2	5	.09603055	9	2	2	.22966001	10	3	4	.12515852
4	1	4	.10732281	7	2	6	.07704676	9	2	3	.15023941	10	3	5	.09981293
4	2	2	.33766765	7	3	3	.20323463	9	2	4	.11184157	10	3	6	.08251154
4	2	3	.21960419	7	3	4	.15158284	9	2	5	.08866188	10	3	7	.06967453
5	1	1	.50785715	7	3	5	.11925885	9	2	6	.07276955	10	3	8	.05949817
5	1	2	.22759566	7	4	4	.19143430	9	2	7	.06085359	10	4	4	.14527354
5	1	3	.14628506	8	1	1	.45809866	9	2	8	.05119466	10	4	5	.11623977
5	1	4	.10517085	8	1	2	.20125494	9	3	3	.17604543	10	4	6	.09631663
5	1	5	.07804404	8	1	3	.13034217	9	3	4	.13174983	10	4	7	.08147598
5	2	2	.29614577	8	1	4	.09626654	9	3	5	.10479769	10	5	5	.13672218
5	2	3	.19375708	8	1	5	.07567814	9	3	6	.08621743	10	5	6	.11361192
5	2	4	.14071473	8	1	6	.06144877	9	3	7	.07222901	11	1	1	.43366606
5	3	3	.26324204	8	1	7	.05056140	9	4	4	.15586806	11	1	2	.18769199
6	1	1	.48625520	8	1	8	.04131341	9	4	5	.12447099	11	1	3	.12141092
6	1	2	.21634964	8	2	2	.23974625	9	4	6	.10269188	11	1	4	.09013698
6	1	3	.13974014	8	2	3	.15701163	9	5	5	.15032736	11	1	5	.07161967
6	1	4	.10197647	8	2	4	.11664921	10	1	1	.44033695	11	1	6	.05917698
6	1	5	.07828541	8	2	5	.09204042	10	1	2	.19144441	11	1	7	.05008910
6	1	6	.06065386	8	2	6	.07492717	10	1	3	.12392329	11	1	8	.04302288
6	2	2	.27047686	8	2	7	.06177340	10	1	4	.09191196	11	1	9	.03722365
6	2	3	.17726814	8	3	3	.18757046	10	1	5	.07286593	11	1	10	.03218799
6	2	4	.13040057	8	3	4	.14021388	10	1	6	.05999046	11	1	11	.02748457
6	2	5	.10063236	8	3	5	.11107481	10	1	7	.05050490	11	2	2	.21495908
6	3	3	.22606222	8	3	6	.09068004	10	1	8	.04302649	11	2	3	.14025685
6	3	4	.16787878	8	4	4	.17030721	10	1	9	.03673310	11	2	4	.10459434
7	1	1	.47035439	8	4	5	.13557243	10	1	10	.03099507	11	2	5	.08333414
7	1	2	.20788929	9	1	1	.44832857	10	2	2	.22158905	11	2	6	.06898368

TABLE II

P = 7.5

N	R	S	COVARIANCE	N	R	S	COVARIANCE	N	R	S	COVARIANCE	N	R	S	COVARIANCE
11	2	7	.05846831	12	2	5	.08120475	13	1	9	.03730890	13	5	8	.07028065
11	2	8	.05027194	12	2	6	.06738232	13	1	10	.03293008	13	5	9	.06179423
11	2	9	.04353201	12	2	7	.05731661	13	1	11	.02913056	13	6	6	.10695905
11	2	10	.03767008	12	2	8	.04954075	13	1	12	.02567834	13	6	7	.09168553
11	3	3	.16004014	12	2	9	.04323750	13	1	13	.02233857	13	6	8	.07985382
11	3	4	.11985325	12	2	10	.03789376	13	2	2	.20465777	13	7	7	.10511521
11	3	5	.09574454	12	2	11	.03313050	13	2	3	.13318039	14	1	1	.41890182
11	3	6	.07940159	12	3	3	.15420432	13	2	4	.09935211	14	1	2	.17922733
11	3	7	.06738865	12	3	4	.11547373	13	2	5	.07934216	14	1	3	.11563426
11	3	8	.05800232	12	3	5	.09235074	13	2	6	.06595082	14	1	4	.08593538
11	3	9	.05026886	12	3	6	.07675741	13	2	7	.05624361	14	1	5	.06851850
11	4	4	.13711248	12	3	7	.06537000	13	2	8	.04879083	14	1	6	.05694076
11	4	5	.10984836	12	3	8	.05655404	13	2	9	.04280396	14	1	7	.04859973
11	4	6	.09128197	12	3	9	.04939532	13	2	10	.03780166	14	1	8	.04223720
11	4	7	.07758830	12	3	10	.04331769	13	2	11	.03345629	14	1	9	.03716560
11	4	8	.06686030	12	4	4	.13059931	13	2	12	.02950436	14	1	10	.03297207
11	5	5	.12676544	12	4	5	.10471513	13	3	3	.14931282	14	1	11	.02938628
11	5	6	.10558928	12	4	6	.08718895	13	3	4	.11178541	14	1	12	.02621153
11	5	7	.08990991	12	4	7	.07435171	13	3	5	.08946905	14	1	13	.02327774
11	6	6	.12372744	12	4	8	.06439024	13	3	6	.07448079	14	1	14	.02040233
12	1	1	.42800556	12	4	9	.05628620	13	3	7	.06358791	14	2	2	.20055849
12	1	2	.18447464	12	5	5	.11911040	13	3	8	.05520833	14	2	3	.13034500
12	1	3	.11923287	12	5	6	.09937609	13	3	9	.04846644	14	2	4	.09722901
12	1	4	.08857126	12	5	7	.08487312	13	3	10	.04282609	14	2	5	.07769799
12	1	5	.07048597	12	5	8	.07358952	13	3	11	.03792120	14	2	6	.06466672
12	1	6	.05838799	12	6	6	.11424008	13	4	4	.12525933	14	2	7	.05525381
12	1	7	.04960443	12	6	7	.09774570	13	4	5	.10048475	14	2	8	.04805948
12	1	8	.04283440	13	1	1	.42313700	13	4	6	.08378461	14	2	9	.04231601
12	1	9	.03735628	13	1	2	.18168044	13	4	7	.07161501	14	2	10	.03756101
12	1	10	.03271885	13	1	3	.11732370	13	4	8	.06223392	14	2	11	.03349092
12	1	11	.02859021	13	1	4	.08718006	13	4	9	.05467376	14	2	12	.02988421
12	1	12	.02465860	13	1	5	.06945589	13	4	10	.04834019	14	2	13	.02654873
12	2	2	.20939907	13	1	6	.05764076	13	5	5	.11301008	14	3	3	.14514289
12	2	3	.13644595	13	1	7	.04910004	13	5	6	.09439589	14	3	4	.10862845
12	2	4	.10178147	13	1	8	.04255660	13	5	7	.08079165	14	3	5	.08698611

TABLE II

P = 7.5

N	R	S	COVARIANCE	N	R	S	COVARIANCE	N	R	S	COVARIANCE	N	R	S	COVARIANCE
14	3	6	.07249825	15	1	9	.03697040	15	4	8	.05867969	16	1	15	.01956205
14	3	7	.06200820	15	1	10	.03291832	15	4	9	.05186532	16	1	16	.01736011
14	3	8	.05397606	15	1	11	.02948546	15	4	10	.04624685	16	2	2	.19380599
14	3	9	.04755456	15	1	12	.02649098	15	4	11	.04147367	16	2	3	.12564927
14	3	10	.04223203	15	1	13	.02379490	15	4	12	.03730016	16	2	4	.09368570
14	3	11	.03767171	15	1	14	.02126764	15	5	5	.10383482	16	2	5	.07492286
14	3	12	.03362722	15	1	15	.01876366	15	5	6	.08685582	16	2	6	.06246201
14	4	4	.12078748	15	2	2	.19697305	15	5	7	.07454191	16	2	7	.05350714
14	4	5	.09692681	15	2	3	.12785563	15	5	8	.06511356	16	2	8	.04670499
14	4	6	.08090055	15	2	4	.09535475	15	5	9	.05758938	16	2	9	.04131757
14	4	7	.06926829	15	2	5	.07623472	15	5	10	.05137820	16	2	10	.03690546
14	4	8	.06034490	15	2	6	.06350964	15	5	11	.04609615	16	2	11	.03318777
14	4	9	.05320025	15	2	7	.05434375	15	6	6	.09642537	16	2	12	.02997283
14	4	10	.04727109	15	2	8	.04736231	15	6	7	.08285286	16	2	13	.02711966
14	4	11	.04218578	15	2	9	.04181390	15	6	8	.07243982	16	2	14	.02451263
14	5	5	.10801441	15	2	10	.03724920	15	6	9	.06411628	16	2	15	.02203760
14	5	6	.09029800	15	2	11	.03337830	15	6	10	.05723597	16	3	3	.13838530
14	5	7	.07740524	15	2	12	.02999897	15	7	7	.09256803	16	3	4	.10348720
14	5	8	.06749480	15	2	13	.02695426	15	7	8	.08102020	16	3	5	.08291152
14	5	9	.05954689	15	2	14	.02409845	15	7	9	.07177241	16	3	6	.06920680
14	5	10	.05294222	15	3	3	.14153815	15	8	8	.09136550	16	3	7	.05933745
14	6	6	.10116494	15	3	4	.10588983	16	1	1	.41188733	16	3	8	.05182884
14	6	7	.08683680	15	3	5	.08482037	16	1	2	.17511225	16	3	9	.04587453
14	6	8	.07579786	15	3	6	.07075438	16	1	3	.11277228	16	3	10	.04099328
14	6	9	.06692856	15	3	7	.06060006	16	1	4	.08379977	16	3	11	.03687688
14	7	7	.09812409	15	3	8	.05285273	16	1	5	.06688000	16	3	12	.03331465
14	7	8	.08575626	15	3	9	.04668747	16	1	6	.05568075	16	3	13	.03015135
15	1	1	.41518190	15	3	10	.04160983	16	1	7	.04765154	16	3	14	.02725942
15	1	2	.17705369	15	3	11	.03730012	16	1	8	.04156328	16	4	4	.11368682
15	1	3	.11412690	15	3	12	.03353484	16	1	9	.03674789	16	4	5	.09124828
15	1	4	.08481469	15	3	13	.03014016	16	1	10	.03280859	16	4	6	.07625986
15	1	5	.06766312	15	4	4	.11697803	16	1	11	.02949232	16	4	7	.06544362
15	1	6	.05628816	15	4	5	.09388479	16	1	12	.02662670	16	4	8	.05720159
15	1	7	.04811511	15	4	6	.07842014	16	1	13	.02408523	16	4	9	.05065751
15	1	8	.04190146	15	4	7	.06723110	16	1	14	.02176433	16	4	10	.04528731

TABLE II

P = 7.5

N	R	S	COVARIANCE	N	R	S	COVARIANCE	N	R	S	COVARIANCE	N	R	S	COVARIANCE
16	4	11	.04075474	17	1	13	.02424397	17	4	6	.07435823	18	1	2	.17178594
16	4	12	.03682952	17	1	14	.02205747	17	4	7	.06386061	18	1	3	.11043296
16	4	13	.03334172	17	1	15	.02003631	17	4	8	.05588067	18	1	4	.08203055
16	5	5	.10027709	17	1	16	.01809814	17	4	9	.04956277	18	1	5	.06549749
16	5	6	.08391579	17	1	17	.01614535	17	4	10	.04439673	18	1	6	.05458867
16	5	7	.07208315	17	2	2	.19098468	17	4	11	.04005707	18	1	7	.04679415
16	5	8	.06305154	17	2	3	.12367767	17	4	12	.03632359	18	1	8	.04090691
16	5	9	.05587100	17	2	4	.09218810	17	4	13	.03303823	18	1	9	.03627257
16	5	10	.04997210	17	2	5	.07373906	17	4	14	.03007882	18	1	10	.03250423
16	5	11	.04498875	17	2	6	.06150898	17	5	5	.09720533	18	1	11	.02935718
16	5	12	.04066979	17	2	7	.05273696	17	5	6	.08136989	18	1	12	.02666777
16	6	6	.09246367	17	2	8	.04608845	17	5	7	.06994447	18	1	13	.02432073
16	6	7	.07951030	17	2	9	.04083689	17	5	8	.06124589	18	1	14	.02223019
16	6	8	.06960528	17	2	10	.03655084	17	5	9	.05435062	18	1	15	.02032720
16	6	9	.06171886	17	2	11	.03295601	17	5	10	.04870685	18	1	16	.01854950
16	6	10	.05523228	17	2	12	.02986739	17	5	11	.04396194	18	1	17	.01682905
16	6	11	.04974692	17	2	13	.02715257	17	5	12	.03987693	18	1	18	.01508434
16	7	7	.08802804	17	2	14	.02470952	17	5	13	.03628002	18	2	2	.18845275
16	7	8	.07713403	17	2	15	.02245016	17	6	6	.08909399	18	2	3	.12190327
16	7	9	.06844597	17	2	16	.02028265	17	6	7	.07665813	18	2	4	.09083537
16	7	10	.06129036	17	3	3	.13560011	17	6	8	.06717460	18	2	5	.07266455
16	8	8	.08600090	17	3	4	.10135898	17	6	9	.05964717	18	2	6	.06063811
16	8	9	.07638125	17	3	5	.08121396	17	6	10	.05347930	18	2	7	.05202639
17	1	1	.40894834	17	3	6	.06782268	17	6	11	.04828905	18	2	8	.04551140
17	1	2	.17336610	17	3	7	.05819882	17	6	12	.04381715	18	2	9	.04037651
17	1	3	.11154718	17	3	8	.05089385	17	7	7	.08423652	18	2	10	.03619697
17	1	4	.08287579	17	3	9	.04511706	17	7	8	.07387707	18	2	11	.03270366
17	1	5	.06616060	17	3	10	.04039792	17	7	9	.06564238	18	2	12	.02971629
17	1	6	.05511540	17	3	11	.03643677	17	7	10	.05888685	18	2	13	.02710769
17	1	7	.04721110	17	3	12	.03303119	17	7	11	.05319629	18	2	14	.02478299
17	1	8	.04123035	17	3	13	.03003610	17	8	8	.08162178	18	2	15	.02266591
17	1	9	.03651242	17	3	14	.02733950	17	8	9	.07257950	18	2	16	.02068741
17	1	10	.03266591	17	3	15	.02484457	17	8	10	.06515130	18	2	17	.01877197
17	1	11	.02944251	17	4	4	.11080957	17	9	9	.08079426	18	3	3	.13311852
17	1	12	.02667502	17	4	5	.08893700	18	1	1	.40630991	18	3	4	.09945813

TABLE II

P = 7.5

N	R	S	COVARIANCE	N	R	S	COVARIANCE	N	R	S	COVARIANCE	N	R	S	COVARIANCE
18	3	5	.07969253	18	6	7	.07419032	19	1	17	.01725784	19	4	4	.10600570
18	3	6	.06657618	18	6	8	.06506276	19	1	18	.01571913	19	4	5	.08506468
18	3	7	.05716641	18	6	9	.05783612	19	1	19	.01415012	19	4	6	.07115626
18	3	8	.05003767	18	6	10	.05193236	19	2	2	.18616576	19	4	7	.06117621
18	3	9	.04441286	18	6	11	.04698285	19	2	3	.12029630	19	4	8	.05361791
18	3	10	.03983051	18	6	12	.04273924	19	2	4	.08960629	19	4	9	.04765915
18	3	11	.03599773	18	6	13	.03902542	19	2	5	.07168411	19	4	10	.04281148
18	3	12	.03271804	18	7	7	.08101395	19	2	6	.05983895	19	4	11	.03876519
18	3	13	.02985268	18	7	8	.07110049	19	2	7	.05136917	19	4	12	.03531314
18	3	14	.02729799	18	7	9	.06324123	19	2	8	.04497156	19	4	13	.03231010
18	3	15	.02497052	18	7	10	.05681368	19	2	9	.03993826	19	4	14	.02964939
18	3	16	.02279463	18	7	11	.05142012	19	2	10	.03585031	19	4	15	.02724826
18	4	4	.10826880	18	7	12	.04679217	19	2	11	.03244282	19	4	16	.02503734
18	4	5	.08689103	18	8	8	.07796767	19	2	12	.02953911	19	5	5	.09215240
18	4	6	.07266890	18	8	9	.06939657	19	2	13	.02701558	19	5	6	.07716671
18	4	7	.06244722	18	8	10	.06237832	19	2	14	.02478166	19	5	7	.06639482
18	4	8	.05469266	18	8	11	.05648304	19	2	15	.02276720	19	5	8	.05822594
18	4	9	.04856742	18	9	9	.07654813	19	2	16	.02091356	19	5	9	.05177902
18	4	10	.04357299	18	9	10	.06885125	19	2	17	.01916534	19	5	10	.04652974
18	4	11	.03939249	19	1	1	.40392803	19	2	18	.01745921	19	5	11	.04214512
18	4	12	.03581306	19	1	2	.17034821	19	3	3	.13089088	19	5	12	.03840216
18	4	13	.03268417	19	1	3	.10941449	19	3	4	.09774805	19	5	13	.03514438
18	4	14	.02989323	19	1	4	.08125398	19	3	5	.07831962	19	5	14	.03225667
18	4	15	.02734948	19	1	5	.06488425	19	3	6	.06544672	19	5	15	.02964964
18	5	5	.09452137	19	1	6	.05409722	19	3	7	.05622562	19	6	6	.08364739
18	5	6	.07913967	19	1	7	.04640011	19	3	8	.04925119	19	6	7	.07203002
18	5	7	.06806390	19	1	8	.04059514	19	3	9	.04375835	19	6	8	.06320758
18	5	8	.05964940	19	1	9	.03603357	19	3	10	.03929342	19	6	9	.05623709
18	5	9	.05299536	19	1	10	.03233229	19	3	11	.03556914	19	6	10	.05055632
18	5	10	.04756477	19	1	11	.02924954	19	3	12	.03239363	19	6	11	.04580768
18	5	11	.04301572	19	1	12	.02662427	19	3	13	.02963253	19	6	12	.04175137
18	5	12	.03911819	19	1	13	.02434401	19	3	14	.02718724	19	6	13	.03821888
18	5	13	.03570932	19	1	14	.02232640	19	3	15	.02498134	19	6	14	.03508614
18	5	14	.03266716	19	1	15	.02050778	19	3	16	.02295088	19	7	7	.07823511
18	6	6	.08618646	19	1	16	.01883498	19	3	17	.02103529	19	7	8	.06869993

TABLE II

P = 7.5

N	R	S	COVARIANCE	N	R	S	COVARIANCE	N	R	S	COVARIANCE	N	R	S	COVARIANCE
19	7	9	.06115722	20	2	3	.11883284	20	4	6	.06979239	20	7	8	.06659990
19	7	10	.05500402	20	2	4	.08848370	20	4	7	.06002599	20	7	9	.05932809
19	7	11	.04985620	20	2	5	.07078527	20	4	8	.05264039	20	7	10	.05340819
19	7	12	.04545579	20	2	6	.05910271	20	4	9	.04682721	20	7	11	.04846754
19	7	13	.04162128	20	2	7	.05075967	20	4	10	.04210684	20	7	12	.04425673
19	8	8	.07486410	20	2	8	.04446627	20	4	11	.03817567	20	7	13	.04060125
19	8	9	.06668542	20	2	9	.03952246	20	4	12	.03483126	20	7	14	.03737385
19	8	10	.06000610	20	2	10	.03551436	20	4	13	.03193241	20	8	8	.07218948
19	8	11	.05441297	20	2	11	.03218072	20	4	14	.02937651	20	8	9	.06434316
19	8	12	.04962814	20	2	12	.02934774	20	4	15	.02708561	20	8	10	.05794917
19	9	9	.07300783	20	2	13	.02689448	20	4	16	.02499719	20	8	11	.05260841
19	9	10	.06573285	20	2	14	.02473320	20	4	17	.02305689	20	8	12	.04805334
19	9	11	.05963460	20	2	15	.02279738	20	5	5	.09004315	20	8	13	.04409655
19	10	10	.07241417	20	2	16	.02103377	20	5	6	.07540650	20	9	9	.07000295
20	1	1	.40176703	20	2	17	.01939618	20	5	7	.06490150	20	9	10	.06307882
20	1	2	.16903372	20	2	18	.01783918	20	5	8	.05694739	20	9	11	.05728986
20	1	3	.10847934	20	2	19	.01630885	20	5	9	.05068055	20	9	12	.05234859
20	1	4	.08053766	20	3	3	.12887807	20	5	10	.04558773	20	10	10	.06896998
20	1	5	.06431537	20	3	4	.09619979	20	5	11	.04134357	20	10	11	.06267118
20	1	6	.05363786	20	3	5	.07707328	20	5	12	.03773085	******	******	******	**********
20	1	7	.04602791	20	3	6	.06441768	20	5	13	.03459794				
20	1	8	.04029607	20	3	7	.05536432	20	5	14	.03183451				
20	1	9	.03579860	20	3	8	.04852638	20	5	15	.02935669				
20	1	10	.03215571	20	3	9	.04314946	20	5	16	.02709712				
20	1	11	.02912810	20	3	10	.03878674	20	6	6	.08140737				
20	1	12	.02655681	20	3	11	.03515577	20	6	7	.07012007				
20	1	13	.02433136	20	3	12	.03206841	20	6	8	.06156240				
20	1	14	.02237168	20	3	13	.02939362	20	6	9	.05481304				
20	1	15	.02061714	20	3	14	.02703622	20	6	10	.04932345				
20	1	16	.01901924	20	3	15	.02492399	20	6	11	.04474542				
20	1	17	.01753601	20	3	16	.02299904	20	6	12	.04084618				
20	1	18	.01612617	20	3	17	.02121113	20	6	13	.03746307				
20	1	19	.01474084	20	3	18	.01951078	20	6	14	.03447762				
20	1	20	.01332162	20	4	4	.10397469	20	6	15	.03179965				
20	2	2	.18408812	20	4	5	.08342242	20	7	7	.07580972				

TABLE II

P = 8.0

N	R	S	COVARIANCE
2	1	1	.68817808
2	1	2	.31182192
3	1	1	.58749664
3	1	2	.26844625
3	1	3	.16470682
3	2	2	.42180807
4	1	1	.53556415
4	1	2	.24364866
4	1	3	.15492461
4	1	4	.10716200
4	2	2	.33933189
4	2	3	.22079542
5	1	1	.50317945
5	1	2	.22741067
5	1	3	.14645705
5	1	4	.10524585
5	1	5	.07778852
5	2	2	.29729197
5	2	3	.19482476
5	2	4	.14139762
5	3	3	.26493148
6	1	1	.48076441
6	1	2	.21583153
6	1	3	.13975796
6	1	4	.10203994
6	1	5	.07824579
6	1	6	.06035649
6	2	2	.27118669
6	2	3	.17813092
6	2	4	.13108050
6	2	5	.10103336
6	3	3	.22750551
6	3	4	.16898027
7	1	1	.46418853
7	1	2	.20708764

N	R	S	COVARIANCE
7	1	3	.13441552
7	1	4	.09894179
7	1	5	.07711749
7	1	6	.06158890
7	1	7	.04896090
7	2	2	.25314075
7	2	3	.16634272
7	2	4	.12325503
7	2	5	.09647293
7	2	6	.07728026
7	3	3	.20442533
7	3	4	.15258220
7	3	5	.12000397
7	4	4	.19278366
8	1	1	.45135688
8	1	2	.20020932
8	1	3	.13006765
8	1	4	.09618763
8	1	5	.07563861
8	1	6	.06138189
8	1	7	.05041703
8	1	8	.04099416
8	2	2	.23979293
8	2	3	.15749490
8	2	4	.11714632
8	2	5	.09245410
8	2	6	.07521795
8	2	7	.06190134
8	3	3	.18853676
8	3	4	.14108729
8	3	5	.11178507
8	3	6	.09119975
8	4	4	.17150525
8	4	5	.13653742
9	1	1	.44108513

N	R	S	COVARIANCE
9	1	2	.19463076
9	1	3	.12645775
9	1	4	.09378172
9	1	5	.07414571
9	1	6	.06071304
9	1	7	.05064501
9	1	8	.04246614
9	1	9	.03514356
9	2	2	.22944883
9	2	3	.15056356
9	2	4	.11223968
9	2	5	.08902479
9	2	6	.07305798
9	2	7	.06104324
9	2	8	.05125292
9	3	3	.17681687
9	3	4	.13250002
9	3	5	.10544442
9	3	6	.08673490
9	3	7	.07259905
9	4	4	.15691569
9	4	5	.12535887
9	4	6	.10340221
9	5	5	.15144107
10	1	1	.43264927
10	1	2	.18999809
10	1	3	.12340762
10	1	4	.09168043
10	1	5	.07273962
10	1	6	.05989835
10	1	7	.05041027
10	1	8	.04290333
10	1	9	.03655157
10	1	10	.03068248
10	2	2	.22115437

N	R	S	COVARIANCE
10	2	3	.14495871
10	2	4	.10820233
10	2	5	.08609877
10	2	6	.07104012
10	2	7	.05987437
10	2	8	.05101609
10	2	9	.04350457
10	3	3	.16775190
10	3	4	.12579488
10	3	5	.10038917
10	3	6	.08299865
10	3	7	.07005862
10	3	8	.05976458
10	4	4	.14618367
10	4	5	.11704177
10	4	6	.09699004
10	4	7	.08201140
10	5	5	.13773412
10	5	6	.11445803
11	1	1	.42557978
11	1	2	.18607779
11	1	3	.12079166
11	1	4	.08983559
11	1	5	.07144667
11	1	6	.05905721
11	1	7	.04998515
11	1	8	.04291161
11	1	9	.03708578
11	1	10	.03199849
11	1	11	.02717874
11	2	2	.21432792
11	2	3	.14031447
11	2	4	.10481347
11	2	5	.08358249
11	2	6	.06921421

TABLE II

P = 8.0

N	R	S	COVARIANCE	N	R	S	COVARIANCE	N	R	S	COVARIANCE	N	R	S	COVARIANCE
11	2	7	.05865936	12	2	5	.08139765	13	1	9	.03718799	13	5	8	.07078275
11	2	8	.05040952	12	2	6	.06757638	13	1	10	.03279912	13	5	9	.06221094
11	2	9	.04360163	12	2	7	.05748861	13	1	11	.02897696	13	6	6	.10776008
11	2	10	.03764720	12	2	8	.04967748	13	1	12	.02548298	13	6	7	.09238878
11	3	3	.16049341	12	2	9	.04332855	13	1	13	.02204827	13	6	8	.08045755
11	3	4	.12038619	12	2	10	.03792680	13	2	2	.20369466	13	7	7	.10593544
11	3	5	.09625095	12	2	11	.03308360	13	2	3	.13302348	14	1	1	.40981822
11	3	6	.07984795	12	3	3	.15452597	13	2	4	.09941990	14	1	2	.17719759
11	3	7	.06776099	12	3	4	.11591302	13	2	5	.07948249	14	1	3	.11475505
11	3	8	.05829096	12	3	5	.09279069	13	2	6	.06610789	14	1	4	.08545243
11	3	9	.05046103	12	3	6	.07715982	13	2	7	.05639226	14	1	5	.06821506
11	4	4	.13789946	12	3	7	.06571935	13	2	8	.04891749	14	1	6	.05672956
11	4	5	.11056629	12	3	8	.05684178	13	2	9	.04289949	14	1	7	.04843810
11	4	6	.09190596	12	3	9	.04961366	13	2	10	.03785770	14	1	8	.04210118
11	4	7	.07810932	12	3	10	.04345517	13	2	11	.03346203	14	1	9	.03704001
11	4	8	.06727127	12	4	4	.13127628	13	2	12	.02943995	14	1	10	.03284587
11	5	5	.12767736	12	4	5	.10535415	13	3	3	.14951691	14	1	11	.02924989
11	5	6	.10637656	12	4	6	.08776030	13	3	4	.11213975	14	1	12	.02605421
11	5	7	.09056686	12	4	7	.07484495	13	3	5	.08984677	14	1	13	.02308235
11	6	6	.12467283	12	4	8	.06479982	13	3	6	.07483916	14	1	14	.02011997
12	1	1	.41955778	12	4	9	.05660588	13	3	7	.06390939	14	2	2	.19945287
12	1	2	.18270899	12	5	5	.11992925	13	3	8	.05548412	14	2	3	.13009541
12	1	3	.11851932	12	5	6	.10010087	13	3	9	.04869035	14	2	4	.09723014
12	1	4	.08820487	12	5	7	.08549695	13	3	10	.04299140	14	2	5	.07778890
12	1	5	.07026747	12	5	8	.07410893	13	3	11	.03801745	14	2	6	.06478751
12	1	6	.05823800	12	6	6	.11511251	13	4	4	.12583772	14	2	7	.05537732
12	1	7	.04948399	12	6	7	.09849491	13	4	5	.10105080	14	2	8	.04817135
12	1	8	.04272079	13	1	1	.41435841	13	4	6	.08430390	14	2	9	.04240730
12	1	9	.03723336	13	1	2	.17977699	13	4	7	.07207505	14	2	10	.03762481
12	1	10	.03257137	13	1	3	.11652387	13	4	8	.06262905	14	2	11	.03352016
12	1	11	.02839654	13	1	4	.08675328	13	4	9	.05499971	14	2	12	.02986922
12	1	12	.02436037	13	1	5	.06919381	13	4	10	.04859115	14	2	13	.02647142
12	2	2	.20859310	13	1	6	.05745996	13	5	5	.11374384	14	3	3	.14524108
12	2	3	.13639088	13	1	7	.04895973	13	5	6	.09505947	14	3	4	.10890544
12	2	4	.10192171	13	1	8	.04243363	13	5	7	.08137618	14	3	5	.08730595

TABLE II

P = 8.0

N	R	S	COVARIANCE	N	R	S	COVARIANCE	N	R	S	COVARIANCE	N	R	S	COVARIANCE
14	3	6	.07281389	15	1	9	.03683642	15	4	8	.05903062	16	1	15	.01936956
14	3	7	.06229999	15	1	10	.03279060	15	4	9	.05217292	16	1	16	.01709312
14	3	8	.05423440	15	1	11	.02935558	15	4	10	.04650814	16	2	2	.19245035
14	3	9	.04777359	15	1	12	.02635106	15	4	11	.04168524	16	2	3	.12523669
14	3	10	.04240678	15	1	13	.02363557	15	4	12	.03745683	16	2	4	.09356828
14	3	11	.03779618	15	1	14	.02107333	15	5	5	.10442052	16	2	5	.07492399
14	3	12	.03369180	15	1	15	.01848911	15	5	6	.08740698	16	2	6	.06251437
14	4	4	.12127705	15	2	2	.19573720	15	5	7	.07504518	16	2	7	.05357969
14	4	5	.09742572	15	2	3	.12752119	15	5	8	.06556378	16	2	8	.04678138
14	4	6	.08136992	15	2	4	.09529434	15	5	9	.05798413	16	2	9	.04138816
14	4	7	.06969342	15	2	5	.07627928	15	5	10	.05171551	16	2	10	.03696384
14	4	8	.06071936	15	2	6	.06359545	15	5	11	.04637315	16	2	11	.03322898
14	4	9	.05352022	15	2	7	.05444166	15	6	6	.09709679	16	2	12	.02999218
14	4	10	.04753286	15	2	8	.04745700	15	6	7	.08346189	16	2	13	.02711172
14	4	11	.04238387	15	2	9	.04189618	15	6	8	.07298308	16	2	14	.02446936
14	5	5	.10867070	15	2	10	.03731261	15	6	9	.06459256	16	2	15	.02194361
14	5	6	.09090352	15	2	11	.03341729	15	6	10	.05764404	16	3	3	.13829957
14	5	7	.07794889	15	2	12	.03000740	15	7	7	.09327989	16	3	4	.10362843
14	5	8	.06797247	15	2	13	.02692336	15	7	8	.08165398	16	3	5	.08312760
14	5	9	.05995627	15	2	14	.02401157	15	7	9	.07232781	16	3	6	.06944291
14	5	10	.05328044	15	3	3	.14154022	15	8	8	.09208937	16	3	7	.05956985
14	6	6	.10189874	15	3	4	.10609608	16	1	1	.40225666	16	3	8	.05204581
14	6	7	.08749246	15	3	5	.08508644	16	1	2	.17285796	16	3	9	.04606955
14	6	8	.07637301	15	3	6	.07102919	16	1	3	.11175150	16	3	10	.04116208
14	6	9	.06742200	15	3	7	.06086189	16	1	4	.08321556	16	3	11	.03701595
14	7	7	.09888951	15	3	8	.05309095	16	1	5	.06650047	16	3	12	.03342023
14	7	8	.08642678	15	3	9	.04689602	16	1	6	.05541164	16	3	13	.03021841
15	1	1	.40581529	15	3	10	.04178434	16	1	7	.04744653	16	3	14	.02727970
15	1	2	.17490746	15	3	11	.03743635	16	1	8	.04139652	16	4	4	.11402254
15	1	3	.11317430	15	3	12	.03362732	16	1	9	.03660345	16	4	5	.09162849
15	1	4	.08427939	15	3	13	.03018001	16	1	10	.03267575	16	4	6	.07663764
15	1	5	.06732055	15	4	4	.11738711	16	1	11	.02936310	16	4	7	.06579998
15	1	6	.05604745	15	4	5	.09432193	16	1	12	.02649440	16	4	8	.05752791
15	1	7	.04793173	15	4	6	.07884229	16	1	13	.02394318	16	4	9	.05094937
15	1	8	.04175052	15	4	7	.06762138	16	1	14	.02160419	16	4	10	.04554186

TABLE II

P = 8.0

N	R	S	COVARIANCE	N	R	S	COVARIANCE	N	R	S	COVARIANCE	N	R	S	COVARIANCE
16	4	11	.04096943	17	1	13	.02411021	17	4	6	.07469440	18	1	2	.16933657
16	4	12	.03700116	17	1	14	.02191430	17	4	7	.06418436	18	1	3	.10928900
16	4	13	.03346540	17	1	15	.01987623	17	4	8	.05618226	18	1	4	.08135743
16	5	5	.10079830	17	1	16	.01790796	17	4	9	.04983719	18	1	5	.06505000
16	5	6	.08441633	17	1	17	.01588563	17	4	10	.04464099	18	1	6	.05426636
16	5	7	.07254748	17	2	2	.18951821	17	4	11	.04026893	18	1	7	.04654735
16	5	8	.06347335	17	2	3	.12319278	17	4	12	.03650075	18	1	8	.04070787
16	5	9	.05624744	17	2	4	.09201766	17	4	13	.03317767	18	1	9	.03610433
16	5	10	.05030139	17	2	5	.07369944	17	4	14	.03017590	18	1	10	.03235558
16	5	11	.04526904	17	2	6	.06152951	17	5	5	.09766739	18	1	11	.02922024
16	5	12	.04089823	17	2	7	.05278477	17	5	6	.08182334	18	1	12	.02653653
16	6	6	.09307750	17	2	8	.04614617	17	5	7	.07037170	18	1	13	.02419020
16	6	7	.08007476	17	2	9	.04089434	17	5	8	.06163941	18	1	14	.02209571
16	6	8	.07011584	17	2	10	.03660144	17	5	9	.05470699	18	1	15	.02018368
16	6	9	.06217389	17	2	11	.03299499	17	5	10	.04902419	18	1	16	.01839009
16	6	10	.05563086	17	2	12	.02989072	17	5	11	.04423876	18	1	17	.01664151
16	6	11	.05008768	17	2	13	.02715623	17	5	12	.04011140	18	1	18	.01483158
16	7	7	.08868904	17	2	14	.02468857	17	5	13	.03646935	18	2	2	.18688321
16	7	8	.07773032	17	2	15	.02239720	17	6	6	.08965467	18	2	3	.12135114
16	7	9	.06897689	17	2	16	.02018337	17	6	7	.07718048	18	2	4	.09061547
16	7	10	.06175561	17	3	3	.13543367	17	6	8	.06765288	18	2	5	.07258665
16	8	8	.08668211	17	3	4	.10144019	17	6	9	.06007917	18	2	6	.06062840
16	8	9	.07698724	17	3	5	.08138351	17	6	10	.05386401	18	2	7	.05205027
17	1	1	.39907011	17	3	6	.06802222	17	6	11	.04862556	18	2	8	.04555050
17	1	2	.17101106	17	3	7	.05840270	17	6	12	.04410399	18	2	9	.04042008
17	1	3	.11046279	17	3	8	.05108928	17	7	7	.08484982	18	2	10	.03623816
17	1	4	.08224574	17	3	9	.04529694	17	7	8	.07443662	18	2	11	.03273769
17	1	5	.06574615	17	3	10	.04055783	17	7	9	.06614694	18	2	12	.02973940
17	1	6	.05481908	17	3	11	.03657345	17	7	10	.05933611	18	2	13	.02711652
17	1	7	.04698490	17	3	12	.03314166	17	7	11	.05358980	18	2	14	.02477397
17	1	8	.04104745	17	3	13	.03011697	17	8	8	.08226132	18	2	15	.02263454
17	1	9	.03635637	17	3	14	.02738609	17	8	9	.07315552	18	2	16	.02062683
17	1	10	.03252580	17	3	15	.02484920	17	8	10	.06566400	18	2	17	.01866880
17	1	11	.02931049	17	4	4	.11107804	17	9	9	.08144179	18	3	3	.13287751
17	1	12	.02654489	17	4	5	.08926461	18	1	1	.39619863	18	3	4	.09948371

TABLE II

P = 8.0

N	R	S	COVARIANCE	N	R	S	COVARIANCE	N	R	S	COVARIANCE	N	R	S	COVARIANCE
18	3	5	.07981864	18	6	7	.07467309	19	1	17	.01709956	19	4	4	.10615493
18	3	6	.06674121	18	6	8	.06550979	19	1	18	.01553443	19	4	5	.08529816
18	3	7	.05734289	18	6	9	.05824454	19	1	19	.01390400	19	4	6	.07141686
18	3	8	.05021174	18	6	10	.05230090	19	2	2	.18449993	19	4	7	.06143929
18	3	9	.04457684	18	6	11	.04731071	19	2	3	.11968139	19	4	8	.05387154
18	3	10	.03997969	18	6	12	.04302546	19	2	4	.08934009	19	4	9	.04789704
18	3	11	.03612886	18	6	13	.03926845	19	2	5	.07157019	19	4	10	.04303005
18	3	12	.03282848	18	7	7	.08158274	19	2	6	.05980052	19	4	11	.03896211
18	3	13	.02993982	18	7	8	.07162472	19	2	7	.05137001	19	4	12	.03548663
18	3	14	.02735872	18	7	9	.06371902	19	2	8	.04499234	19	4	13	.03245847
18	3	15	.02500046	18	7	10	.05724447	19	2	9	.03996765	19	4	14	.02977065
18	3	16	.02278660	18	7	11	.05180361	19	2	10	.03588114	19	4	15	.02733969
18	4	4	.10847535	18	7	12	.04712772	19	2	11	.03247019	19	4	16	.02509477
18	4	5	.08716987	18	8	8	.07856739	19	2	12	.02955934	19	5	5	.09250966
18	4	6	.07296609	18	8	9	.06994227	19	2	13	.02702563	19	5	6	.07753550
18	4	7	.06273987	18	8	10	.06287002	19	2	14	.02477861	19	5	7	.06675373
18	4	8	.05496993	18	8	11	.05692083	19	2	15	.02274786	19	5	8	.05856522
18	4	9	.04882366	18	9	9	.07716154	19	2	16	.02087379	19	5	9	.05209392
18	4	10	.04380495	18	9	10	.06940361	19	2	17	.01909872	19	5	10	.04681775
18	4	11	.03959796	19	1	1	.39359652	19	2	18	.01735320	19	5	11	.04240471
18	4	12	.03599013	19	1	2	.16781016	19	3	3	.13058063	19	5	12	.03863214
18	4	13	.03283070	19	1	3	.10821454	19	3	4	.09772186	19	5	13	.03534344
18	4	14	.03000637	19	1	4	.08054026	19	3	5	.07840510	19	5	14	.03242305
18	4	15	.02742483	19	1	5	.06440548	19	3	6	.06557918	19	5	15	.02978070
18	5	5	.09492898	19	1	6	.05375007	19	3	7	.05637587	19	6	6	.08411350
18	5	6	.07954930	19	1	7	.04613339	19	3	8	.04940437	19	6	7	.07247565
18	5	7	.06845599	19	1	8	.04038021	19	3	9	.04390616	19	6	8	.06362467
18	5	8	.06001532	19	1	9	.03585289	19	3	10	.03943082	19	6	9	.05662208
18	5	9	.05333101	19	1	10	.03217437	19	3	11	.03569277	19	6	10	.05090759
18	5	10	.04786800	19	1	11	.02910634	19	3	12	.03250096	19	6	11	.04612434
18	5	11	.04328509	19	1	12	.02648985	19	3	13	.02972131	19	6	12	.04203267
18	5	12	.03935235	19	1	13	.02421361	19	3	14	.02725513	19	6	13	.03846382
18	5	13	.03590647	19	1	14	.02219590	19	3	15	.02502547	19	6	14	.03529313
18	5	14	.03282457	19	1	15	.02037315	19	3	16	.02296715	19	7	7	.07876240
18	6	6	.08669804	19	1	16	.01869167	19	3	17	.02101696	19	7	8	.06919056

TABLE II

P = 8.0

N	R	S	COVARIANCE	N	R	S	COVARIANCE	N	R	S	COVARIANCE	N	R	S	COVARIANCE
19	7	9	.06160860	20	2	3	.11815909	20	4	6	.07001863	20	7	8	.06705870
19	7	10	.05541524	20	2	4	.08817402	20	4	7	.06026099	20	7	9	.05975382
19	7	11	.05022689	20	2	5	.07063736	20	4	8	.05287120	20	7	10	.05379952
19	7	12	.04578557	20	2	6	.05903700	20	4	9	.04704692	20	7	11	.04882395
19	7	13	.04190940	20	2	7	.05073840	20	4	10	.04231148	20	7	12	.04457788
19	8	8	.07542615	20	2	8	.04446917	20	4	11	.03836275	20	7	13	.04088667
19	8	9	.06720140	20	2	9	.03953765	20	4	12	.03499899	20	7	14	.03762268
19	8	10	.06047568	20	2	10	.03553432	20	4	13	.03207929	20	8	8	.07271608
19	8	11	.05483621	20	2	11	.03220037	20	4	14	.02950099	20	8	9	.06483045
19	8	12	.05000495	20	2	12	.02936338	20	4	15	.02718581	20	8	10	.05839642
19	9	9	.07358788	20	2	13	.02690320	20	4	16	.02507052	20	8	11	.05301552
19	9	10	.06626032	20	2	14	.02473242	20	4	17	.02309942	20	8	12	.04842031
19	9	11	.06010988	20	2	15	.02278456	20	5	5	.09035369	20	8	13	.04442312
19	10	10	.07299980	20	2	16	.02100599	20	5	6	.07573712	20	9	9	.07055096
20	1	1	.39122672	20	2	17	.01934959	20	5	7	.06522908	20	9	10	.06358127
20	1	2	.16641201	20	2	18	.01776778	20	5	8	.05726112	20	9	11	.05774701
20	1	3	.10722659	20	2	19	.01620083	20	5	9	.05097502	20	9	12	.05276069
20	1	4	.07978559	20	3	3	.12850326	20	5	10	.04586003	20	10	10	.06952770
20	1	5	.06380691	20	3	4	.09612524	20	5	11	.04159201	20	10	11	.06317840
20	1	6	.05326701	20	3	5	.07712063	20	5	12	.03795424	*****	*****	*****	**********
20	1	7	.04574200	20	3	6	.06451938	20	5	13	.03479520				
20	1	8	.04006558	20	3	7	.05548955	20	5	14	.03200438				
20	1	9	.03560546	20	3	8	.04865929	20	5	15	.02949746				
20	1	10	.03198810	20	3	9	.04328116	20	5	16	.02720622				
20	1	11	.02897782	20	3	10	.03891183	20	6	6	.08183123				
20	1	12	.02641786	20	3	11	.03527071	20	6	7	.07053085				
20	1	13	.02419908	20	3	12	.03217066	20	6	8	.06195095				
20	1	14	.02224221	20	3	13	.02948109	20	6	9	.05517513				
20	1	15	.02048699	20	3	14	.02710696	20	6	10	.04965704				
20	1	16	.01888489	20	3	15	.02497587	20	6	11	.04504945				
20	1	17	.01739332	20	3	16	.02302938	20	6	12	.04111993				
20	1	18	.01596933	20	3	17	.02121609	20	6	13	.03770580				
20	1	19	.01455909	20	3	18	.01948399	20	6	14	.03468829				
20	1	20	.01308182	20	4	4	.10407066	20	6	15	.03197666				
20	2	2	.18233199	20	4	5	.08361357	20	7	7	.07629833				

TABLE II

P = 8.5

N	R	S	COVARIANCE	N	R	S	COVARIANCE	N	R	S	COVARIANCE	N	R	S	COVARIANCE
2	1	1	.68769306	7	1	3	.13429372	9	1	2	.19353789	10	2	3	.14511097
2	1	2	.31230694	7	1	4	.09893648	9	1	3	.12610502	10	2	4	.10846256
3	1	1	.58549144	7	1	5	.07710105	9	1	4	.09364149	10	2	5	.08636089
3	1	2	.26896992	7	1	6	.06149463	9	1	5	.07407267	10	2	6	.07126645
3	1	3	.16475078	7	1	7	.04868743	9	1	6	.06064840	10	2	7	.06004607
3	2	2	.42363589	7	2	2	.25343454	9	1	7	.05055516	10	2	8	.05111741
4	1	1	.53242443	7	2	3	.16691311	9	1	8	.04231891	10	2	9	.04351129
4	1	2	.24381875	7	2	4	.12376986	9	1	9	.03486840	10	3	3	.16826851
4	1	3	.15516223	7	2	5	.09685402	9	2	2	.22925630	10	3	4	.12634413
4	1	4	.10701887	7	2	6	.07747968	9	2	3	.15083855	10	3	5	.10088746
4	2	2	.34076979	7	3	3	.20545577	9	2	4	.11258111	10	3	6	.08341996
4	2	3	.22182496	7	3	4	.15344831	9	2	5	.08933674	10	3	7	.07039038
5	1	1	.49917330	7	3	5	.12064909	9	2	6	.07330569	10	3	8	.05999380
5	1	2	.22724375	7	4	4	.19395431	9	2	7	.06120520	10	4	4	.14697189
5	1	3	.14660103	8	1	1	.44559994	9	2	8	.05130074	10	4	5	.11773729
5	1	4	.10530669	8	1	2	.19929934	9	3	3	.17748142	10	4	6	.09757411
5	1	5	.07756452	8	1	3	.12982451	9	3	4	.13314874	10	4	7	.08247528
5	2	2	.29827932	8	1	4	.09611506	9	3	5	.10600427	10	5	5	.13861269
5	2	3	.19574783	8	1	5	.07560096	9	3	6	.08718257	10	5	6	.11519266
5	2	4	.14198661	8	1	6	.06132104	9	3	7	.07291828	11	1	1	.41869233
5	3	3	.26639531	8	1	7	.05028956	9	4	4	.15782397	11	1	2	.18467995
6	1	1	.47606656	8	1	8	.04071712	9	4	5	.12612926	11	1	3	.12025085
6	1	2	.21537646	8	2	2	.23982407	9	4	6	.10401819	11	1	4	.08957041
6	1	3	.13976816	8	2	3	.15790856	9	5	5	.15240800	11	1	5	.07129335
6	1	4	.10209088	8	2	4	.11757414	10	1	1	.42609608	11	1	6	.05895053
6	1	5	.07820800	8	2	5	.09281032	10	1	2	.18874400	11	1	7	.04989258
6	1	6	.06009714	8	2	6	.07546755	10	1	3	.12295612	11	1	8	.04281302
6	2	2	.27179427	8	2	7	.06200938	10	1	4	.09147565	11	1	9	.03696448
6	2	3	.17887555	8	3	3	.18937139	10	1	5	.07262673	11	1	10	.03183303
6	2	4	.13166760	8	3	4	.14184361	10	1	6	.05981562	11	1	11	.02691446
6	2	5	.10137791	8	3	5	.11240024	10	1	7	.05032577	11	2	2	.21377187
6	3	3	.22875603	8	3	6	.09164903	10	1	8	.04279448	11	2	3	.14035750
6	3	4	.16993480	8	4	4	.17254462	10	1	9	.03639270	11	2	4	.10499879
7	1	1	.45891832	8	4	5	.13737467	10	1	10	.03041203	11	2	5	.08379446
7	1	2	.20638782	9	1	1	.43490538	10	2	2	.22076824	11	2	6	.06941146

TABLE II

P = 8.5

N	R	S	COVARIANCE	N	R	S	COVARIANCE	N	R	S	COVARIANCE	N	R	S	COVARIANCE
11	2	7	.05882269	12	2	5	.08156125	13	1	9	.03708131	13	5	8	.07121836
11	2	8	.05052656	12	2	6	.06774179	13	1	10	.03268399	13	5	9	.06257210
11	2	9	.04365972	12	2	7	.05763534	13	1	11	.02884253	13	6	6	.10845564
11	2	10	.03762510	12	2	8	.04979383	13	1	12	.02531296	13	6	7	.09299966
11	3	3	.16088035	12	2	9	.04340535	13	1	13	.02179790	13	6	8	.08098185
11	3	4	.12084507	12	2	10	.03795330	13	2	2	.20285216	13	7	7	.10664813
11	3	5	.09668820	12	2	11	.03304088	13	2	3	.13288022	14	1	1	.40209777
11	3	6	.08023371	12	3	3	.15479821	13	2	4	.09947343	14	1	2	.17544519
11	3	7	.06808263	12	3	4	.11629010	13	2	5	.07960028	14	1	3	.11399083
11	3	8	.05853980	12	3	5	.09316984	13	2	6	.06624103	14	1	4	.08503063
11	3	9	.05062581	12	3	6	.07750715	13	2	7	.05651862	14	1	5	.06794897
11	4	4	.13858006	12	3	7	.06602096	13	2	8	.04902508	14	1	6	.05654376
11	4	5	.11118838	12	3	8	.05708994	13	2	9	.04298023	14	1	7	.04829559
11	4	6	.09244698	12	3	9	.04980144	13	2	10	.03790425	14	1	8	.04198115
11	4	7	.07856089	12	3	10	.04357251	13	2	11	.03346499	14	1	9	.03692926
11	4	8	.06762690	12	4	4	.13186074	13	2	12	.02938224	14	1	10	.03273484
11	5	5	.12846874	12	4	5	.10590723	13	3	3	.14968676	14	1	11	.02913027
11	5	6	.10706013	12	4	6	.08825535	13	3	4	.11244266	14	1	12	.02591677
11	5	7	.09113705	12	4	7	.07527236	13	3	5	.09017154	14	1	13	.02291251
11	6	6	.12549401	12	4	8	.06515446	13	3	6	.07514798	14	1	14	.01987667
12	1	1	.41236782	12	4	9	.05688211	13	3	7	.06418665	14	2	2	.19848766
12	1	2	.18118172	12	5	5	.12063937	13	3	8	.05572188	14	2	3	.12987183
12	1	3	.11789736	12	5	6	.10072995	13	3	9	.04888307	14	2	4	.09722570
12	1	4	.08788355	12	5	7	.08603844	13	3	10	.04313314	14	2	5	.07786363
12	1	5	.07007476	12	5	8	.07455946	13	3	11	.03809907	14	2	6	.06488903
12	1	6	.05810513	12	6	6	.11587030	13	4	4	.12633602	14	2	7	.05548176
12	1	7	.04937713	12	6	7	.09914570	13	4	5	.10154007	14	2	8	.04826605
12	1	8	.04262018	13	1	1	.40689219	13	4	6	.08475342	14	2	9	.04248436
12	1	9	.03712496	13	1	2	.17813216	13	4	7	.07247349	14	2	10	.03767813
12	1	10	.03244202	13	1	3	.11582777	13	4	8	.06297118	14	2	11	.03354358
12	1	11	.02822774	13	1	4	.08637986	13	4	9	.05528160	14	2	12	.02985436
12	1	12	.02410293	13	1	5	.06896344	13	4	10	.04880763	14	2	13	.02640272
12	2	2	.20788610	13	1	6	.05730044	13	5	5	.11437957	14	3	3	.14531878
12	2	3	.13633601	13	1	7	.04883567	13	5	6	.09563509	14	3	4	.10914085
12	2	4	.10203829	13	1	8	.04232488	13	5	7	.08188344	14	3	5	.08758013

TABLE II

P = 8.5

N	R	S	COVARIANCE	N	R	S	COVARIANCE	N	R	S	COVARIANCE	N	R	S	COVARIANCE
14	3	6	.07308536	15	1	9	.03671842	15	4	8	.05933416	16	1	15	.01920258
14	3	7	.06255130	15	1	10	.03267824	15	4	9	.05243892	16	1	16	.01686342
14	3	8	.05445694	15	1	11	.02924156	15	4	10	.04673388	16	2	2	.19127030
14	3	9	.04796210	15	1	12	.02622857	15	4	11	.04186767	16	2	3	.12487207
14	3	10	.04255684	15	1	13	.02349657	15	4	12	.03759137	16	2	4	.09346088
14	3	11	.03790249	15	1	14	.02090462	15	5	5	.10492677	16	2	5	.07492059
14	3	12	.03374601	15	1	15	.01825273	15	5	6	.08788428	16	2	6	.06255627
14	4	4	.12169775	15	2	2	.19466001	15	5	7	.07548142	16	2	7	.05363970
14	4	5	.09785626	15	2	3	.12722414	15	5	8	.06595417	16	2	8	.04684514
14	4	6	.08177576	15	2	4	.09523643	15	5	9	.05832636	16	2	9	.04144718
14	4	7	.07006134	15	2	5	.07631366	15	5	10	.05200772	16	2	10	.03701246
14	4	8	.06104346	15	2	6	.06366648	15	5	11	.04661273	16	2	11	.03326283
14	4	9	.05379700	15	2	7	.05452378	15	6	6	.09767913	16	2	12	.03000718
14	4	10	.04775894	15	2	8	.04753671	15	6	7	.08399056	16	2	13	.02710310
14	4	11	.04255440	15	2	9	.04196538	15	6	8	.07345480	16	2	14	.02443022
14	5	5	.10923868	15	2	10	.03736561	15	6	9	.06500603	16	2	15	.02186076
14	5	6	.09142838	15	2	11	.03344921	15	6	10	.05799804	16	3	3	.13821746
14	5	7	.07842045	15	2	12	.03001287	15	7	7	.09389828	16	3	4	.10374550
14	5	8	.06838683	15	2	13	.02689481	15	7	8	.08220470	16	3	5	.08331119
14	5	9	.06031119	15	2	14	.02393474	15	7	9	.07281030	16	3	6	.06964488
14	5	10	.05357329	15	3	3	.14153438	15	8	8	.09271849	16	3	7	.05976923
14	6	6	.10253561	15	3	4	.10626980	16	1	1	.39408134	16	3	8	.05223220
14	6	7	.08806185	15	3	5	.08531365	16	1	2	.17091473	16	3	9	.04623713
14	6	8	.07687252	15	3	6	.07126498	16	1	3	.11086604	16	3	10	.04130700
14	6	9	.06785031	15	3	7	.06108701	16	1	4	.08270668	16	3	11	.03713510
14	7	7	.09955457	15	3	8	.05329592	16	1	5	.06616880	16	3	12	.03351029
14	7	8	.08700941	15	3	9	.04707540	16	1	6	.05517584	16	3	13	.03027498
15	1	1	.39785934	15	3	10	.04193423	16	1	7	.04726652	16	3	14	.02729563
15	1	2	.17305599	15	3	11	.03755299	16	1	8	.04124990	16	4	4	.11430890
15	1	3	.11234718	15	3	12	.03370591	16	1	9	.03647640	16	4	5	.09195525
15	1	4	.08381255	15	3	13	.03021284	16	1	10	.03255895	16	4	6	.07696334
15	1	5	.06702072	15	4	4	.11773751	16	1	11	.02924963	16	4	7	.06610771
15	1	6	.05583615	15	4	5	.09469845	16	1	12	.02637846	16	4	8	.05780993
15	1	7	.04777041	15	4	6	.07920682	16	1	13	.02381900	16	4	9	.05120162
15	1	6	.04161758	15	4	7	.06795880	16	1	14	.02146466	16	4	10	.04576177

TABLE II

P = 8.5

N	R	S	COVARIANCE	N	R	S	COVARIANCE	N	R	S	COVARIANCE	N	R	S	COVARIANCE
16	4	11	.04115467	17	1	13	.02399316	17	4	6	.07498372	18	1	2	.16722803
16	4	12	.03714889	17	1	14	.02178932	17	4	7	.06446357	18	1	3	.10829832
16	4	13	.03357132	17	1	15	.01973690	17	4	8	.05644263	18	1	4	.08077227
16	5	5	.10124814	17	1	16	.01774314	17	4	9	.05007419	18	1	5	.06465987
16	5	6	.08484933	17	1	17	.01566234	17	4	10	.04485192	18	1	6	.05398470
16	5	7	.07294965	17	2	2	.18824311	17	4	11	.04045174	18	1	7	.04633130
16	5	8	.06383891	17	2	3	.12276570	17	4	12	.03665337	18	1	8	.04053339
16	5	9	.05657369	17	2	4	.09186430	17	4	13	.03329742	18	1	9	.03595671
16	5	10	.05058667	17	2	5	.07366063	17	4	14	.03025873	18	1	10	.03222512
16	5	11	.04551162	17	2	6	.06154370	17	5	5	.09806552	18	1	11	.02910007
16	5	12	.04109555	17	2	7	.05282323	17	5	6	.08221513	18	1	12	.02642146
16	6	6	.09360947	17	2	8	.04619365	17	5	7	.07074140	18	1	13	.02407591
16	6	7	.08056446	17	2	9	.04094191	17	5	8	.06198022	18	1	14	.02197817
16	6	8	.07055900	17	2	10	.03664329	17	5	9	.05501571	18	1	15	.02005852
16	6	9	.06256888	17	2	11	.03302691	17	5	10	.04929906	18	1	16	.01825148
16	6	10	.05597670	17	2	12	.02990919	17	5	11	.04447837	18	1	17	.01647910
16	6	11	.05038309	17	2	13	.02715771	17	5	12	.04031410	18	1	18	.01461441
16	7	7	.08926304	17	2	14	.02466879	17	5	13	.03663265	18	2	2	.18551984
16	7	8	.07824836	17	2	15	.02234978	17	6	6	.09014014	18	2	3	.12086604
16	7	9	.06943815	17	2	16	.02009608	17	6	7	.07763334	18	2	4	.09041926
16	7	10	.06215966	17	3	3	.13528154	17	6	8	.06806782	18	2	5	.07251460
16	8	8	.08727416	17	3	4	.10150501	17	6	9	.06045405	18	2	6	.06061629
16	8	9	.07751392	17	3	5	.08152649	17	6	10	.05419781	18	2	7	.05206789
17	1	1	.39068945	17	3	6	.06819225	17	6	11	.04891737	18	2	8	.04558176
17	1	2	.16898240	17	3	7	.05857716	17	6	12	.04435246	18	2	9	.04045555
17	1	3	.10952296	17	3	8	.05125683	17	7	7	.08538212	18	2	10	.03627181
17	1	4	.08169751	17	3	9	.04545129	17	7	8	.07492257	18	2	11	.03276529
17	1	5	.06538442	17	3	10	.04069501	17	7	9	.06658524	18	2	12	.02975767
17	1	6	.05455982	17	3	11	.03669054	17	7	10	.05972631	18	2	13	.02712250
17	1	7	.04678660	17	3	12	.03323603	17	7	11	.05393139	18	2	14	.02476455
17	1	8	.04088688	17	3	13	.03018562	17	8	8	.08281705	18	2	15	.02260582
17	1	9	.03621928	17	3	14	.02742496	17	8	9	.07365615	18	2	16	.02057291
17	1	10	.03240272	17	3	15	.02485163	17	8	10	.06610953	18	2	17	.01857826
17	1	11	.02919459	17	4	4	.11130575	17	9	9	.08200467	18	3	3	.13266073
17	1	12	.02643079	17	4	5	.08954539	18	1	1	.38762492	18	3	4	.09950014

TABLE II

P = 0.5

N	R	S	COVARIANCE	N	R	S	COVARIANCE	N	R	S	COVARIANCE	N	R	S	COVARIANCE
18	3	5	.07992376	18	6	7	.07509130	19	1	17	.01696205	19	4	4	.10627871
18	3	6	.06688112	18	6	8	.06589739	19	1	18	.01537460	19	4	5	.08549674
18	3	7	.05749341	18	6	9	.05859882	19	1	19	.01369266	19	4	6	.07164011
18	3	8	.05036064	18	6	10	.05262060	19	2	2	.18305419	19	4	7	.06166541
18	3	9	.04471729	18	6	11	.04759502	19	2	3	.11914216	19	4	8	.0540899[illegible]
18	3	10	.04010750	18	6	12	.04327349	19	2	4	.08910381	19	4	9	.04810210
18	3	11	.03624112	18	6	13	.03947877	19	2	5	.07146688	19	4	10	.04321851
18	3	12	.03292284	18	7	7	.08207610	19	2	6	.05976346	19	4	11	.03913189
18	3	13	.03001396	18	7	8	.07207980	19	2	7	.05136759	19	4	12	.03563611
18	3	14	.02740993	18	7	9	.06413395	19	2	8	.04500766	19	4	13	.03258612
18	3	15	.02502494	18	7	10	.05761859	19	2	9	.03999077	19	4	14	.02987471
18	3	16	.02277814	18	7	11	.05213654	19	2	10	.03590576	19	4	15	.02741777
18	4	4	.10864906	18	7	12	.04741884	19	2	11	.03249199	19	4	16	.02514325
18	4	5	.08740805	18	8	8	.07908836	19	2	12	.02957511	19	5	5	.09281618
18	4	6	.07322131	18	8	9	.07041649	19	2	13	.02703267	19	5	6	.07785320
18	4	7	.06299186	18	8	10	.06329732	19	2	14	.02477439	19	5	7	.06706362
18	4	8	.05520901	18	8	11	.05730116	19	2	15	.02272959	19	5	8	.05885854
18	4	9	.04904477	18	9	9	.07769476	19	2	16	.02083788	19	5	9	.05236635
18	4	10	.04400513	18	9	10	.06988377	19	2	17	.01903968	19	5	10	.04706698
18	4	11	.03977520	19	1	1	.38484063	19	2	18	.01726033	19	5	11	.04262933
18	4	12	.03614270	19	1	2	.16562663	19	3	3	.13030390	19	5	12	.03883103
18	4	13	.03295671	19	1	3	.10717613	19	3	4	.09769328	19	5	13	.03551540
18	4	14	.03010329	19	1	4	.07992034	19	3	5	.07847481	19	5	14	.03256653
18	4	15	.02748883	19	1	5	.06398848	19	3	6	.06569066	19	5	15	.02989335
18	5	5	.09527949	19	1	6	.05344706	19	3	7	.05650349	19	6	6	.08451620
18	5	6	.07990275	19	1	7	.04590018	19	3	8	.04953502	19	6	7	.07286135
18	5	7	.06879493	19	1	8	.04019201	19	3	9	.04403248	19	6	8	.06398605
18	5	8	.06033196	19	1	9	.03569454	19	3	10	.03954834	19	6	9	.05695586
18	5	9	.05362161	19	1	10	.03203589	19	3	11	.03579849	19	6	10	.05121219
18	5	10	.04813055	19	1	11	.02898078	19	3	12	.03259261	19	6	11	.04639890
18	5	11	.04351825	19	1	12	.02637202	19	3	13	.02979691	19	6	12	.04227645
18	5	12	.03955485	19	1	13	.02409941	19	3	14	.02731262	19	6	13	.03867590
18	5	13	.03607671	19	1	14	.02208176	19	3	15	.02506233	19	6	14	.03547208
18	5	14	.03296012	19	1	15	.02025561	19	3	16	.02297982	19	7	7	.07921946
18	6	6	.08714054	19	1	16	.01856682	19	3	17	.02099964	19	7	8	.06961624

TABLE II

P = 8.5

N	R	S	COVARIANCE
19	7	9	.06200044
19	7	10	.05577227
19	7	11	.05054870
19	7	12	.04607175
19	7	13	.04215920
19	8	8	.07591421
19	8	9	.06764967
19	8	10	.06088370
19	8	11	.05520392
19	8	12	.05033219
19	9	9	.07409203
19	9	10	.06671884
19	9	11	.06052299
19	10	10	.07350895
20	1	1	.38229834
20	1	2	.16415782
20	1	3	.10614321
20	1	4	.07913286
20	1	5	.06336445
20	1	6	.05294362
20	1	7	.04549226
20	1	8	.03986397
20	1	9	.03543635
20	1	10	.03184126
20	1	11	.02884614
20	1	12	.02629613
20	1	13	.02408326
20	1	14	.02212896
20	1	15	.02037329
20	1	16	.01876770
20	1	17	.01726913
20	1	18	.01583318
20	1	19	.01440192
20	1	20	.01287603
20	2	2	.18080908

N	R	S	COVARIANCE
20	2	3	.11756920
20	2	4	.08790014
20	2	5	.07050460
20	2	6	.05897625
20	2	7	.05071676
20	2	8	.04446894
20	2	9	.03954841
20	2	10	.03554947
20	2	11	.03221545
20	2	12	.02937515
20	2	13	.02690908
20	2	14	.02473017
20	2	15	.02277194
20	2	16	.02098047
20	2	17	.01930787
20	2	18	.01770469
20	2	19	.01610633
20	3	3	.12817067
20	3	4	.09605465
20	3	5	.07715716
20	3	6	.06460403
20	3	7	.05559532
20	3	8	.04877223
20	3	9	.04339340
20	3	10	.03901859
20	3	11	.03536884
20	3	12	.03225787
20	3	13	.02955556
20	3	14	.02716695
20	3	15	.02501951
20	3	16	.02305433
20	3	17	.02121901
20	3	18	.01945942
20	4	4	.10414807
20	4	5	.08377522

N	R	S	COVARIANCE
20	4	6	.07021186
20	4	7	.06046256
20	4	8	.05306964
20	4	9	.04723607
20	4	10	.04248777
20	4	11	.03852392
20	4	12	.03514344
20	4	13	.03220565
20	4	14	.02960789
20	4	15	.02727159
20	4	16	.02513291
20	4	17	.02313499
20	5	5	.09061939
20	5	6	.07602145
20	5	7	.06551155
20	5	8	.05753207
20	5	9	.05122956
20	5	10	.04609553
20	5	11	.04180688
20	5	12	.03814738
20	5	13	.03496562
20	5	14	.03215095
20	5	15	.02961865
20	5	16	.02729979
20	6	6	.08219698
20	6	7	.07088605
20	6	8	.06228735
20	6	9	.05548884
20	6	10	.04994617
20	6	11	.04531298
20	6	12	.04135715
20	6	13	.03791601
20	6	14	.03487055
20	6	15	.03212953
20	7	7	.07672152

N	R	S	COVARIANCE
20	7	8	.06745653
20	7	9	.06012321
20	7	10	.05413918
20	7	11	.04913330
20	7	12	.04485656
20	7	13	.04113421
20	7	14	.03783827
20	8	8	.07317313
20	8	9	.06525364
20	8	10	.05878495
20	8	11	.05336919
20	8	12	.04873903
20	8	13	.04470660
20	9	9	.07102715
20	9	10	.06401799
20	9	11	.05814437
20	9	12	.05311880
20	10	10	.07001258
20	10	11	.06361937

****

TABLE II

P = 9.0

N	R	S	COVARIANCE	N	R	S	COVARIANCE	N	R	S	COVARIANCE	N	R	S	COVARIANCE
2	1	1	.68727509	7	1	3	.13418316	9	1	2	.19258010	10	2	3	.14523924
2	1	2	.31272491	7	1	4	.09892874	9	1	3	.12579269	10	2	4	.10868711
3	1	1	.58375346	7	1	5	.07708410	9	1	4	.09351566	10	2	5	.08658802
3	1	2	.26942296	7	1	6	.06141001	9	1	5	.07400619	10	2	6	.07146268
3	1	3	.16478511	7	1	7	.04844749	9	1	6	.06058968	10	2	7	.06019459
3	2	2	.42523101	7	2	2	.25368503	9	1	7	.05047469	10	2	8	.05120428
4	1	1	.52970346	7	2	3	.16740899	9	1	8	.04218865	10	2	9	.04351530
4	1	2	.24396253	7	2	4	.12421847	9	1	9	.03462775	10	3	3	.16871669
4	1	3	.15536625	7	2	5	.09718569	9	2	2	.22908067	10	3	4	.12682298
4	1	4	.10689081	7	2	6	.07765191	9	2	3	.15107463	10	3	5	.10132261
4	2	2	.34202453	7	3	3	.20635625	9	2	4	.11287710	10	3	6	.08378794
4	2	3	.22272362	7	3	4	.15420613	9	2	5	.08960773	10	3	7	.07067982
5	1	1	.49570405	7	3	5	.12121306	9	2	6	.07352068	10	3	8	.06019309
5	1	2	.22709276	7	4	4	.19497953	9	2	7	.06134508	10	4	4	.14766113
5	1	3	.14672319	8	1	1	.44062734	9	2	8	.05134056	10	4	5	.11834621
5	1	4	.10535678	8	1	2	.19850042	9	3	3	.17805981	10	4	6	.09808550
5	1	5	.07736662	8	1	3	.12960786	9	3	4	.13371525	10	4	7	.08288105
5	2	2	.29913867	8	1	4	.09604842	9	3	5	.10649362	10	5	5	.13938259
5	2	3	.19655376	8	1	5	.07556546	9	3	6	.08757367	10	5	6	.11583645
5	2	4	.14249977	8	1	6	.06126563	9	3	7	.07319647	11	1	1	.41275656
5	3	3	.26767586	8	1	7	.05017626	9	4	4	.15861894	11	1	2	.18345794
6	1	1	.47200183	8	1	8	.04047450	9	4	5	.12680401	11	1	3	.11977464
6	1	2	.21497380	8	2	2	.23984429	9	4	6	.10455743	11	1	4	.08933540
6	1	3	.13977315	8	2	3	.15826655	9	5	5	.15325535	11	1	5	.07115665
6	1	4	.10213240	8	2	4	.11794617	10	1	1	.42044443	11	1	6	.05885502
6	1	5	.07817234	8	2	5	.09312024	10	1	2	.18764642	11	1	7	.04980971
6	1	6	.05986901	8	2	6	.07568412	10	1	3	.12255767	11	1	8	.04272514
6	2	2	.27232014	8	2	7	.06210175	10	1	4	.09129339	11	1	9	.03685700
6	2	3	.17952469	8	3	3	.19009953	10	1	5	.07252538	11	1	10	.03168735
6	2	4	.13217965	8	3	4	.14250490	10	1	6	.05974107	11	1	11	.02668385
6	2	5	.10167712	8	3	5	.11293821	10	1	7	.05024996	11	2	2	.21327857
6	3	3	.22984995	8	3	6	.09204127	10	1	8	.04269765	11	2	3	.14038991
6	3	4	.17076991	8	4	4	.17345486	10	1	9	.03625253	11	2	4	.10515745
7	1	1	.45436235	8	4	5	.13810796	10	1	10	.03017580	11	2	5	.08397746
7	1	2	.20577177	9	1	1	.42957181	10	2	2	.22042333	11	2	6	.06958209

TABLE II

P = 9.0

N	R	S	COVARIANCE	N	R	S	COVARIANCE	N	R	S	COVARIANCE	N	R	S	COVARIANCE
11	2	7	.05896388	12	2	5	.08170169	13	1	9	.03698655	13	5	8	.07159986
11	2	8	.05062731	12	2	6	.06788442	13	1	10	.03258204	13	5	9	.06288812
11	2	9	.04370884	12	2	7	.05776194	13	1	11	.02872394	13	6	6	.10906528
11	2	10	.03760406	12	2	8	.04989399	13	1	12	.02516369	13	6	7	.09353524
11	3	3	.16121444	12	2	9	.04347094	13	1	13	.02157981	13	6	8	.08144140
11	3	4	.12124429	12	2	10	.03797487	13	2	2	.20210919	13	7	7	.10727312
11	3	5	.09706953	12	2	11	.03300193	13	2	3	.13274940	14	1	1	.39545652
11	3	6	.08057039	12	3	3	.15503145	13	2	4	.09951631	14	1	2	.17391720
11	3	7	.06836325	12	3	4	.11661725	13	2	5	.07970042	14	1	3	.11332058
11	3	8	.05875651	12	3	5	.09349996	13	2	6	.06635526	14	1	4	.08465917
11	3	9	.05076864	12	3	6	.07780998	13	2	7	.05662730	14	1	5	.06771385
11	4	4	.13917447	12	3	7	.06628396	13	2	8	.04911755	14	1	6	.05637913
11	4	5	.11173260	12	3	8	.05730615	13	2	9	.04304931	14	1	7	.04816908
11	4	6	.09292053	12	3	9	.04996463	13	2	10	.03794344	14	1	8	.04187452
11	4	7	.07895602	12	3	10	.04367381	13	2	11	.03346605	14	1	9	.03683094
11	4	8	.06793767	12	4	4	.13237040	13	2	12	.02933033	14	1	10	.03263644
11	5	5	.12916198	12	4	5	.10639062	13	3	3	.14983004	14	1	11	.02902455
11	5	6	.10765919	12	4	6	.08868840	13	3	4	.11270450	14	1	12	.02579571
11	5	7	.09163659	12	4	7	.07564628	13	3	5	.09045371	14	1	13	.02276356
11	6	6	.12621393	12	4	8	.06546451	13	3	6	.07541686	14	1	14	.01966490
12	1	1	.40617542	12	4	9	.05712319	13	3	7	.06442821	14	2	2	.19763796
12	1	2	.17984785	12	5	5	.12126104	13	3	8	.05592897	14	2	3	.12967072
12	1	3	.11735058	12	5	6	.10128110	13	3	9	.04905069	14	2	4	.09721771
12	1	4	.08759958	12	5	7	.08651288	13	3	10	.04325599	14	2	5	.07792593
12	1	5	.06990366	12	5	8	.07495397	13	3	11	.03816912	14	2	6	.06497543
12	1	6	.05798674	12	6	6	.11653465	13	4	4	.12676975	14	2	7	.05557115
12	1	7	.04928177	12	6	7	.09971625	13	4	5	.10196716	14	2	8	.04834718
12	1	8	.04253052	13	1	1	.40046586	13	4	6	.08514633	14	2	9	.04255021
12	1	9	.03702870	13	1	2	.17669685	13	4	7	.07282191	14	2	10	.03772329
12	1	10	.03232769	13	1	3	.11521659	13	4	8	.06327028	14	2	11	.03356261
12	1	11	.02807935	13	1	4	.08605051	13	4	9	.05552780	14	2	12	.02983992
12	1	12	.02387849	13	1	5	.06875946	13	4	10	.04899627	14	2	13	.02634132
12	2	2	.20726115	13	1	6	.05715876	13	5	5	.11493567	14	3	3	.14538118
12	2	3	.13628259	13	1	7	.04872527	13	5	6	.09613915	14	3	4	.10934329
12	2	4	.10213654	13	1	8	.04222812	13	5	7	.08232780	14	3	5	.08781773

TABLE II

P = 9.0

N	R	S	COVARIANCE	N	R	S	COVARIANCE	N	R	S	COVARIANCE	N	R	S	COVARIANCE
14	3	6	.07332129	15	1	9	.03661376	15	4	8	.05959928	16	1	15	.01905640
14	3	7	.06276998	15	1	10	.03257867	15	4	9	.05267121	16	1	16	.01666374
14	3	8	.05465063	15	1	11	.02914070	15	4	10	.04693086	16	2	2	.19023406
14	3	9	.04812605	15	1	12	.02612049	15	4	11	.04202658	16	2	3	.12454775
14	3	10	.04268708	15	1	13	.02337429	15	4	12	.03770816	16	2	4	.09336267
14	3	11	.03799432	15	1	14	.02075678	15	5	5	.10536868	16	2	5	.07491430
14	3	12	.03379209	15	1	15	.01804711	15	5	6	.08830162	16	2	6	.06259028
14	4	4	.12206311	15	2	2	.19371303	15	5	7	.07586319	16	2	7	.05369003
14	4	5	.09823155	15	2	3	.12695882	15	5	8	.06629591	16	2	8	.04689907
14	4	6	.08213014	15	2	4	.09518157	15	5	9	.05862589	16	2	9	.04149718
14	4	7	.07038285	15	2	5	.07634053	15	5	10	.05226329	16	2	10	.03705351
14	4	8	.06132670	15	2	6	.06372608	15	5	11	.04682199	16	2	11	.03329104
14	4	9	.05403875	15	2	7	.05459354	15	6	6	.09818901	16	2	12	.03001895
14	4	10	.04795615	15	2	8	.04760467	15	6	7	.08445378	16	2	13	.02709426
14	4	11	.04270274	15	2	9	.04202433	15	6	8	.07386823	16	2	14	.02439473
14	5	5	.10973504	15	2	10	.03741050	15	6	9	.06536835	16	2	15	.02178723
14	5	6	.09188768	15	2	11	.03347571	15	6	10	.05830805	16	3	3	.13813972
14	5	7	.07883336	15	2	12	.03001628	15	7	7	.09444047	16	3	4	.10384389
14	5	8	.06874967	15	2	13	.02686850	15	7	8	.08268765	16	3	5	.08346901
14	5	9	.06062185	15	2	14	.02386636	15	7	9	.07323336	16	3	6	.06981957
14	5	10	.05382932	15	3	3	.14152348	15	8	8	.09327031	16	3	7	.05994213
14	6	6	.10309355	15	3	4	.10641797	16	1	1	.38705644	16	3	8	.05239403
14	6	7	.08856094	15	3	5	.08550988	16	1	2	.16922265	16	3	9	.04638264
14	6	8	.07731036	15	3	6	.07146946	16	1	3	.11009082	16	3	10	.04143277
14	6	9	.06822560	15	3	7	.06128260	16	1	4	.08225957	16	3	11	.03723830
14	7	7	.10013780	15	3	8	.05347413	16	1	5	.06587657	16	3	12	.03358800
14	7	8	.08752035	15	3	9	.04723132	16	1	6	.05496761	16	3	13	.03032330
15	1	1	.39101932	15	3	10	.04206435	16	1	7	.04710729	16	3	14	.02730832
15	1	2	.17144276	15	3	11	.03765396	16	1	8	.04112006	16	4	4	.11455597
15	1	3	.11162245	15	3	12	.03377348	16	1	9	.03636384	16	4	5	.09223905
15	1	4	.08340194	15	3	13	.03024026	16	1	10	.03245550	16	4	6	.07724702
15	1	5	.06675620	15	4	4	.11804096	16	1	11	.02914924	16	4	7	.06637613
15	1	6	.05564929	15	4	5	.09502612	16	1	12	.02627604	16	4	8	.05805607
15	1	7	.04762747	15	4	6	.07952475	16	1	13	.02370957	16	4	9	.05142181
15	1	8	.04149968	15	4	7	.06825342	16	1	14	.02134204	16	4	10	.04595363

TABLE II

P = 9.0

N	R	S	COVARIANCE	N	R	S	COVARIANCE	N	R	S	COVARIANCE	N	R	S	COVARIANCE
16	4	11	.04131611	17	1	13	.02388990	17	4	6	.07523530	18	1	2	.16539417
16	4	12	.03727736	17	1	14	.02167929	17	4	7	.06470682	18	1	3	.10743223
16	4	13	.03366301	17	1	15	.01961456	17	4	8	.05666969	18	1	4	.08025902
16	5	5	.10164031	17	1	16	.01759895	17	4	9	.05028094	18	1	5	.06431684
16	5	6	.08522758	17	1	17	.01546835	17	4	10	.04503589	18	1	6	.05373657
16	5	7	.07330135	17	2	2	.18712448	17	4	11	.04061107	18	1	7	.04614065
16	5	8	.06415875	17	2	3	.12238688	17	4	12	.03678619	18	1	8	.04037925
16	5	9	.05685917	17	2	4	.09172586	17	4	13	.03340137	18	1	9	.03582620
16	5	10	.05083620	17	2	5	.07362329	17	4	14	.03033021	18	1	10	.03210975
16	5	11	.04572361	17	2	6	.06155339	17	5	5	.09841208	18	1	11	.02899383
16	5	12	.04126771	17	2	7	.05285461	17	5	6	.08255701	18	1	12	.02631980
16	6	6	.09407491	17	2	8	.04623327	17	5	7	.07106445	18	1	13	.02397504
16	6	7	.08099333	17	2	9	.04098183	17	5	8	.06227823	18	1	14	.02187460
16	6	8	.07094728	17	2	10	.03667838	17	5	9	.05528574	18	1	15	.01994845
16	6	9	.06291496	17	2	11	.03305342	17	5	10	.04953944	18	1	16	.01812988
16	6	10	.05627962	17	2	12	.02992400	17	5	11	.04468781	18	1	17	.01633712
16	6	11	.05064163	17	2	13	.02715772	17	5	12	.04049108	18	1	18	.01442585
16	7	7	.08976613	17	2	14	.02465025	17	5	13	.03677494	18	2	2	.18432477
16	7	8	.07870258	17	2	15	.02230714	17	6	6	.09056457	18	2	3	.12043665
16	7	9	.06984260	17	2	16	.02001877	17	6	7	.07802971	18	2	4	.09024335
16	7	10	.06251383	17	3	3	.13514248	17	6	8	.06843122	18	2	5	.07244813
16	8	8	.08779345	17	3	4	.10155751	17	6	9	.06078244	18	2	6	.06060289
16	8	9	.07797589	17	3	5	.08164859	17	6	10	.05449016	18	2	7	.05208098
17	1	1	.38349171	17	3	6	.06833881	17	6	11	.04917283	18	2	8	.04560713
17	1	2	.16721701	17	3	7	.05872810	17	6	12	.04456975	18	2	9	.04048484
17	1	3	.10870077	17	3	8	.05140206	17	7	7	.08584847	18	2	10	.03629969
17	1	4	.08121627	17	3	9	.04558515	17	7	8	.07534854	18	2	11	.03278801
17	1	5	.06506606	17	3	10	.04081395	17	7	9	.06696951	18	2	12	.02977232
17	1	6	.05433117	17	3	11	.03679195	17	7	10	.06006836	18	2	13	.02712646
17	1	7	.04661142	17	3	12	.03331754	17	7	11	.05423069	18	2	14	.02475509
17	1	8	.04074488	17	3	13	.03024460	17	8	8	.08330442	18	2	15	.02257954
17	1	9	.03609796	17	3	14	.02745781	17	8	9	.07409527	18	2	16	.02052467
17	1	10	.03229378	17	3	15	.02485255	17	8	10	.06650027	18	2	17	.01849821
17	1	11	.02909207	17	4	4	.11150120	17	9	9	.08249848	18	3	3	.13246510
17	1	12	.02632999	17	4	5	.08978867	18	1	1	.38026491	18	3	4	.09951015

TABLE II

P = 9.0

N	R	S	COVARIANCE	N	R	S	COVARIANCE	N	R	S	COVARIANCE	N	R	S	COVARIANCE
18	3	5	.08001256	18	6	7	.07545709	19	1	17	.01684149	19	4	4	.10638279
18	3	6	.06700115	18	6	8	.06623666	19	1	18	.01523495	19	4	5	.08566762
18	3	7	.05762326	18	6	9	.05890903	19	1	19	.01350927	19	4	6	.07183344
18	3	8	.05048942	18	6	10	.05290055	19	2	2	.18178789	19	4	7	.06186183
18	3	9	.04483890	18	6	11	.04784392	19	2	3	.11866566	19	4	8	.05427995
18	3	10	.04021818	18	6	12	.04349047	19	2	4	.08889288	19	4	9	.04828066
18	3	11	.03633828	18	6	13	.03966257	19	2	5	.07137303	19	4	10	.04338268
18	3	12	.03300436	18	7	7	.08250811	19	2	6	.05972817	19	4	11	.03927976
18	3	13	.03007779	18	7	8	.07247856	19	2	7	.05136309	19	4	12	.03576621
18	3	14	.02745366	18	7	9	.06449764	19	2	8	.04501902	19	4	13	.03269709
18	3	15	.02504524	18	7	10	.05794652	19	2	9	.04000922	19	4	14	.02996497
18	3	16	.02276960	18	7	11	.05242829	19	2	10	.03592571	19	4	15	.02748520
18	4	4	.10879703	18	7	12	.04767380	19	2	11	.03250962	19	4	16	.02518468
18	4	5	.08761381	18	8	8	.07954512	19	2	12	.02958756	19	5	5	.09308199
18	4	6	.07344283	18	8	9	.07083239	19	2	13	.02703756	19	5	6	.07812973
18	4	7	.06321110	18	8	10	.06367207	19	2	14	.02476949	19	5	7	.06733387
18	4	8	.05541728	18	8	11	.05763464	19	2	15	.02271245	19	5	8	.05911464
18	4	9	.04923749	18	9	9	.07816254	19	2	16	.02080538	19	5	9	.05260436
18	4	10	.04417963	18	9	10	.07030500	19	2	17	.01898704	19	5	10	.04728477
18	4	11	.03992964	19	1	1	.37732768	19	2	18	.01717833	19	5	11	.04282559
18	4	12	.03627551	19	1	2	.16372856	19	3	3	.13005584	19	5	12	.03900472
18	4	13	.03306620	19	1	3	.10626886	19	3	4	.09766381	19	5	13	.03566544
18	4	14	.03018722	19	1	4	.07937700	19	3	5	.07853250	19	5	14	.03269152
18	4	15	.02754383	19	1	5	.06362213	19	3	6	.06578569	19	5	15	.02999121
18	5	5	.09558408	19	1	6	.05318035	19	3	7	.05661318	19	6	6	.08486761
18	5	6	.08021080	19	1	7	.04569460	19	3	8	.04964772	19	6	7	.07319844
18	5	7	.06909083	19	1	8	.04002592	19	3	9	.04414165	19	6	8	.06430218
18	5	8	.06060865	19	1	9	.03555467	19	3	10	.03964997	19	6	9	.05724799
18	5	9	.05387565	19	1	10	.03191353	19	3	11	.03588989	19	6	10	.05147884
18	5	10	.04836007	19	1	11	.02886982	19	3	12	.03267176	19	6	11	.04663922
18	5	11	.04372202	19	1	12	.02626795	19	3	13	.02986203	19	6	12	.04248975
18	5	12	.03973169	19	1	13	.02399861	19	3	14	.02736189	19	6	13	.03886132
18	5	13	.03622518	19	1	14	.02198113	19	3	15	.02509352	19	6	14	.03562831
18	5	14	.03307806	19	1	15	.02015212	19	3	16	.02298983	19	7	7	.07961943
18	6	6	.08752706	19	1	16	.01845710	19	3	17	.02098340	19	7	8	.06998906

TABLE II

P = 9.0

N	R	S	COVARIANCE
19	7	9	.06234377
19	7	10	.05608517
19	7	11	.05083069
19	7	12	.04632242
19	7	13	.04237786
19	8	8	.07634197
19	8	9	.06804272
19	8	10	.06124152
19	8	11	.05552635
19	8	12	.05061902
19	9	9	.07453426
19	9	10	.06712109
19	9	11	.06088536
19	10	10	.07395568
20	1	1	.37464074
20	1	2	.16219932
20	1	3	.10519722
20	1	4	.07856113
20	1	5	.06297602
20	1	6	.05265920
20	1	7	.04527230
20	1	8	.03968620
20	1	9	.03528713
20	1	10	.03171161
20	1	11	.02872985
20	1	12	.02618864
20	1	13	.02398104
20	1	14	.02202908
20	1	15	.02027313
20	1	16	.01866462
20	1	17	.01716007
20	1	18	.01571390
20	1	19	.01426469
20	1	20	.01269754
20	2	2	.17947610

N	R	S	COVARIANCE
20	2	3	.11704862
20	2	4	.08765638
20	2	5	.07038497
20	2	6	.05892021
20	2	7	.05069539
20	2	8	.04446665
20	2	9	.03955599
20	2	10	.03556111
20	2	11	.03222718
20	2	12	.02938409
20	2	13	.02691295
20	2	14	.02472699
20	2	15	.02275975
20	2	16	.02095706
20	2	17	.01927034
20	2	18	.01764859
20	2	19	.01602299
20	3	3	.12787381
20	3	4	.09598836
20	3	5	.07718569
20	3	6	.06467546
20	3	7	.05568577
20	3	8	.04886934
20	3	9	.04349016
20	3	10	.03911073
20	3	11	.03545355
20	3	12	.03233312
20	3	13	.02961969
20	3	14	.02721843
20	3	15	.02505668
20	3	16	.02307512
20	3	17	.02122052
20	3	18	.01943689
20	4	4	.10421141
20	4	5	.08391360

N	R	S	COVARIANCE
20	4	6	.07037874
20	4	7	.06063733
20	4	8	.05324205
20	4	9	.04740060
20	4	10	.04264120
20	4	11	.03866421
20	4	12	.03526912
20	4	13	.03231551
20	4	14	.02970068
20	4	15	.02734583
20	4	16	.02518660
20	4	17	.02316514
20	5	5	.09084924
20	5	6	.07626854
20	5	7	.06575761
20	5	8	.05776842
20	5	9	.05145178
20	5	10	.04630120
20	5	11	.04199455
20	5	12	.03831603
20	5	13	.03511433
20	5	14	.03227871
20	5	15	.02972409
20	5	16	.02738093
20	6	6	.08251578
20	6	7	.07119621
20	6	8	.06258143
20	6	9	.05576327
20	6	10	.05019918
20	6	11	.04554358
20	6	12	.04156469
20	6	13	.03809982
20	6	14	.03502976
20	6	15	.03226286
20	7	7	.07709161

N	R	S	COVARIANCE
20	7	8	.06780477
20	7	9	.06044675
20	7	10	.05443675
20	7	11	.04940433
20	7	12	.04510067
20	7	13	.04135093
20	7	14	.03802686
20	8	8	.07357355
20	8	9	.06562459
20	8	10	.05912560
20	8	11	.05367929
20	8	12	.04901842
20	8	13	.04495499
20	9	9	.07144477
20	9	10	.06440108
20	9	11	.05849293
20	9	12	.05343287
20	10	10	.07043801
20	10	11	.06400628

****

TABLE II

P = 9.5

N	R	S	COVARIANCE
2	1	1	.68691123
2	1	2	.31308877
3	1	1	.58223268
3	1	2	.26981874
3	1	3	.16481225
3	2	2	.42663518
4	1	1	.52732285
4	1	2	.24408552
4	1	3	.15554330
4	1	4	.10677567
4	2	2	.34312901
4	2	3	.22351483
5	1	1	.49267073
5	1	2	.22695577
5	1	3	.14682802
5	1	4	.10539856
5	1	5	.07719058
5	2	2	.29989337
5	2	3	.19726350
5	2	4	.14295085
5	3	3	.26880551
6	1	1	.46845058
6	1	2	.21461514
6	1	3	.13977452
6	1	4	.10216668
6	1	5	.07813888
6	1	6	.05966683
6	2	2	.27277969
6	2	3	.18009560
6	2	4	.13263014
6	2	5	.10193935
6	3	3	.23081493
6	3	4	.17150668
7	1	1	.45038498
7	1	2	.20522544

N	R	S	COVARIANCE
7	1	3	.13408257
7	1	4	.09891954
7	1	5	.07706718
7	1	6	.06133370
7	1	7	.04823531
7	2	2	.25390099
7	2	3	.16784403
7	2	4	.12461283
7	2	5	.09747695
7	2	6	.07780213
7	3	3	.20714988
7	3	4	.15487477
7	3	5	.12171027
7	4	4	.19588482
8	1	1	.43628946
8	1	2	.19779351
8	1	3	.12941374
8	1	4	.09598723
8	1	5	.07553218
8	1	6	.06121507
8	1	7	.05007495
8	1	8	.04026027
8	2	2	.23985674
8	2	3	.15857937
8	2	4	.11827264
8	2	5	.09339231
8	2	6	.07587378
8	2	7	.06218159
8	3	3	.19074031
8	3	4	.14308800
8	3	5	.11341264
8	3	6	.09238667
8	4	4	.17425861
8	4	5	.13875550
9	1	1	.42492228

N	R	S	COVARIANCE
9	1	2	.19173393
9	1	3	.12551430
9	1	4	.09340224
9	1	5	.07394557
9	1	6	.06053620
9	1	7	.05040227
9	1	8	.04207264
9	1	9	.03441555
9	2	2	.22892022
9	2	3	.15127944
9	2	4	.11313613
9	2	5	.08984531
9	2	6	.07370903
9	2	7	.06146707
9	2	8	.05137413
9	3	3	.17856775
9	3	4	.13421424
9	3	5	.10692500
9	3	6	.08791826
9	3	7	.07344104
9	4	4	.15932055
9	4	5	.12739989
9	4	6	.10503343
9	5	5	.15400400
10	1	1	.41552086
10	1	2	.18667791
10	1	3	.12220353
10	1	4	.09113023
10	1	5	.07243400
10	1	6	.05967364
10	1	7	.05018165
10	1	8	.04261101
10	1	9	.03612799
10	1	10	.02996769
10	2	2	.22011361

N	R	S	COVARIANCE
10	2	3	.14534861
10	2	4	.10888278
10	2	5	.08678670
10	2	6	.07163441
10	2	7	.06032431
10	2	8	.05127955
10	2	9	.04351740
10	3	3	.16910914
10	3	4	.12724414
10	3	5	.10170588
10	3	6	.08411210
10	3	7	.07093454
10	3	8	.06036793
10	4	4	.14826892
10	4	5	.11888373
10	4	6	.09853698
10	4	7	.08323898
10	5	5	.14006281
10	5	6	.11640528
11	1	1	.40758866
11	1	2	.18238069
11	1	3	.11935219
11	1	4	.08912579
11	1	5	.07103410
11	1	6	.05876909
11	1	7	.04973517
11	1	8	.04264637
11	1	9	.03676114
11	1	10	.03155813
11	1	11	.02648088
11	2	2	.21283817
11	2	3	.14041449
11	2	4	.10529474
11	2	5	.08413700
11	2	6	.06973114

TABLE II

P = 9.5

N	R	S	COVARIANCE	N	R	S	COVARIANCE	N	R	S	COVARIANCE	N	R	S	COVARIANCE
11	2	7	.05908713	12	2	5	.08182350	13	1	9	.03690186	13	5	8	.07193674
11	2	8	.05071491	12	2	6	.06800863	13	1	10	.03249115	13	5	9	.06316696
11	2	9	.04375085	12	2	7	.05787225	13	1	11	.02861858	13	6	6	.10960397
11	2	10	.03758416	12	2	8	.04998110	13	1	12	.02503160	13	6	7	.09400861
11	3	3	.16150576	12	2	9	.04352757	13	1	13	.02138816	13	6	8	.08184750
11	3	4	.12159474	12	2	10	.03799264	13	2	2	.20144928	13	7	7	.10782566
11	3	5	.09740502	12	2	11	.03296637	13	2	3	.13262976	14	1	1	.38968396
11	3	6	.08086679	12	3	3	.15523340	13	2	4	.09955108	14	1	2	.17257329
11	3	7	.06861021	12	3	4	.11690374	13	2	5	.07978653	14	1	3	.11272808
11	3	8	.05894693	12	3	5	.09378996	13	2	6	.06645429	14	1	4	.08432963
11	3	9	.05089361	12	3	6	.07807632	13	2	7	.05672173	14	1	5	.06750466
11	4	4	.13969807	12	3	7	.06651531	13	2	8	.04919784	14	1	6	.05623231
11	4	5	.11221270	12	3	8	.05749618	13	2	9	.04310904	14	1	7	.04805607
11	4	6	.09333849	12	3	9	.05010775	13	2	10	.03797683	14	1	8	.04177922
11	4	7	.07930466	12	3	10	.04376211	13	2	11	.03346581	14	1	9	.03674311
11	4	8	.06821153	12	4	4	.13281875	13	2	12	.02928344	14	1	10	.03254868
11	5	5	.12977427	12	4	5	.10681668	13	3	3	.14995235	14	1	11	.02893047
11	5	6	.10818850	12	4	6	.08907041	13	3	4	.11293305	14	1	12	.02568829
11	5	7	.09207784	12	4	7	.07597616	13	3	5	.09070114	14	1	13	.02263188
11	6	6	.12685022	12	4	8	.06573788	13	3	6	.07565305	14	1	14	.01947892
12	1	1	.40078725	12	4	9	.05733540	13	3	7	.06464053	14	2	2	.19688436
12	1	2	.17867299	12	5	5	.12180981	13	3	8	.05611095	14	2	3	.12948908
12	1	3	.11686622	12	5	6	.10176795	13	3	9	.04919780	14	2	4	.09720749
12	1	4	.08734690	12	5	7	.08693199	13	3	10	.04336348	14	2	5	.07797852
12	1	5	.06975079	12	5	8	.07530227	13	3	11	.03822986	14	2	6	.06504981
12	1	6	.05788063	12	6	6	.11712183	13	4	4	.12715068	14	2	7	.05564848
12	1	7	.04919621	12	6	7	.10022053	13	4	5	.10234322	14	2	8	.04841743
12	1	8	.04245017	13	1	1	.39487720	13	4	6	.08549269	14	2	9	.04260709
12	1	9	.03694270	13	1	2	.17543360	13	4	7	.07312916	14	2	10	.03776198
12	1	10	.03222594	13	1	3	.11467581	13	4	8	.06353399	14	2	11	.03357826
12	1	11	.02794790	13	1	4	.08575794	13	4	9	.05574466	14	2	12	.02982608
12	1	12	.02368111	13	1	5	.06857766	13	4	10	.04916210	14	2	13	.02628616
12	2	2	.20670493	13	1	6	.05703213	13	5	5	.11542621	14	3	3	.14543190
12	2	3	.13623131	13	1	7	.04862646	13	5	6	.09658421	14	3	4	.10951916
12	2	4	.10222032	13	1	8	.04214151	13	5	7	.08272025	14	3	5	.08802558

TABLE II

P = 9.5

N	R	S	COVARIANCE	N	R	S	COVARIANCE	N	R	S	COVARIANCE	N	R	S	COVARIANCE
14	3	6	.07352823	15	1	9	.03652034	15	4	8	.05983285	16	1	15	.01892737
14	3	7	.06296199	15	1	10	.03248987	15	4	9	.05287582	16	1	16	.01648860
14	3	8	.05482072	15	1	11	.02905088	15	4	10	.04710423	16	2	2	.18931700
14	3	9	.04826992	15	1	12	.02602443	15	4	11	.04216622	16	2	3	.12425755
14	3	10	.04280116	15	1	13	.02326590	15	4	12	.03781046	16	2	4	.09327274
14	3	11	.03807442	15	1	14	.02062620	15	5	5	.10575777	16	2	5	.07490619
14	3	12	.03383171	15	1	15	.01786665	15	5	6	.08866962	16	2	6	.06261825
14	4	4	.12238333	15	2	2	.19287414	15	5	7	.07620007	16	2	7	.05373275
14	4	5	.09856159	15	2	3	.12672058	15	5	8	.06659756	16	2	8	.04694521
14	4	6	.08244225	15	2	4	.09512992	15	5	9	.05889024	16	2	9	.04154002
14	4	7	.07066620	15	2	5	.07636175	15	5	10	.05248871	16	2	10	.03708856
14	4	8	.06157635	15	2	6	.06377671	15	5	11	.04700632	16	2	11	.03331484
14	4	9	.05425173	15	2	7	.05465347	15	6	6	.09863914	16	2	12	.03002829
14	4	10	.04812967	15	2	8	.04766324	15	6	7	.08486299	16	2	13	.02708545
14	4	11	.04283294	15	2	9	.04207510	15	6	8	.07423353	16	2	14	.02436247
14	5	5	.11017250	15	2	10	.03744896	15	6	9	.06568845	16	2	15	.02172156
14	5	6	.09229296	15	2	11	.03349799	15	6	10	.05858179	16	3	3	.13806659
14	5	7	.07919790	15	2	12	.03001822	15	7	7	.09491972	16	3	4	.10392756
14	5	8	.06907003	15	2	13	.02684427	15	7	8	.08311462	16	3	5	.08360607
14	5	9	.06089602	15	2	14	.02380513	15	7	9	.07360732	16	3	6	.06997212
14	5	10	.05405505	15	3	3	.14150941	15	8	8	.09375824	16	3	7	.06009348
14	6	6	.10358637	15	3	4	.10654575	16	1	1	.38095627	16	3	8	.05253583
14	6	7	.08900199	15	3	5	.08568102	16	1	2	.16773619	16	3	9	.04651017
14	6	8	.07769730	15	3	6	.07164847	16	1	3	.10940658	16	3	10	.04154292
14	6	9	.06855712	15	3	7	.06145410	16	1	4	.08186370	16	3	11	.03732855
14	7	7	.10065341	15	3	8	.05363047	16	1	5	.06561722	16	3	12	.03365570
14	7	8	.08797205	15	3	9	.04736808	16	1	6	.05478244	16	3	13	.03036502
15	1	1	.38507692	15	3	10	.04217836	16	1	7	.04696548	16	3	14	.02731855
15	1	2	.17002476	15	3	11	.03774220	16	1	8	.04100431	16	4	4	.11477124
15	1	3	.11098231	15	3	12	.03383217	16	1	9	.03626346	16	4	5	.09248780
15	1	4	.08303806	15	3	13	.03026344	16	1	10	.03236328	16	4	6	.07749631
15	1	5	.06652118	15	4	4	.11830624	16	1	11	.02905982	16	4	7	.06661230
15	1	6	.05548290	15	4	5	.09531384	16	1	12	.02618496	16	4	8	.05827277
15	1	7	.04750001	15	4	6	.07980447	16	1	13	.02361243	16	4	9	.05161568
15	1	8	.04139444	15	4	7	.06851287	16	1	14	.02123344	16	4	10	.04612249

TABLE II

P = 9.5

N	R	S	COVARIANCE	N	R	S	COVARIANCE	N	R	S	COVARIANCE	N	R	S	COVARIANCE
16	4	11	.04145805	17	1	13	.02379815	17	4	6	.07545607	18	1	2	.16378483
16	4	12	.03739010	17	1	14	.02158170	17	4	7	.06492062	18	1	3	.10666873
16	4	13	.03374316	17	1	15	.01950631	17	4	8	.05686942	18	1	4	.07980527
16	5	5	.10198521	17	1	16	.01747176	17	4	9	.05046287	18	1	5	.06401293
16	5	6	.08556083	17	1	17	.01529829	17	4	10	.04519776	18	1	6	.05351635
16	5	7	.07361152	17	2	2	.18613534	17	4	11	.04075116	18	1	7	.04597122
16	5	8	.06444095	17	2	3	.12204873	17	4	12	.03690283	18	1	8	.04024213
16	5	9	.05711106	17	2	4	.09160043	17	4	13	.03349242	18	1	9	.03571003
16	5	10	.05105630	17	2	5	.07358772	17	4	14	.03039250	18	1	10	.03200703
16	5	11	.04591044	17	2	6	.06155983	17	5	5	.09871647	18	1	11	.02889926
16	5	12	.04141921	17	2	7	.05288056	17	5	6	.08285795	18	1	12	.02622935
16	6	6	.09448557	17	2	8	.04626674	17	5	7	.07134913	18	1	13	.02388539
16	6	7	.08137203	17	2	9	.04101575	17	5	8	.06254102	18	1	14	.02178266
16	6	8	.07129027	17	2	10	.03670815	17	5	9	.05552391	18	1	15	.01985090
16	6	9	.06322068	17	2	11	.03307572	17	5	10	.04975144	18	1	16	.01802235
16	6	10	.05654714	17	2	12	.02993604	17	5	11	.04487242	18	1	17	.01621195
16	6	11	.05086978	17	2	13	.02715673	17	5	12	.04064693	18	1	18	.01426062
16	7	7	.09021069	17	2	14	.02463294	17	5	13	.03690002	18	2	2	.18326883
16	7	8	.07910410	17	2	15	.02226864	17	6	6	.09093878	18	2	3	.12005405
16	7	9	.07020012	17	2	16	.01994985	17	6	7	.07837952	18	2	4	.09008488
16	7	10	.06282682	17	3	3	.13501520	17	6	8	.06875210	18	2	5	.07238683
16	8	8	.08825264	17	3	4	.10160055	17	6	9	.06107246	18	2	6	.06058891
16	8	9	.07838438	17	3	5	.08175398	17	6	10	.05474834	18	2	7	.05209073
17	1	1	.37724424	17	3	6	.06846639	17	6	11	.04939832	18	2	8	.04562798
17	1	2	.16566699	17	3	7	.05885995	17	6	12	.04476139	18	2	9	.04050935
17	1	3	.10797553	17	3	8	.05152912	17	7	7	.08626039	18	2	10	.03632308
17	1	4	.08079052	17	3	9	.04570233	17	7	8	.07572499	18	2	11	.03280696
17	1	5	.06478377	17	3	10	.04091805	17	7	9	.06730915	18	2	12	.02978420
17	1	6	.05412806	17	3	11	.03688061	17	7	10	.06037066	18	2	13	.02712897
17	1	7	.04645559	17	3	12	.03338865	17	7	11	.05449509	18	2	14	.02474581
17	1	8	.04061843	17	3	13	.03029579	17	8	8	.08373532	18	2	15	.02255545
17	1	9	.03598988	17	3	14	.02748591	17	8	9	.07448356	18	2	16	.02048129
17	1	10	.03219673	17	3	15	.02485244	17	8	10	.06684574	18	2	17	.01842695
17	1	11	.02900077	17	4	4	.11167072	17	9	9	.08293517	18	3	3	.13228792
17	1	12	.02624030	17	4	5	.09000146	18	1	1	.37387930	18	3	4	.09951561

TABLE II

P = 9.5

N	R	S	COVARIANCE	N	R	S	COVARIANCE	N	R	S	COVARIANCE	N	R	S	COVARIANCE
18	3	5	.08008844	18	6	7	.07577972	19	1	17	.01673496	19	4	4	.10647135
18	3	6	.06710520	18	6	8	.06653611	19	1	18	.01511192	19	4	5	.08581617
18	3	7	.05773639	18	6	9	.05918292	19	1	19	.01334864	19	4	6	.07200247
18	3	8	.05060187	18	6	10	.05314772	19	2	2	.18066973	19	4	7	.06203401
18	3	9	.04494521	18	6	11	.04806362	19	2	3	.11824167	19	4	8	.05444677
18	3	10	.04031496	18	6	12	.04368190	19	2	4	.08870355	19	4	9	.04843754
18	3	11	.03642318	18	6	13	.03982454	19	2	5	.07128757	19	4	10	.04352696
18	3	12	.03307548	18	7	7	.08288952	19	2	6	.05969485	19	4	11	.03940969
18	3	13	.03013330	18	7	8	.07283083	19	2	7	.05135727	19	4	12	.03588048
18	3	14	.02749140	18	7	9	.06481902	19	2	8	.04502746	19	4	13	.03279444
18	3	15	.02506228	18	7	10	.05823631	19	2	9	.04002411	19	4	14	.03004398
18	3	16	.02276117	18	7	11	.05268606	19	2	10	.03594208	19	4	15	.02754401
18	4	4	.10892445	18	7	12	.04789894	19	2	11	.03252406	19	4	16	.02522046
18	4	5	.08779329	18	8	8	.07994886	19	2	12	.02959750	19	5	5	.09331467
18	4	6	.07363688	18	8	9	.07120011	19	2	13	.02704090	19	5	6	.07837257
18	4	7	.06340356	18	8	10	.06400342	19	2	14	.02476423	19	5	7	.06757162
18	4	8	.05560032	18	8	11	.05792942	19	2	15	.02269643	19	5	8	.05934016
18	4	9	.04940696	18	9	9	.07857622	19	2	16	.02077586	19	5	9	.05281407
18	4	10	.04433308	18	9	10	.07067753	19	2	17	.01893984	19	5	10	.04747671
18	4	11	.04006541	19	1	1	.37081201	19	2	18	.01710542	19	5	11	.04299853
18	4	12	.03639217	19	1	2	.16206366	19	3	3	.12983244	19	5	12	.03915771
18	4	13	.03316222	19	1	3	.10546951	19	3	4	.09763437	19	5	13	.03579749
18	4	14	.03026060	19	1	4	.07889695	19	3	5	.07858083	19	5	14	.03280136
18	4	15	.02759158	19	1	5	.06329779	19	3	6	.06586757	19	5	15	.03007700
18	5	5	.09585119	19	1	6	.05294384	19	3	7	.05670841	19	6	6	.08517691
18	5	6	.08048166	19	1	7	.04551206	19	3	8	.04974590	19	6	7	.07349555
18	5	7	.06935137	19	1	8	.03987830	19	3	9	.04423691	19	6	8	.06458106
18	5	8	.06085248	19	1	9	.03543028	19	3	10	.03973870	19	6	9	.05750580
18	5	9	.05409961	19	1	10	.03180466	19	3	11	.03596968	19	6	10	.05171421
18	5	10	.04856244	19	1	11	.02877110	19	3	12	.03274078	19	6	11	.04685132
18	5	11	.04390162	19	1	12	.02617537	19	3	13	.02991870	19	6	12	.04267794
18	5	12	.03988746	19	1	13	.02390902	19	3	14	.02740456	19	6	13	.03902479
18	5	13	.03635580	19	1	14	.02189176	19	3	15	.02512023	19	6	14	.03576590
18	5	14	.03318160	19	1	15	.02006034	19	3	16	.02299782	19	7	7	.07997237
18	6	6	.08786757	19	1	16	.01835995	19	3	17	.02096823	19	7	8	.07031828

TABLE II

P = 9.5

N	R	S	COVARIANCE
19	7	9	.06264707
19	7	10	.05636162
19	7	11	.05107983
19	7	12	.04654380
19	7	13	.04257084
19	8	8	.07671995
19	8	9	.06839017
19	8	10	.06155786
19	8	11	.05581138
19	8	12	.05087250
19	9	9	.07492530
19	9	10	.06747683
19	9	11	.06120580
19	10	10	.07435079
20	1	1	.36800219
20	1	2	.16048219
20	1	3	.10436416
20	1	4	.07805630
20	1	5	.06263235
20	1	6	.05240718
20	1	7	.04507715
20	1	8	.03952833
20	1	9	.03515450
20	1	10	.03159634
20	1	11	.02862644
20	1	12	.02609307
20	1	13	.02389019
20	1	14	.02194037
20	1	15	.02018425
20	1	16	.01857325
20	1	17	.01706356
20	1	18	.01560855
20	1	19	.01414384
20	1	20	.01254128
20	2	2	.17829978

N	R	S	COVARIANCE
20	2	3	.11658594
20	2	4	.08743815
20	2	5	.07027677
20	2	6	.05886855
20	2	7	.05067466
20	2	8	.04446301
20	2	9	.03956127
20	2	10	.03557011
20	2	11	.03223638
20	2	12	.02939092
20	2	13	.02691539
20	2	14	.02472326
20	2	15	.02274811
20	2	16	.02093556
20	2	17	.01923644
20	2	18	.01759839
20	2	19	.01594896
20	3	3	.12760742
20	3	4	.09592638
20	3	5	.07720822
20	3	6	.06473643
20	3	7	.05576394
20	3	8	.04895368
20	3	9	.04357441
20	3	10	.03919105
20	3	11	.03552741
20	3	12	.03239868
20	3	13	.02967547
20	3	14	.02726306
20	3	15	.02508868
20	3	16	.02309267
20	3	17	.02122105
20	3	18	.01941622
20	4	4	.10426391
20	4	5	.08403332

N	R	S	COVARIANCE
20	4	6	.07052428
20	4	7	.06079028
20	4	8	.05339323
20	4	9	.04754501
20	4	10	.04277593
20	4	11	.03878741
20	4	12	.03537947
20	4	13	.03241188
20	4	14	.02978196
20	4	15	.02741070
20	4	16	.02523329
20	4	17	.02319098
20	5	5	.09105000
20	5	6	.07648524
20	5	7	.06597385
20	5	8	.05797639
20	5	9	.05164746
20	5	10	.04648236
20	5	11	.04215987
20	5	12	.03846456
20	5	13	.03524523
20	5	14	.03239104
20	5	15	.02981664
20	5	16	.02745193
20	6	6	.08279611
20	6	7	.07146938
20	6	8	.06284068
20	6	9	.05600535
20	6	10	.05042242
20	6	11	.04574707
20	6	12	.04174778
20	6	13	.03826190
20	6	14	.03517003
20	6	15	.03238017
20	7	7	.07741799

N	R	S	COVARIANCE
20	7	8	.06811214
20	7	9	.06073246
20	7	10	.05469961
20	7	11	.04964375
20	7	12	.04531625
20	7	13	.04154224
20	7	14	.03819322
20	8	8	.07392725
20	8	9	.06595240
20	8	10	.05942572
20	8	11	.05395340
20	8	12	.04926534
20	8	13	.04517442
20	9	9	.07181397
20	9	10	.06473984
20	9	11	.05880116
20	9	12	.05371055
20	10	10	.07081428
20	10	11	.06434849

****

TABLE II

P = 10.0

N	R	S	COVARIANCE	N	R	S	COVARIANCE	N	R	S	COVARIANCE	N	R	S	COVARIANCE
2	1	1	.68659165	7	1	3	.13399080	9	1	2	.19098104	10	2	3	.14544283
2	1	2	.31340835	7	1	4	.09890951	9	1	3	.12526469	10	2	4	.10905479
3	1	1	.58089085	7	1	5	.07705062	9	1	4	.09329957	10	2	5	.08696194
3	1	2	.27016743	7	1	6	.06126460	9	1	5	.07389015	10	2	6	.07178595
3	1	3	.16483391	7	1	7	.04804636	9	1	6	.06048738	10	2	7	.06043857
3	2	2	.42788073	7	2	2	.25408898	9	1	7	.05033679	10	2	8	.05134536
4	1	1	.52522258	7	2	3	.16822875	9	1	8	.04196867	10	2	9	.04351812
4	1	2	.24419184	7	2	4	.12496222	9	1	9	.03422704	10	3	3	.16945562
4	1	3	.15569833	7	2	5	.09773476	9	2	2	.22877332	10	3	4	.12761743
4	1	4	.10667165	7	2	6	.07793428	9	2	3	.15145875	10	3	5	.10204602
4	2	2	.34410867	7	3	3	.20785459	9	2	4	.11336469	10	3	6	.08439983
4	2	3	.22421676	7	3	4	.15546908	9	2	5	.09005528	10	3	7	.07116041
5	1	1	.48999617	7	3	5	.12215190	9	2	6	.07387538	10	3	8	.06052255
5	1	2	.22683111	7	4	4	.19669006	9	2	7	.06157439	10	4	4	.14880888
5	1	3	.14691891	8	1	1	.43247235	9	2	8	.05140274	10	4	5	.11936172
5	1	4	.10543380	8	1	2	.19716368	9	3	3	.17901735	10	4	6	.09893847
5	1	5	.07703299	8	1	3	.12923889	9	3	4	.13465708	10	4	7	.08355705
5	2	2	.30056143	8	1	4	.09593097	9	3	5	.10730813	10	5	5	.14066815
5	2	3	.19789330	8	1	5	.07550107	9	3	6	.08822417	10	5	6	.11691150
5	2	4	.14335045	8	1	6	.06116884	9	3	7	.07365773	11	1	1	.40304918
5	3	3	.26980943	8	1	7	.04998386	9	4	4	.15994432	11	1	2	.18142402
6	1	1	.46532148	8	1	8	.04006976	9	4	5	.12792994	11	1	3	.11897497
6	1	2	.21429373	8	2	2	.23986360	9	4	6	.10545668	11	1	4	.08893774
6	1	3	.13977340	8	2	3	.15885503	9	5	5	.15467023	11	1	5	.07092366
6	1	4	.10219531	8	2	4	.11856142	10	1	1	.41119359	11	1	6	.05869143
6	1	5	.07810759	8	2	5	.09363307	10	1	2	.18581705	11	1	7	.04966781
6	1	6	.05948644	8	2	6	.07604123	10	1	3	.12188678	11	1	8	.04257539
6	2	2	.27318469	8	2	7	.06225123	10	1	4	.09098339	11	1	9	.03667514
6	2	3	.18060160	8	3	3	.19130855	10	1	5	.07235124	11	1	10	.03144275
6	2	4	.13302955	8	3	4	.14360598	10	1	6	.05961241	11	1	11	.02630088
6	2	5	.10217105	8	3	5	.11383414	10	1	7	.05011983	11	2	2	.21244271
6	3	3	.23167249	8	3	6	.09269313	10	1	8	.04253305	11	2	3	.14043317
6	3	4	.17216151	8	4	4	.17497352	10	1	9	.03601663	11	2	4	.10541464
7	1	1	.44688279	8	4	5	.13933150	10	1	10	.02978300	11	2	5	.08427729
7	1	2	.20473771	9	1	1	.42083348	10	2	2	.21983412	11	2	6	.06986242

TABLE II

P = 10.0

N	R	S	COVARIANCE	N	R	S	COVARIANCE	N	R	S	COVARIANCE	N	R	S	COVARIANCE
11	2	7	.05919563	12	2	5	.08193011	13	1	9	.03682574	13	5	8	.07223638
11	2	8	.05079176	12	2	6	.06811774	13	1	10	.03240966	13	5	9	.06341481
11	2	9	.04378715	12	2	7	.05796922	13	1	11	.02852437	13	6	6	.11008341
11	2	10	.03756544	12	2	8	.05005752	13	1	12	.02491392	13	6	7	.09443003
11	3	3	.16176197	12	2	9	.04357692	13	1	13	.02121843	13	6	8	.08220895
11	3	4	.12190484	12	2	10	.03800746	13	2	2	.20085933	13	7	7	.10831763
11	3	5	.09770245	12	2	11	.03293382	13	2	3	.13252013	14	1	1	.38462079
11	3	6	.08112973	12	3	3	.15540989	13	2	4	.09957959	14	1	2	.17138221
11	3	7	.06882922	12	3	4	.11715668	13	2	5	.07986128	14	1	3	.11220061
11	3	8	.05911556	12	3	5	.09404672	13	2	6	.06654093	14	1	4	.08403534
11	3	9	.05100385	12	3	6	.07831239	13	2	7	.05680452	14	1	5	.06731738
11	4	4	.14016278	12	3	7	.06672039	13	2	8	.04926820	14	1	6	.05610059
11	4	5	.11263937	12	3	8	.05766452	13	2	9	.04316118	14	1	7	.04795455
11	4	6	.09371009	12	3	9	.05023428	13	2	10	.03800558	14	1	8	.04169356
11	4	7	.07961455	12	3	10	.04383976	13	2	11	.03346467	14	1	9	.03666421
11	4	8	.06845470	12	4	4	.13321620	13	2	12	.02924091	14	1	10	.03246993
11	5	5	.13031899	12	4	5	.10719504	13	3	3	.15005782	14	1	11	.02884623
11	5	6	.10865957	12	4	6	.08940990	13	3	4	.11313424	14	1	12	.02559234
11	5	7	.09247043	12	4	7	.07626934	13	3	5	.09091984	14	1	13	.02251466
11	6	6	.12741665	12	4	8	.06598070	13	3	6	.07586216	14	1	14	.01931431
12	1	1	.39605672	12	4	9	.05752363	13	3	7	.06482861	14	2	2	.19621155
12	1	2	.17763041	12	5	5	.12229778	13	3	8	.05627211	14	2	3	.12932435
12	1	3	.11643424	12	5	6	.10220113	13	3	9	.04932793	14	2	4	.09719593
12	1	4	.08712065	12	5	7	.08730491	13	3	10	.04345830	14	2	5	.07802339
12	1	5	.06961345	12	5	8	.07561203	13	3	11	.03828301	14	2	6	.06511444
12	1	6	.05778506	12	6	6	.11764454	13	4	4	.12748787	14	2	7	.05571601
12	1	7	.04911906	12	6	7	.10066946	13	4	5	.10267685	14	2	8	.04847882
12	1	8	.04237779	13	1	1	.38997307	13	4	6	.08580029	14	2	9	.04265670
12	1	9	.03686544	13	1	2	.17431333	13	4	7	.07340213	14	2	10	.03779546
12	1	10	.03213484	13	1	3	.11419397	13	4	8	.06376822	14	2	11	.03359129
12	1	11	.02783067	13	1	4	.08549637	13	4	9	.05593714	14	2	12	.02981291
12	1	12	.02350620	13	1	5	.06841465	13	4	10	.04930903	14	2	13	.02623637
12	2	2	.20620681	13	1	6	.05691834	13	5	5	.11586214	14	3	3	.14547360
12	2	3	.13618247	13	1	7	.04853754	13	5	6	.09698003	14	3	4	.10967333
12	2	4	.10229251	13	1	8	.04206357	13	5	7	.08306940	14	3	5	.08820892

TABLE II

P = 10.0

N	R	S	COVARIANCE	N	R	S	COVARIANCE	N	R	S	COVARIANCE	N	R	S	COVARIANCE
14	3	6	.07371118	15	1	9	.03643648	15	4	8	.06004016	16	1	15	.01881266
14	3	7	.06313191	15	1	10	.03241021	15	4	9	.05305741	16	1	16	.01633374
14	3	8	.05497126	15	1	11	.02897040	15	4	10	.04725800	16	2	2	.18849979
14	3	9	.04839719	15	1	12	.02593852	15	4	11	.04228991	16	2	3	.12399646
14	3	10	.04290191	15	1	13	.02316918	15	4	12	.03790082	16	2	4	.09319026
14	3	11	.03814489	15	1	14	.02051004	15	5	5	.10610298	16	2	5	.07489698
14	3	12	.03386611	15	1	15	.01770701	15	5	6	.08899654	16	2	6	.06264151
14	4	4	.12266629	15	2	2	.19212596	15	5	7	.07649955	16	2	7	.05376940
14	4	5	.09885407	15	2	3	.12650559	15	5	8	.06686577	16	2	8	.04698509
14	4	6	.08271922	15	2	4	.09508149	15	5	9	.05912525	16	2	9	.04157709
14	4	7	.07091780	15	2	5	.07637867	15	5	10	.05268900	16	2	10	.03711880
14	4	8	.06179805	15	2	6	.06382017	15	5	11	.04716993	16	2	11	.03333514
14	4	9	.05444078	15	2	7	.05470547	15	6	6	.09903945	16	2	12	.03003576
14	4	10	.04828353	15	2	8	.04771421	15	6	7	.08522712	16	2	13	.02707683
14	4	11	.04294812	15	2	9	.04211925	15	6	8	.07455865	16	2	14	.02433304
14	5	5	.11056097	15	2	10	.03748223	15	6	9	.06597330	16	2	15	.02166257
14	5	6	.09265323	15	2	11	.03351693	15	6	10	.05882525	16	3	3	.13799803
14	5	7	.07952211	15	2	12	.03001910	15	7	7	.09534638	16	3	4	.10399948
14	5	8	.06935496	15	2	13	.02682192	15	7	8	.08349481	16	3	5	.08372619
14	5	9	.06113978	15	2	14	.02375003	15	7	9	.07394026	16	3	6	.07010647
14	5	10	.05425555	15	3	3	.14149340	15	8	8	.09419277	16	3	7	.06022706
14	6	6	.10402483	15	3	4	.10665700	16	1	1	.37561032	16	3	8	.05266109
14	6	7	.08939455	15	3	5	.08583156	16	1	2	.16642017	16	3	9	.04662285
14	6	8	.07804172	15	3	6	.07180646	16	1	3	.10879827	16	3	10	.04164018
14	6	9	.06885210	15	3	7	.06160568	16	1	4	.08151080	16	3	11	.03740812
14	7	7	.10111251	15	3	8	.05376873	16	1	5	.06538552	16	3	12	.03371521
14	7	8	.08837426	15	3	9	.04748900	16	1	6	.05461675	16	3	13	.03040138
15	1	1	.37986707	15	3	10	.04227905	16	1	7	.04683842	16	3	14	.02732687
15	1	2	.16876871	15	3	11	.03781996	16	1	8	.04090052	16	4	4	.11496046
15	1	3	.11041283	15	3	12	.03388361	16	1	9	.03617342	16	4	5	.09270761
15	1	4	.08271341	15	3	13	.03028325	16	1	10	.03228058	16	4	6	.07771708
15	1	5	.06631102	15	4	4	.11854012	16	1	11	.02897969	16	4	7	.06682170
15	1	6	.05533384	15	4	5	.09556848	16	1	12	.02610344	16	4	8	.05846500
15	1	7	.04738567	15	4	6	.08005247	16	1	13	.02352563	16	4	9	.05178767
15	1	8	.04129997	15	4	7	.06874310	16	1	14	.02113662	16	4	10	.04627224

TABLE II

P = 10.0

N	R	S	COVARIANCE	N	R	S	COVARIANCE	N	R	S	COVARIANCE	N	R	S	COVARIANCE
16	4	11	.04158382	17	1	13	.02371612	17	4	6	.07565133	18	1	2	.16236133
16	4	12	.03748982	17	1	14	.02149458	17	4	7	.06511000	18	1	3	.10599069
16	4	13	.03381380	17	1	15	.01940986	17	4	8	.05704648	18	1	4	.07940131
16	5	5	.10229089	17	1	16	.01735875	17	4	9	.05062419	18	1	5	.06374186
16	5	6	.08585666	17	1	17	.01514799	17	4	10	.04534126	18	1	6	.05331964
16	5	7	.07388709	17	2	2	.18525455	17	4	11	.04087530	18	1	7	.04581970
16	5	8	.06469177	17	2	3	.12174513	17	4	12	.03700607	18	1	8	.04011940
16	5	9	.05733495	17	2	4	.09148639	17	4	13	.03357284	18	1	9	.03560599
16	5	10	.05125187	17	2	5	.07355407	17	4	14	.03044726	18	1	10	.03191531
16	5	11	.04607635	17	2	6	.06156389	17	5	5	.09898592	18	1	11	.02881458
16	5	12	.04155356	17	2	7	.05290225	17	5	6	.08312486	18	1	12	.02614838
16	6	6	.09485058	17	2	8	.04629532	17	5	7	.07160191	18	1	13	.02380520
16	6	7	.08170887	17	2	9	.04104487	17	5	8	.06277448	18	1	14	.02170052
16	6	8	.07159545	17	2	10	.03673369	17	5	9	.05573554	18	1	15	.01976388
16	6	9	.06349272	17	2	11	.03309467	17	5	10	.04993979	18	1	16	.01792660
16	6	10	.05678511	17	2	12	.02994592	17	5	11	.04503637	18	1	17	.01610080
16	6	11	.05107261	17	2	13	.02715508	17	5	12	.04078521	18	1	18	.01411467
16	7	7	.09060636	17	2	14	.02461681	17	5	13	.03701083	18	2	2	.18232916
16	7	8	.07946157	17	2	15	.02223373	17	6	6	.09127117	18	2	3	.11971106
16	7	9	.07051844	17	2	16	.01988804	17	6	7	.07869052	18	2	4	.08994148
16	7	10	.06310541	17	3	3	.13489849	17	6	8	.06903752	18	2	5	.07233028
16	8	8	.08866157	17	3	4	.10163625	17	6	9	.06133047	18	2	6	.06057480
16	8	9	.07874817	17	3	5	.08184583	17	6	10	.05497799	18	2	7	.05209798
17	1	1	.37177138	17	3	6	.06857844	17	6	11	.04959882	18	2	8	.04564530
17	1	2	.16429534	17	3	7	.05897610	17	6	12	.04493167	18	2	9	.04053007
17	1	3	.10733114	17	3	6	.05164121	17	7	7	.08662689	18	2	10	.03634293
17	1	4	.08041124	17	3	9	.04580576	17	7	8	.07606006	18	2	11	.03282293
17	1	5	.06453181	17	3	10	.04100992	17	7	9	.06761152	18	2	12	.02979396
17	1	6	.05394648	17	3	11	.03695878	17	7	10	.06063975	18	2	13	.02713843
17	1	7	.04631611	17	3	12	.03345120	17	7	11	.05473036	18	2	14	.02473682
17	1	8	.04050514	17	3	13	.03034062	17	8	8	.08411901	18	2	15	.02253335
17	1	9	.03589300	17	3	14	.02751017	17	8	9	.07482936	18	2	16	.02044209
17	1	10	.03210974	17	3	15	.02485162	17	8	10	.06715338	18	2	17	.01836312
17	1	11	.02891898	17	4	4	.11181908	17	9	9	.08332412	18	3	3	.13212686
17	1	12	.02616001	17	4	5	.09018912	18	1	1	.36828753	18	3	4	.09951783

TABLE II

P = 10.0

N	R	S	COVARIANCE	N	R	S	COVARIANCE	N	R	S	COVARIANCE	N	R	S	COVARIANCE
18	3	5	.08015396	18	6	7	.07606639	19	1	17	.01664016	19	4	4	.10654750
18	3	6	.06719624	18	6	8	.06680234	19	1	18	.01500272	19	4	5	.08594644
18	3	7	.05783581	18	6	9	.05942651	19	1	19	.01320681	19	4	6	.07215149
18	3	8	.05070091	18	6	10	.05336756	19	2	2	.17967529	19	4	7	.06218617
18	3	9	.04503892	18	6	11	.04825898	19	2	3	.11786204	19	4	8	.05459438
18	3	10	.04040029	18	6	12	.04385203	19	2	4	.08853276	19	4	9	.04857646
18	3	11	.03649799	18	6	13	.03996837	19	2	5	.07120953	19	4	10	.04365475
18	3	12	.03313806	18	7	7	.08322873	19	2	6	.05966351	19	4	11	.03952477
18	3	13	.03018200	18	7	8	.07314429	19	2	7	.05135066	19	4	12	.03598162
18	3	14	.02752430	18	7	9	.06510507	19	2	8	.04503373	19	4	13	.03288052
18	3	15	.02507674	18	7	10	.05849425	19	2	9	.04003627	19	4	14	.03011373
18	3	16	.02275299	18	7	11	.05291544	19	2	10	.03595566	19	4	15	.02759573
18	4	4	.10903528	18	7	12	.04809919	19	2	11	.03253600	19	4	16	.02525165
18	4	5	.08795119	18	8	8	.08030830	19	2	12	.02960552	19	5	5	.09352002
18	4	6	.07380827	18	8	9	.07152755	19	2	13	.02704310	19	5	6	.07858753
18	4	7	.06357385	18	8	10	.06429848	19	2	14	.02475883	19	5	7	.06778239
18	4	8	.05576244	18	8	11	.05819186	19	2	15	.02268151	19	5	8	.05954028
18	4	9	.04955713	18	9	9	.07894467	19	2	16	.02074897	19	5	9	.05300024
18	4	10	.04446908	18	9	10	.07100933	19	2	17	.01889729	19	5	10	.04764714
18	4	11	.04018569	19	1	1	.36510842	19	2	18	.01704018	19	5	11	.04315208
18	4	12	.03649544	19	1	2	.16059163	19	3	3	.12963032	19	5	12	.03929349
18	4	13	.03324709	19	1	3	.10475998	19	3	4	.09760551	19	5	13	.03591460
18	4	14	.03032528	19	1	4	.07846980	19	3	5	.07862177	19	5	14	.03289865
18	4	15	.02763341	19	1	5	.06300868	19	3	6	.06593881	19	5	15	.03015281
18	5	5	.09608732	19	1	6	.05273272	19	3	7	.05679185	19	6	6	.08545124
18	5	6	.08072167	19	1	7	.04534893	19	3	8	.04983219	19	6	7	.07375939
18	5	7	.06958254	19	1	8	.03974626	19	3	9	.04432074	19	6	8	.06482888
18	5	8	.06106897	19	1	9	.03531895	19	3	10	.03981684	19	6	9	.05773501
18	5	9	.05429854	19	1	10	.03170719	19	3	11	.03603992	19	6	10	.05192349
18	5	10	.04874219	19	1	11	.02868272	19	3	12	.03280149	19	6	11	.04703991
18	5	11	.04406111	19	1	12	.02609252	19	3	13	.02996843	19	6	12	.04284520
18	5	12	.04002571	19	1	13	.02382887	19	3	14	.02744185	19	6	13	.03916999
18	5	13	.03647160	19	1	14	.02181188	19	3	15	.02514332	19	6	14	.03588798
18	5	14	.03327323	19	1	15	.01997839	19	3	16	.02300427	19	7	7	.08028611
18	6	6	.08816981	19	1	16	.01827333	19	3	17	.02095409	19	7	8	.07061113

TABLE II

P = 10.0

N	R	S	COVARIANCE	N	R	S	COVARIANCE	N	R	S	COVARIANCE	N	R	S	COVARIANCE
19	7	9	.06291696	20	2	3	.11617210	20	4	6	.07065231	20	7	8	.06838543
19	7	10	.05660765	20	2	4	.08724172	20	4	7	.06092525	20	7	9	.06098661
19	7	11	.05130152	20	2	5	.07017851	20	4	8	.05352684	20	7	10	.05493347
19	7	12	.04674074	20	2	6	.05882089	20	4	9	.04767277	20	7	11	.04985677
19	7	13	.04274242	20	2	7	.05065477	20	4	10	.04289518	20	7	12	.04550804
19	8	8	.07705636	20	2	8	.04445852	20	4	11	.03889647	20	7	13	.04171237
19	8	9	.06869950	20	2	9	.03956486	20	4	12	.03547711	20	7	14	.03834105
19	8	10	.06183953	20	2	10	.03557713	20	4	13	.03249710	20	8	8	.07424193
19	8	11	.05606516	20	2	11	.03224367	20	4	14	.02985374	20	8	9	.06624419
19	8	12	.05109812	20	2	12	.02939617	20	4	15	.02746785	20	8	10	.05969479
19	9	9	.07527356	20	2	13	.02691679	20	4	16	.02527423	20	8	11	.05419744
19	9	10	.06779368	20	2	14	.02471922	20	4	17	.02321335	20	8	12	.04948514
19	9	11	.06149118	20	2	15	.02273708	20	5	5	.09122684	20	8	13	.04536967
19	10	10	.07470273	20	2	16	.02091580	20	5	6	.07667681	20	9	9	.07214273
20	1	1	.36219305	20	2	17	.01920570	20	5	7	.06616538	20	9	10	.06504154
20	1	2	.15896456	20	2	18	.01755324	20	5	8	.05816079	20	9	11	.05907567
20	1	3	.10362505	20	2	19	.01588277	20	5	9	.05182107	20	9	12	.05395781
20	1	4	.07760734	20	3	3	.12736715	20	5	10	.04664315	20	10	10	.07114945
20	1	5	.06232619	20	3	4	.09586857	20	5	11	.04230660	20	10	11	.06465331
20	1	6	.05218235	20	3	5	.07722616	20	5	12	.03859635	******			**************
20	1	7	.04490286	20	3	6	.06478901	20	5	13	.03536132				
20	1	8	.03938722	20	3	7	.05583215	20	5	14	.03249058				
20	1	9	.03503589	20	3	8	.04902760	20	5	15	.02989853				
20	1	10	.03149320	20	3	9	.04364841	20	5	16	.02751459				
20	1	11	.02853390	20	3	10	.03926167	20	6	6	.08304452				
20	1	12	.02600755	20	3	11	.03559235	20	6	7	.07171181				
20	1	13	.02380892	20	3	12	.03245631	20	6	8	.06307095				
20	1	14	.02186107	20	3	13	.02972443	20	6	9	.05622048				
20	1	15	.02010486	20	3	14	.02730212	20	6	10	.05062086				
20	1	16	.01849173	20	3	15	.02511651	20	6	11	.04592796				
20	1	17	.01697757	20	3	16	.02310765	20	6	12	.04191051				
20	1	18	.01551486	20	3	17	.02122088	20	6	13	.03840588				
20	1	19	.01403663	20	3	18	.01939722	20	6	14	.03529455				
20	1	20	.01240336	20	4	4	.10430792	20	6	15	.03248418				
20	2	2	.17725416	20	4	5	.08413787	20	7	7	.07770796				